Perspectives on the African Environment, Science and Technology

Perspectives on AFRICAN Environment, Science and Technology

Edited by

Toyin Falola and Maurice Amutabi

AFRICA WORLD PRESS

TRENTON | LONDON | CAPE TOWN | NAIROBI | ADDIS ABABA | ASMARA | IBADAN | NEW DELHI

AFRICA WORLD PRESS
541 West Ingham Avenue | Suite B
Trenton, New Jersey 08638

First Printing 2012

Book and cover design: Saverance Publishing Services

Library of Congress Cataloging-in-Publication Data

Perspectives on the African environment, science and technology / edited by Toyin Falola and Maurice Amutabi.

p. cm.

Includes bibliographical references and index.

ISBN 978-1-59221-883-7 (hard cover) -- ISBN 978-1-59221-884-4 (pbk.) 1. Science--Social aspects--Africa. 2. Technology--Social aspects--Africa. 3. Technology and civilization. 4. Environmental protection--Africa. I. Falola, Toyin. II. Amutabi, M. N. (Maurice Nyamanga)

Q175.52.A4P47 2011

303.48'3096--dc23

2011039533

To

Professor A. B. Assensoh,

to mark his retirement from the University of Indiana

TABLE OF CONTENTS

PREFACE

For many years now, Africa has been grappling with issues of science, technology and the environment. The three are intimately and intricately intertwined because sound scientific and technological approaches ensure that the environment is preserved and protected. Since the Earth Summit at Rio de Janeiro in 1992 and the promulgation of Agenda 21 on Global Environment and Kyoto Protocol of 2005, many African governments have enacted many environmental laws. Issues of environment have become very sensitive at national and global levels, largely as a result of the current environment-related calamities across the globe. The chapters in this volume were first presented at a conference held at the University of Texas at Austin (March 27-29, 2009) on the theme: "Science, Technology and the Environment in Africa." The conference brought together scholars from different disciplinary backgrounds who exchanged views and ideas on many matters related to the environment in Africa. This volume emanated from the conference: it attends to issues that have not received significant attention in Africa.

The authors come from different parts of the world, making this a truly international endeavor. The scholars have used their expertise to discuss, clarify, elucidate and interrogate issues that are germane to understanding how Africa has dealt with science, technology and environment. The chapters are based on original research, rigorous analyses and interpretations. The authors have deployed exciting and compelling approaches to organize their data. The chapters recognize the importance of the environment, which is emerging as an important theme in African studies. Technology is also emerging as a new frontier in African studies.

From indigenous knowledge to gender and architecture, the chapters deal with science and technology and how they relate to the environment. Some of the authors have pointed to areas of weakness and shown how these can be addressed. For example, it is clear from many chapters that African societies have had environmental ethos, embedded in their cultures. Some authors have also noted that indigenous knowledge is disappearing and call for the preservation of local knowledge. They also see the need for an inventory to record what has happened in the past and where Africa is going. This book provides a unique opportunity for looking at the African environment and technology from the earliest period to the present, and using fresh insights, interpretations and analyses to understand the process of change.

African environment and technology are more than a mere catalogue of environmental degradation and technological shortcomings, but a holistic evaluation of what had happened in the past and the implications for the present and future of Africa. The book acknowledges that the concepts of environment and technology are facing fundamental changes in the context of global knowledge economy. There have been many changes that have occurred at an accelerating speed of scientific and technological advancement that are addressed in the book, and understanding these dynamics is necessary in appreciating the ways in which African governments and other actors in science, technology and environment have functioned in the past. How has Africa dealt with changes in science, technology, and environment? This book covers a wide range of topics that are presented by various interpretations by scholars from as diverse fields as engineering, political science, history, anthropology, education and women and gender studies. This is in recognition of the fact that changes in environment and technology cannot be captured by one discipline.

We would like to thank all the contributors to this volume: Elinami V. Swai, Eesuola Olukayode Segun, Aribidesi Usman, Nelson O. Fashina, Monsuru Muritala, Mike O. Odey, Hauwau Evelyn Yusuf, Ahmad Bawa Abdul-Qadir, Ann Albuyeh, Shadrack Wanjala Nasong'o, Bukola Adeyemi Oyeniyi, Jolanda Morkel, Maria Leus, Maurice N. Amutabi, Soji Oyeranmi, Ogundiran Soumonni, I. A. Ademiluyi, R.A. Asiyanbola, Adam Carpinelli, Nnenna George-Kalu and George Kalu. These scholars travelled to Ausitn, Texas to attend and present their papers at the conference and were able to follow through by revising them.We would like to salute the conference organizers under Toyin Falola, for without their sacrifices and commitments this book would not have seen the light of day. We are grateful to all the graduate students, most especially Emily Brownell and Adam Paddock who worked tirelessly to ensure a successful conference.

Toyin Falola, Austin, 2011
Maurice N. Amutabi, Nairobi, 2011

INTRODUCTION

Maurice Amutabi and Toyin Falola

After witnessing some of the worst environmental disasters, including the Sahel droughts of the 1970s and 1980s, rapid and unplanned urbanization accompanied by unprecedented urban sprawl, and the AIDS pandemic, Africa needs an environmental inventory of its problems and challenges. Experts agree that Africa needs more scientific and technological innovations for dealing with its environmental problems. This book fulfills a major objective in African studies, in part because environmental issues have been taken for granted and also because science and technology have often been relegated to the rear burner. Societies need information regarding changes that have taken place in science, technology, and environment over the years and how they are impacting human lives on the continent. The overwhelming attendance by the majority of world leaders of the climate change conference held in Copenhagen, Denmark in 2009 illuminated the sense of concern that we all do have over the environment.

Since the Earth Summit at Rio de Janeiro in 1992 and the promulgation of Agenda 21 on Global Environment and Kyoto Protocol, there has been more attention on interdisciplinary and multidisciplinary approaches in the study of the environment. This is because environment is a broad phenomenon that can only be well understood by encompassing all disciplines. As African countries move from one environmental problem to another, and deal with various technological challenges, policy makers and implementers need to be informed by research on how these changes occur and affect the continent. They can only intervene by using research produced by scholars. This anthology recognizes the need to understand environmental and technological issues as they relate to Africa.

In response, this book is an interdisciplinary and multidisciplinary project, which produces an examination of some issues and challenges. The chapters provide us with the views and interpretations of a diverse range of scholars who cover many themes in Africa's environment, and contribute to our understanding of the African past and present. The volume demonstrates how research on the environmental practices of indigenous African societies in the pre-colonial era can contribute to our understanding of these societies more broadly. From Nigeria and Cameroon in West Africa to Kenya and Tanzania in East Africa, the volume provides a wide geography of examples. Furthermore, authors use different theoretical approaches and deploy interdisciplinary methods in examining African environments.

Connections are made by the contributors on the intimate relationship between environment, science, and technology. Numerous African countries have had agencies dealing with environmental policy design and implementation since the 1960s, but many only acquired significant momentum after the Stockholm Conference on the Environment in 1972. Most African countries also have national forests, national parks and reserves, and the regulations to use and preserve them. Following the Earth Summit at Rio de Janeiro in 1992, the promulgation of Agenda 21 on Global Environment, and the Kyoto Protocol, environment has become a big issue in global politics. These international agreements changed public opinion about the environment, creating a global environmental consciousness. On February 16, 2005, the Kyoto Protocol came into effect after many years of campaigning by environmentalists and climatologists. The Protocol was adopted in Kyoto, Japan, on December 11, 1997, following Agenda 21 and Earth Summit in Rio de Janeiro in 1992. The Kyoto Protocol is an international agreement linked to the United Nations Framework Convention on Climate Change. It aims at preventing climate changes and global warming; however, it runs out in 2012, which has created apprehension among scientists, environmentalists, and climatologists who are concerned about the decay of environment and fearful that things will return to the old ways.

There is a realization in the chapters of this volume that the fate of the world is connected and only collective efforts can bring positive results. Africa faces a myriad of environmental problems: desertification; deforestation; soil erosion; siltation and sedimentation of streams and rivers; pollution with pesticides; drought; flooding; air pollution; sanitation; urban sprawl; reduced natural biodiversity; pandemic diseases such as AIDS and malaria; and many other calamities related to the environment. The solutions for these problems are expected to come from science and technology.

Over the years, many countries in West, Central, East, and Southern Africa have suffered forest loss, and many are now suffering from land degradation, reduction in carbon absorption, soil erosion, flooding, acid rain, a loss of biodiversity, soil acidity, and the loss of rare trees. In particular, the countries near the Sahara have been affected more by problems of desertification, especially visible with the drying up of Lake Chad, evidence that the Sahara is expanding. Even though African countries have some of the best environmental laws, good structures, and institutional arrangements on paper, as well as external funding,

African governments have not, for the most part, been successful in managing the environment. This can be attributed to inadequate budgetary allocation for environmental issues as well as a lack of political will. The fact that many African societies are attached to land is partly to blame, as well as the traditional disconnection between official rhetoric and actual practice in African states. Environmental issues are also related to problems of poor leadership, issues of governance, a lack of transparency and accountability, corruption, fiscal crises, and ignorance on suitable tradeoffs between the environment and development.

The loss of forest cover has come at a great disadvantage. Many river basins are without enough water and Africa is losing biodiversity as a result of the loss of habitat for indigenous flora and fauna. In many parts of Africa, environmental degradation has been caused by clearing forests for farming and the overexploitation of natural resources. There has been exploitation of forests for harvest of all kinds of wood for local use and export. Local tree species have been replaced by exotic species, which has also created problems. Some of the forests have been cleared to pave way for plantations of cash crops such as coffee, tea, and sugarcane, among others. In some areas, charcoal burning and fuel use have been the cause of loss of forest cover and land degradation. Overgrazing and poor land use patterns by farmers can also be attributed to the rapid deterioration of ecosystems in African countries.

Many of Africa's environmental problems have been blamed on rapid population growth, which has put a lot of pressure on the land. Overpopulation has also stimulated settlement in fragile lands in both rural and urban areas, and weakened environmental protection. Desertification in particular has been a major problem. Countries bordering the Sahara are losing natural vegetation through overgrazing and rampant clearing for farming and fuel. Evidence also suggests that areas around the Sahara are running the danger of becoming more arid. Scholars have prescribed some solutions to the pressure on environmental degradation, promoting smaller families or population control and planning and scientifically and technologically sound farming practices, especially the use of machines, chemical fertilizers, and genetically modified foods. Scholars have observed that scientific and technological innovations such as the use of chemical fertilizers, better seeds, and genetically modified seeds have played an important role in the switch from horizontal expansion of agriculture to rising output on farms that account for reduction in deforestation. It is a fact that because of rapid population increase in the 1970s and 1980s, Africa lost significant forest cover because individuals cleared forests to create room for new settlements.

Between 1950 and 1990 Africa lost over 80 percent of its forest cover. The Food and Agricultural Association of the United Nations (FAO) estimates that about 13 million hectares of forests have been lost all over the world due to deforestation, mainly in tropical Africa.[1] African forests specifically are lost due to unscrupulous exporters from Asia (mainly China and India) and the North (mainly countries from the European Union) who are the main consumers of forest products from Africa.[2] Recent practices have not been without additional problems. For example, chemical fertilizers have been blamed for the destruction of soil texture and composition, while genetically modified foods are tied to child obesity. A few of these issues are addressed by some chapters in this book.

The twenty one chapters in this book are divided into four parts, which broadly spell out the salient themes in environment, science, and technology in Africa. Part I provides a background to the rest of the book. The first chapter in this section is by Aribidesi Usman who provides an insightful historical overview of iron working metallurgy in pre-colonial Africa. He uses archaeological, historical, ethnographical, and oral sources to understand the processes of the spread of iron, the technological practices associated with iron production, socio-cultural and economic context, and the decline of traditional iron working. Usman utilizes information from his research in north central Nigeria as well as from other works carried out in different parts of Africa and deploys materials from his two decades of personal research on iron metallurgy in north central Nigeria. Usman shows how this area—known as northern Yoruba and home to groups such as the Igbomina, Ilorin, Ibolo, Ekiti, and Okun—is rich in pre-colonial metallurgical evidence. The chapter reveals that Africa is rich in various kinds of metals such as gold, copper, tin, and iron and that metals were important commodities of interregional and intercontinental trade during the pre-colonial period. This chapter allows us to have glimpses of the rich history of metallurgy in pre-colonial Africa.

In Chapter two, Elinami Swai provides a reexamination of the notion of science and technology in Africa, using the hill furrows in Kilimanjaro, Tanzania, among the Chagga people as an example. The thrust of Swai's argument is that, hill furrows—an irrigation technology that has been used by various ethnic groups in East Africa, specifically in the regions along the Rift Valley such as the Engaruka and Sonjo in Arusha and Taita and Marakwet in Kenya—as used by the Chagga people, provide one of the excellent examples through which we can see that African societies had strong environmental ethos and practices. The main argument is that the discourse of hill furrows has not received enough attention for its contribution to environmental protection and preservation. Swai faults accounts of early western travelers and colonialists who saw the hill furrow system as backward and presented simply as "traditional water manage-

1 Food and Agricultural Organization (FAO) of the United Nation, http://faostat.fao.org/, accessed March 18, 2010.

2 E. B. Barbier and J. C. Burgess, "The Economics of Tropical Deforestation," *Journal of Economic Survey*, Vol.15, No 3 (2001):412-433.

ment" without recognizing or understanding the complex ways in which the system operated. The chapter shows how hill furrows are used for the supply of water and for irrigation and argues that Europeans did not give credit to the scientific and technological inventiveness present in the system. Swai sees this as part of "othering" and "distancing" process present in the European representation of Africa, where African knowledge was often regarded as 'traditional' in order to be marginalized.

Chapter three, by Olukayode Segun Eesuola, focuses on pre-colonial scientific and technological innovations among the Yoruba. Eesuola argues that "the Yoruba people of West Africa, like other peoples of the world, had developed and were still developing the science and technology relevant to the challenges of their environment before they were interrupted by colonialism." The chapter draws evidence from Yoruba social structures, the fabrication of tools, medicine, pharmaceuticals, communications, arts, and equipment. The author asserts that for Africa to rank alongside the West in the current wind of globalization, it must revisit the significant science and technology she left behind in the face of colonialism. Using compelling examples from among the Yoruba, the chapter maintains that it was colonialism that halted the development of scientific and technological processes in Africa. Eesuola points out that Europeans replaced African methods with new ones from Europe, and that "this is the historical scenario that is responsible for the present claim that Africa is synonymous with technological backwardness." It is a compelling chapter that builds on the deconstruction, postmodern, and postcolonial discourses on Africa that are engaged in reenacting and recovering knowledge from the African past.

Nelson O. Fashina, in Chapter four, discusses the Yoruba counting system. The chapter examines the nature, properties, and characteristic features of Yoruba counting system and its origin in Ifa divination wisdom, as well as how these "numeric" resources provide models or frameworks of science and technology in the Yoruba language. Fashina assesses the Ifa oracular source of Yoruba counting system and provides grounds on which this can be compared with experiences from elsewhere. He examines the rationale of numerical language among the Yoruba, which allows us to appreciate the level of scientific prowess in pre-colonial Africa, looks at how the Yoruba deployed numeral figures or symbols other than words expressing figures, and compares the Yoruba numerology with English, Roman, and Arabic alternatives. The chapter shows that the Yoruba "formulaic alternation of additive and subtractive counting system, which invariably underscores early Yoruba knowledge and existence of a proto-arithmetic and mathematical science." Furthermore, it demonstrates that "the sequential use of two types of numeric counting codes, namely the five digit count numeric node and the ten digit count numeric node." Fashina explains that the numeric code "manifests in two ways, namely the Yoruba cosmic symbols or signs of the Odu Ifa, and the proto numeric bonded pairs as seen in [Yoruba] Ifa divination verse." The chapter, like Chapter three, also argues that the colonial moment was disruptive and interfered with the natural development of science and technology in Africa.

Elinami Swai provides in Chapter five a broad interrogation of the various processes that have shaped perception and the presentation of knowledge in African societies. Using examples from Tanzania, Swai faults ways in which indigenous knowledge systems were pushed aside during the colonial period. She explains that indigenous knowledge was replaced by Western knowledge, which was privileged and became almost exclusive. She argues that the methods that were used for this emasculation of African knowledge included the introduction of certificates, formal classrooms, and schools, as well as western languages in Africa. With the acceptance of western knowledge as representing science and technology, the take-over was final. Swai asserts that the process of the encroachment of European knowledge hegemony on Africa, which she refers to as "eurocentricization," continues through post-colonial processes such as globalization.

Chapters in Part II deal with a cross-section of environmental issues and threats in Africa. In Chapter six, Monsuru Muritala highlights recurring calamities associated with oil and gas in Nigeria, in particular the politics of oil and implications on the environment in Nigeria and the politics of revenue allocation and resource control. Muritala uses a historical approach that goes back to the Mineral Act of 1912. He argues that the liberalization of the oil exploration in Nigeria following the amendment of the Mineral Act of 1958 in 1960 had greater negative consequences for the environment than when Shell was the only oil producing company in the country. He demonstrates that oil remains the mainstay of Nigeria's economy and also its major problem. The argument in this chapter seems to be that the weak technological base of Nigeria at this period puts it at a disadvantage as far as the control and protection of the environment are concerned. He says that due to mismanagement, corruption, environmental degradation, and poor leadership, oil and gas have become a curse for Nigeria.

Chapter seven, by T. O. Adejumo, is entitled "Potential of microbial strains for reclamation and clean-up of polluted Nigerian soils." It provides an interesting analysis of the use of microbial strains, which is one of the recent problems in environmental management. The chapter discusses the mechanisms, impacts, and the extent of the use of microorganisms, especially fungi and bacteria, in the breaking down of environmental pollutants in Southern Nigeria. In this chapter, Adejumo points out that fungi and bacteria are capable of breaking down hydrocarbons and organic contaminants in oil-polluted soils and are an effective means of disposing hazardous wastes. Adejemo says this involves the transformation of simple/complex chemical compounds into non hazardous, water-soluble products. He points out that the lignin-degrading white-rot fungi have been discovered to degrade varied toxic, stubborn, and persistent environmental pollutants. Citing various studies, he confirms the high potency of these microorganisms and the possibility of using them in bioremediation of petroleum refinery and petrochemical waste waters. Adejumo gives some of the advantages of the use of bioremediation, which include cost, effectiveness, and simplicity. He suggests that these advantages can be put to good use in Nigeria given that crude oil contamination of soils in the Niger Delta is a perennial problem. Adejumo suggests that the use of these fungi is better than the traditional methods of remediation, including digging up the contaminated soil and removing it to a landfill; the cap and contain method; or the use of high-temperature incineration and chemical decomposition such as de-chlorination and oxidation, which tend to generate many negative effects. In closing his arguments, Adejumo recommends the use of microbial strains in removing pollutants from the soil in a rapid, cost-effective, ecologically responsible, and sustainable manner. As a final point, Adejumo recommends that more studies into oil pollution problems are needed in order to help in correcting the situation.

In Chapter eight, Mike Odey analyses the processes of food production and the impact on environment in central Nigeria, which he refers to as the food basket of the country. The chapter provides a historical analysis of the implications of perverse agricultural production under traditional methods on the Benue environment over time. Odey argues that accelerated food production constitutes the main reason for the undue pressure on arable land. The chapter further seeks to demonstrate how farmlands can be used more efficiently to achieve food security without jeopardizing the environment. The most important question Odey answers is how to keep a fair balance between food production and the improvement of environmental resources in the region to meet demands of sustainable development. The chapter concludes by pointing out that environmental insecurity in Central Nigeria cannot be isolated from similar instances in the country or be taken for granted, but rather carefully interrogated to avoid a larger ecological disaster like what is currently going on in the Niger Delta.

Chapter nine, by Hauwau Evelyn Yusuf and Ahmad Bawa Abdul-Qadir, is a study of water sanitation needs in slums, using Kaduna State located in the North Western region of Nigeria as a case study. The chapter discusses the relationship between environmental factors and poor sanitation habits and sustainable development. It also examines water supply and sanitation problems in Kaduna. The authors' main argument is that sanitation habits as impediment to development has always been assessed based on economic advancement and technological development, with little attention given to the impact on environment. The principal objective of the research on which this chapter is based was to find out the implications of lack of portable water and poor sanitation in inhibiting development. Yusuf and Abdul-Qadir argue that there is need to refocus on these two needs—water and sanitation—in order to provide lasting solutions in rapidly urbanizing Nigeria where the need for portable and proper sanitation are becoming ever more important. Because of rapid urbanization in Africa, this chapter provides an interesting dissection of what African cities need to do in dealing with sanitation problems.

In the tenth chapter, Ann Albuyeh deals with the link between language use and the environment. She shows that regardless of the language they use, a number of African writers celebrate nature and voice their particular environmental concerns. Albueyeh cites extensively from the writings of Igbo writer and activist Christopher Okigbo as well as Ogoni writer and activist Ken Saro-Wiwa. The statement, "The land is a god and is worshipped as such," provides a fitting background to Albueyeh's analysis of the role of Saro-Wiwa. Saro-Wiwa was an environmental activist who gained international fame and support for his Ogoni people during his fight against the Royal Dutch Shell Oil Company and a complicit Nigerian government. Albuyeh makes the point that Saro-Wiwa wrote in English, arguing that "however relevant language choice is to environmental concerns in the context of either artistic expression or protest, central to real environmental progress in Africa are questions regarding language and the debate over whose science and technology will have what role in Africa's future development." Does the language in which knowledge is conveyed matter? She suggests that it does, saying that "the goal of developing African environmental resources from North to South, East to West, will ultimately require the participation of all Africans highlights the fact that the language choices made will constitute a complex and critical component in the direction of Africa's environmental future." Her position is in consonance with governments and international environmental activists in Africa who have been involved in producing environmental brochures in all languages in order to reach the majority on the African continent. The use of indigenous language stations to propagate environmental information makes this recognition more explicit.

Ademola O. Dasylva in Chapter eleven continues the literary discourse on environment in Africa. Using a combination of Anthony Vital's suggested "African Ecocriticism" and interpretive approach, Dasylva examines, among other things, how African writers respond to the serious issue of environmental degradation and its implications on the people's psyche. Specifically, he examines the works of selected African writers such as Niyi Osundare, Tanure Ojaide, Ayi Kwei Armah, and Ken Saro Wiwa, and their discussion of environment. He analyzes the extent to which the writers' understanding, and subsequent interpretation, of the complexity and contradictions in African environment have been examined. He also examines the variety in African writers' responses, including protestation, and subversive droppings in his chapter. Dasylva suggests that these writers seem to strengthen Africans' historical self-understanding for sustainable and healthy ways of inhabiting their place and argues that through an analysis of these writers, one can detect that Africans have been aware of their environment and the changes that affected them. He interrogates how the selected African writers subordinate their postcolonial discourse to a literal and non-literal celebration of nature. The chapter further proposes an understanding of the textuality of the different forms of violence centered by ecological issues of degradation and death in the selected works.

Chapters in Part III of this book address issues of ecology, science, technology, and conflicts in the postcolonial era in Africa. In Chapter twelve, Shadrack Wanjala Nasong'o argues that there is an environmental dimension to conflicts in Kenya in particular and in Africa in general that has either not been understood or effectively incorporated into the development of conflict prevention and management strategies. Deploying powerful political science approaches, he contends that the causal variables that help explain the occurrence of social conflict—defined as antagonism and tension that escalate into open hostilities, fighting, and deadly clashes between social groups whether they are based on class, ethnic, racial, or religious factors—can be divided into three: underlying causes, intermediate causes, and precipitating causes. He argues that whereas the precipitating and intermediate causes of social conflict in Africa and elsewhere have been adequately covered in available literatures, the underlying causes have not. In regard to Kenya and the East African region in general, Nasong'o contends that most social conflicts have the ecological dynamic as their underlying causal variable. The chapter elaborates on the ecological roots of conflict in the "White Highlands," the Shifta war in Northern Kenya in the 1960s, the ethnic clashes of the 1990s, and the Kikuyu-Maasai conflict in Narok. The chapter concludes that unless this ecological dimension is taken into account in designing tools and techniques for conflict prevention, the management of social conflict will remain overly problematic.

Chapter thirteen, by Bukola Adeyemi Oyeniyi, provides a critical assessment of Nigeria's efforts at ensuring the promotion of science and technology since independence in 1960. His chapter seeks to understand what factors have militated against the development and implementation of sustainable science and technology policy in Nigeria. Oyeniyi faults successive Nigerian governments for their half-hearted attempts at promoting science and technology. He contends that, from independence, efforts at promoting science and technology have been lackadaisical and replete with organizational ineptitude and institutional corruption. To him, even the formation of many commissions of inquiry, seminars, symposia, and lectures have not helped matters. Oyeniyi suggests that the government has not recognized the importance of science and technology in development, saying that "the nation lacks the understanding, the political will and leadership to do what would bring about this noble ideal." He adds further that Nigeria lacks "the culture that is capable of ensuring the development of science and technology." The chapter concludes that in order to ensure that scientific development attains sustainable growth in Nigeria, it is necessary to tie it more directly with indigenous technological development and traditional science.

In the fourteenth chapter, Jolanda Morkel and Maria Leus provide an architectural interpretation of environmental issues in post-apartheid South Africa. They show how South Africa has battled with the effects of policy and legislation as well as the new democratic dispensation over the last two decades, as well as examine the relationship between social justice, education, and ecological sustainability. The authors provide answers to "post-apartheid" tensions and negotiations and look at how the post-apartheid governments have sought to redress imbalances of the past. The chapter interrogates "how place can be defined so that the emotional value of the past can be positively reinterpreted from a future-focused perspective." In what way does architecture, considering the meaning of inter cultural exchange, serves to contribute to cultural diversity? How can architecture make a contribution towards empowering and transforming communities by presenting sustainable and democratic learning environments? These are the questions that Morkel and Leus interrogate and seek to answer in this chapter.

In Chapter fifteen, Maurice N. Amutabi uses architecture to interrogate the environmental ethos among the Abaluyia people of western Kenya. He argues that there is overwhelming architectural evidence that suggests the Abaluyia have been conscious about their environment for generations. The chapter answers many question, including: How did indigenous architecture function in regard to environmental preservation and sustainable development? Was indigenous architecture user friendly? How does modern architecture affect the environment? Can we draw comparisons between indigenous and modern architecture with the view of acquiring more environmentally friendly approaches? Using important landmarks such as Chitambe 'Fort' near Webuye town, as well as burial sites of the *Nabongos* (kings) of Wanga kingdom at Elureko, and shrines of rainmakers in Ebusiekwe, Amutabi argues that these sites reveal a deep sense of knowledge of environmental ethos and careful planning. He argues that these archaeological artifacts provide useful information on how diverse architectural fundamentals have evolved among the Abaluyia and helped in preserving and conserving the environment. The chapter also examines worship sites (*olusambwa*), sacrificial platforms, irrigation terraces and canals, clay structures, pottery, huts made of wood and grass, mud mounds, and a mixture of numerous building materials to interrogate the relationship between architecture and environment. By looking at how these features were conceived and produced, and examining the architectural motifs, the chapter also reveals aspects of the cultural history of the Abaluyia people, which are of great use to conservationists.

Chapter sixteen, by Soji Oyeranmi, argues that the poorest of the poor Nigerians, as in other parts of sub-Saharan Africa, are the ones who are mostly affected by environmental degradation due to geographical locations, fragile economies, low incomes, a heavy reliance on climate sensitive sectors such as extractive industries and agriculture, and a lack of adequate modern conservation skills and facilities. Oyeranmi argues that the poor will still suffer more from environmental degradation compared to their rich counterparts unless urgent steps are taken to abate the crumbling of the environment. He asks and answers many questions in this chapter, which include: Is there any nexus between climate and economy? Is the cynicism about climate change real? What are the causes? Can the impact be positively reversed? Is there any link between poverty and climate change? Do they reinforce each other? What are the ways out of the impending climate doom for poor countries in Africa (especially Nigeria)? What are the implications of climate change on Nigerian economy? Oyeranmi asserts that dangers posed by climate change are nearly as dire as those posed by nuclear weapons. He argues that while climate change actually threatens all our futures, the threat appears to differ from place to place.

Part IV covers many themes. The first chapter of this section, by Ogundiran Soumonni, is Chapter seventeen. The chapter provides an interesting examination of the challenge of biofuels development in Africa, using an institutional analysis and development (IAD) perspective. He says that although biofuels have been promoted as having the potential to reduce the demand for fossil fuels, promote economic development, and mitigate the impact of global climate change, there is a lot more to these claims that needs to be thoroughly interrogated. Soumonni explains that this is because the development of biofuels has been marred by controversies and challenges, especially their viability as sustainable energy options for the future. One of the effects of use of biofuels is that they lead to scarcity of food as well as soaring food costs due to increased demand for ethanol. Soumonni argues that growing demand for ethanol is leading to higher crop prices, which could lead farmers to till more acres and remove environmentally sensitive land from conservation programs, thereby negatively impacting soil and water resources. He also asserts that the clearing of forests and savannas to be replaced with plantations of oil palm, soybean, jatropha, and other biofuel crops is causing a huge impact on the environment, especially the "carbon debt." Specifically, he says, that more carbon dioxide is released in this process than the reductions that can be obtained from using the biofuels. He seems to favor the use of cellulosic ethanol from crop wastes (stalks, cobs, leaves) or non-edible plants, which he says have the potential to reduce the destruction of habitats, reverse carbon debts, and lessen the competition with food crops. To use or not to use biofuels or fossil oil is currently very controversial and this chapter will no doubt contribute to the growing debate.

In Chapter eighteen, I. A. Ademiluyi and R. A. Asiyanbola examine the role of folklore and folk life in environmental education and sanitation in Yorubaland. They argue that indigenous methods of educating and creating awareness on environmental issues in Yorubaland before the advent and widespread of western education was largely through folklore and taboos embedded in culture. Ademiluyi and Asiyanbola assert that the advent of the modern science, with its perception of the environment, has rendered many of these taboos irrelevant in environmental considerations. They suggest that the indigenous ways of knowing are capable of raising awareness about environmental sanitation, which is a part of the general environmental education process. The chapter demonstrates that traditional education is pragmatic in nature and helps in solving everyday problems such as knowledge of important plants for treating ailments like stomach ache, headache, and malaria, as well as plants to keep snakes away from homes and those used to ward-off mosquitoes, among other uses. Ademiluyi and Asiyanbola argue that the approach of indigenous knowledge of learning by doing makes the transfer of knowledge and ideas much easier and effective and they call for a collective approach to solving environmental issues, saying that no "single learning system—formal, non-formal, and informal—can effectively cover the range of necessary awareness-raising." They show that indigenous knowledge provides a thorough knowledge of plant species, and that the medicinal uses of plants, soil, and water, and other uses of the environment, are contained in the local or indigenous education and, therefore, should be appreciated. Thus, they suggest that "the methods of indigenous environmental learning—folklore, folksong, folk drama, etc.—be integrated into the national education curriculum to supplement formal Environmental Education System." Ademiluyi and Asiyanbola's chapter will certainly help in stimulating interest in an important debate because indigenous knowledge systems (IKS) and indigenous education (IE) are being revisited all over Africa as alternative ways of knowing and learning.

Adam Carpinelli, in Chapter nineteen, examines land rights among the Saramaka people of Suriname. The chapter uses field interviews conducted among the Saramaka in August 2006 regarding their land rights case brought to the Organization of American States. He shows how the Saramaka, who are people of African descent living in the interior of the Amazon rainforest, have engaged many forces that they have encountered. Carpinelli explains how these people, whose ancestors were enslaved Africans, settled in the forest after fleeing from plantation slavery and formed their own autonomous communities are an enduring example of survival and taming of the environment. He shows how, after over two hundred years of a fairly harmonious existence, they have been threatened by a new colonialism in the late twentieth century, that is, neoliberal globalization, and how they are dealing with this new threat. The chapter addresses how socio-economic and political changes impact human memory among the Saramaka since the era of the slave trade, with an emphasis on cultural links between Africa and the Americas. He interrogates the way neoliberal globalization is affecting the Saramaka people and their natural environment, especially deforest by new global demands, particularly by Chinese loggers.

In Chapter twenty, Nnenna George-Kalu and George Kalu analyze the current environmental protection efforts by governmental and non-governmental bodies in Nigeria in the light of current global awareness of ecological issues in the Niger Delta and other parts of Nigeria. They argue that although the advent of oil discovery and prospecting in Nigeria was heralded as a welcome development with great prospects of economic benefits, this has not translated into positive development for everyone in the country. Oil exploration has created environmental disasters. The authors point out that civil society organizations, government agencies, and the oil-prospecting companies have not succeeded in ending oil-related environmental crises. Kalu and Kalu suggest that a precautionary approach, as opposed to a reactionary system, would enable both the remediation of existent abuse and the prevention of future damage with sustainability as the goal. They advocate a shift from the present emphasis on industrial and large-scale polluting activities to environmental control that focuses on the quality of life of Nigerian masses.

Chapter twenty one, the final chapter in the book by 'Lai Olurode, examines issues of gender, work, and technology in Africa, focusing on rural areas. The chapter pays attention to ways in which the introduction of technology has altered the gender balance in rural Africa, thereby fostering collaborative activities between men and women. The chapter uses case studies from Kenya and Nigeria. Olurode argues that gender stereotypes and discrimination have been with society for a long time and have persisted in spite of social change dynamics. Olurode justifies his focus on rural areas saying that "There cannot be a better platform for assessing the gender differentials in access to technology in Africa than in rural production since this sector remains the most resilient of Africa's political economy by being a major provider of employment for its teeming population." He explains that this is also the sector in which more women are found and have been "able to sustain their visibility and thus warded off marginalization." This is a good point, for it recognizes the gains that women have made in African countries in recent past. For example, there are now ministers of gender and women issues. He argues that technology has been altering gender perception about work and that it has served as a bridge in the worlds of work of men and women.

All of these chapters collectively have made three main contributions. First, the book fills an existing academic lacuna in the area of environment in Africa. There are a few books that have used interdisciplinary approaches in studying environmental issues in Africa. This is what gives this book special status, because it seeks to fill an important scholarly vacuum in African studies covering issues of environment. Second, since 1992, after the Earth Summit in Rio de Janeiro, there has been tremendous amount of research on environment, but mainly by professional environmentalists interested in issues such as global warming and depletion of the ozone layer. Thus, the study of environment has not enjoyed research interest among many social scientists in the past, largely because it was seen as a preserve of biological and physical scientists. This is a change that this volume contributes to, with various chapters analyzing environmental changes in Africa using data based on empirical research drawn from different fields. Citing relevant examples, the contributors demonstrate that environmental issues are not a monopoly of biological and physical scientists, but social scientists as well. Finally, the book has presented a coherent and logical account of environmental changes in Africa, especially those that have previously been ignored in the past. The fact that the contributors to this volume come from different disciplines, such as engineering, political science, history, anthropology, education, and women and gender studies, allows them to provide a holistic study of African environment.

Part I

REVISITING AFRICAN SCIENCE AND TECHNOLOGY

• Chapter 1 •

PRECOLONIAL AFRICAN IRON METALLURGY: METHODOLOGY, SOCIO-CULTURAL CONTEXT, AND DECLINE

Aribidesi Usman

Introduction

Africa, in early European writings, was portrayed as "backward," contributing nothing to the technological innovations of the world.[1] As Clark put it, Africa was treated as a "cultural museum in which archaic cultural traditions . . . continued . . . without contributing to the main course of human progress."[2] For a long time, there was no interest in the study of Africa's past. And when Europeans finally began to do so, the interest was mainly in topics related to early hominids, a direct consequence of the discoveries of Australopithecus species by Raymond Dart and Robert Broom in South Africa between 1924 and 1947, and by Louis Leakey in East Africa after 1931.[3] However, the focus in the Stone Age of Africa discouraged archaeologists from taking an interest in the later period or Iron Age of Africa. The belief in the relatively recent introduction of metals to Africa,[4] may have also contributed to the delay in studying the later period of Africa's prehistory. However, since the 1960s, following the independence of African countries from colonialism, there have been increasing interest and renewed efforts to research the later period of Africa's past.

Africa is rich in various kinds of metals such as gold, copper, tin, and iron. All of these have been exploited in significant quantities by Africans for thousands of years before the arrival of Arabs and Europeans in Africa. Metals were important commodities of interregional and intercontinental trade during precolonial period. Gold was obtained from the Red Sea hills and Nubia during the Pharaonic, Kushite, and later periods.[5] In the western Sudan, Gold from the Niger River and from modern Ghana reached North Africa and Europe as early as the eight century A.D.

and contributed to the extraordinary wealth and fame of the indigenous African states and empires of Ghana and Mali. Gold mining from Zimbabwe, which was traded as far as the Persian Gulf, Indian, and China contributed to the rise of Great Zimbabwe and other Zambezian states early in the second millennium A.D.

Copper, tin, and iron working was widespread in Africa and their exploitation formed an integral part of the fabric of African societies in the past. Copper in Zaire and Zambia, for example, was used in making ceremonial and utilitarian objects. The bronze work of Ife and Benin in Nigeria demonstrate the vitality of this precolonial African metals. Tin came primarily from Bauchi plateau in central Nigeria as well as from eastern Zaire. Tin were mined and used in bronze casting. The mining of tin in north central Nigeria have led to the discovery of clay figurines popularly known as Nok Culture. Iron working and the making from iron of agricultural tools, weapons, and utensils took place in many parts of Africa in precolonial period. The technology of iron working was practiced in Africa between the ninth century BC and 200 AD.

This chapter examines iron metallurgy in pre-colonial Africa using archaeological, historical, ethnographical, and oral traditions information to understand the processes of spread of iron, the technological practices, socio-cultural and economic context, and the decline of traditional iron working. In this chapter I utilized information from my research in north central Nigeria as well as from other works carried out in different parts of Africa. The almost two decades of my research in north central Nigeria, has produced, among others, substantial evidence of iron metallurgy. This area, known as northern Yoruba, is the home of Yoruba groups such as Igbomina, Ilorin, Ibolo, Ekiti, and Okun. The non-Yoruba groups like Bariba, Nupe, Igala, Ebira, and the Kakanda, share borders with the northern Yoruba (Fig. 1.1).

Archaeological research conducted in various parts of Africa in the last three decades has produced considerable amount of data on iron working. As a result of this new information, the analysis of iron technology in Africa has become less simple and less straightforward. Old assumptions about African iron working are being re-evaluated in the light of new information. It has been claimed that "the cultures antecedent to many modern ethnic groups seem to have begun within that period that archaeologists in Africa termed the Iron Age."[6] The people demonstrated ability for converting iron ores to metal and shaping them into tools and weapons for use in their daily activities. Some scholars even consider the Iron Age as the actual beginning of African history.[7] In other words, "the 'Iron Age' was not some ahistorical descriptive entity but rather a vital part of Africa's past."[8]

Two approaches have been employed in the study of African Iron Age. The first, which is essentially cultural in perspective, focuses on the economic, social and political aspect of iron using communities and the cultural development of

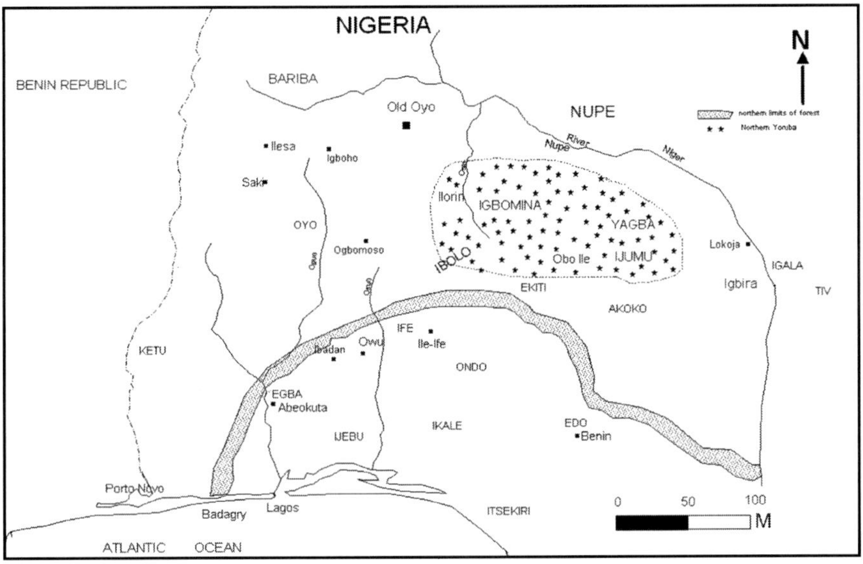

Figure 1.1: Map of Yorubaland showing Northern Yoruba area

Iron Age societies in Africa since about the middle of the first millennium B.C.[9] The second and more recent approach concentrates on the technology of iron production as well other factors (e.g., economic, cultural, social, and environmental factors) that are indirectly related to iron working.[10] This later approach, which has generated more interest and results in recent years,[11] is followed in this chapter.

The Development of Iron Technology and Distribution of Sites

The debate on the development and expansion of African metallurgies is one of the interesting scholarly focus in the history of African archaeology.[12] There is no agreement among scholars on the origins of African iron working. However, recently two schools of thought have emerged, the "monophyletic" and "polyphyletic."[13] These two are discussed below.

The 'monophyletic' argument, which is somewhat diffusionist, recognizes a single center, and beliefs that iron metallurgy occurred between Armenia and the Anatolian plateau during the first half of the second millennium B.C.[14] From this center, knowledge of iron working diffused to nearby and distant communities before arriving finally in sub-Saharan Africa.[15] This argument has often been supported by scholars who did not believe in the idea of parallel and independent invention of iron working in sub-Saharan Africa. Phillipson, for example, wrote that "the associated technology is so complex, and in earlier African societies no other process involved heating material to such high temperatures, we have to consider a northerly source of sub-Saharan iron-working knowledge rather than duplicate independent discovery."[16] He was of the opinion that prior knowledge of metallurgy, and especially a complex and high standard pyrotechnology used in heating material to high temperature, was essential or necessary for iron working technology to take place. Mauny has expressed similar belief when he says "it seems impossible that an invention of iron technology independent from the Mediterranean zone occurred there. This technology is so complex that it could not have emerged from a context totally ignorant of metallurgy."[17]

As the above views suggest, for iron working to be conducted in sub-Saharan Africa, it had to be introduced from outside the continent or from one part of the continent to another. As a result of this, archaeological research in Africa has been aimed at identifying dispersion and diffusion centers of iron metallurgy. The two centers of diffusion often suggested are Meroe located in the upper Nile valley, and Carthaginian North Africa on the Mediterranean. However, the primacy of the Nile Valley for transmitting knowledge of iron technology to sub-Saharan Africa has been refuted.[18] Similarly, the argument for Meroe has been challenged.[19] For those who subscribe to this school of thought, the Carthaginian hypothesis is now the most likely theory for the introduction of knowledge of iron working to sub-Saharan Africa. However, no substantial archaeological evidence has been made available to support this idea.[20]

The second school of thought, the 'polyphyletic' hypotheses, is based on local, indigenous or autochthonous development. Beside the archaeological evidence employed to support this position, the proponents of this hypothesis point to the abundant iron ore sources found throughout Africa and the variations of iron working traditions in Africa.[21] According to this school, there is almost lack of ethnographic record demonstrating iron working skills by the Berbers of the Sahara desert. As a result, it is argued that iron working was an independent invention in sub-Saharan Africa, since many of the practices related to the technology were different from those obtained in the north.[22] Following the review of the theories of the origins of iron technology in Africa, Diop has this to say: "Today, it thus seems proven that in Africa, traditional iron metallurgy is very old, widespread and autochthonous; however, it is still needed to identify the cradles of this metallurgy, to date them more securely, and to highlight the hypothetical routes of iron throughout the continent."[23]

The slag heaps at Meroe in Sudan have become important point of reference for many scholars. In fact, Meroe has been described humorously as the Birmingham of Africa.[24] A trench excavation carried out on one of the slag mounds gave the size of the mounds estimated at a depth of about 12 feet (4 m).[25] The investigation by Shinnie has provided further information on iron working techniques employed and the chronology of iron working. It has been suggested that Meroe was already an urban center before it became the capital of Kush, and slag from iron smelting has produced a radiocarbon date of sixth century B.C.[26] Furthermore a temple, dated to about 246-266 A.D., was built on top of a slag mound, indicating that the slag mound was older than the temple. From the available data, it is estimated that iron smelting producing the large slag mounds occurred within this period.[27]

An excavation at the site of Do Dimmi in Niger has produced evidence of early iron working dated from the mid-ninth century B.C.[28] The Nok Culture sites of Taruga and Samun Dukiya in central Nigeria have provided one of the most elaborate evidence for iron working in Africa.[29] A large series of carbon and thermoluminiscent dates obtained from early furnaces suggest that iron smelting was established between 800 B.C. and 500 B.C.[30] Nok Culture is associated with the distinctive stylized terracotta figurines whose eyes are triangle-shaped with the pupil represented by a pierced hole. The population of the Nok region practiced agriculture and used iron and stone tools. In Opi near Nsukka in southeastern Nigeria, a date of 500 BC was obtained from an iron-smelting furnace, and considered to be very close to the Nok (Taruga) dates.[31]

At Daboya site in northern Ghana, archaeological research has indicated that a Terminal Late Stone Age component (Kintampo) was replaced by a distinct pottery style said to be associated with iron-using culture, even though no iron materials or relics has been found.[32] This tradition has been radiocarbon dated to between the mid-first millennium B.C. and the early first millennium A.D., and represents the first well established Early Iron Age culture in West Africa outside of the Nok culture.[33]

In Gabon, archaeological investigations of smelting furnaces from Otoumbi site have yielded dates from 700 B.C. to 450 B.C. in the area of Moyen-Ogoue.[34] In the Central African Republic, a number of settlements as well as complex system of iron working dating from the ninth or tenth century B.C. has been documented by Zangato.[35] The West-Central Africa, particularly the Zaire Basin region, is important for the several postulations of the origins and movements of the Bantu-speakers and the diffusion of iron technology. A number of sites have been investigated in the region with ceramic types that perhaps suggest Early Iron Age dated to between the fourth century and the first century B.C.[36]

In the Great Lakes region of eastern Africa, iron working sites date from around 1450 B.C. onward at Rwiyange I, Mubuga V, Kabacusi, and Gasiza I and Mirama III in Rwanda and Burundi.[37] One of the most important of the evidence for early iron working is from the region of BuHaya in northwestern Tanzania.[38] These sites have demonstrated a high standard iron technology with series of dates that ranged from the second millennium B.C. through to the early first millennium A.D.[39] The interest in Iron Age societies in South Africa over the last three decades has led to recovery of substantial data that helped in the understanding of the later prehistory of the region. It is observed that iron using population appeared in the northern Transvaal of South Africa by the late third/ mid fourth century A.D.[40] and also on the Natal coast at about the same time.[41]

The rapidity of the spread of iron using societies southward from eastern and central Africa between the late centuries B.C. and early centuries A.D. has been attributed to the Bantu speakers. The Bantu are groups with close cultural and ethnic identity. It has long been recognized the similarity in the language of most of eastern and southern Africa.[42] One theory provides a link between the expansion of the Bantu and the beginnings of the Iron Age. The working of iron encouraged agricultural production and provided effective tools that allowed the Bantu group to dominate the peoples in the areas where they settled. The Bantu "were a dominant minority, specialized to hunting with the spear, constantly attracting new adherents . . . by their fabulous prestige as suppliers of meat, constantly throwing off new bands of migratory adventures, until the whole southern sub-continent was iron-using and Bantu speaking."[43] In a way, Bantu expansion has been associated with the spread of agriculture, iron working technology, and establishment of complex societies in the southern and eastern Africa.

Finally, there was spread of Iron Age communities in areas where sites had already indicated the early presence of iron working. From the first century B.C. to about the ninth century A.D., a number of sites have been found in West, Central and East Africa. Examples include Rim in Upper Volta (Burkina Faso), Begho and Bono Manso in Ghana, and Koro Toro in Chad,[44] and in Cameroon the Tsanga II site at Maroua[45] and Rambura and Kabuya in Rwanda.[46] In northern Yoruba (north central Nigeria) archaeological investigations in the Ookun area indicate a continuous stratified sequence of occupation from the Late Stone Age deposits ca. the fourth century B.C. into the Iron Age period in the ninth century A.D.[47] This suggests that iron-producing agricultural communities were occupying the region at the beginning of the first millennium A.D. The substantial quantity of iron slags and tuyerès, and iron arrowheads from the lower iron-bearing stratum at Oluwaju rock shelter in Iffe-Ijumu, indicates that iron-working and iron-using activities started about the same time. Similar evidence has been reported at Obo Aiyegunle where radiocarbon dates clustering around the ninth century A.D. have been obtained for iron smelting.[48]

Iron Working Processes, Materials and Features

While there are variations in the working of iron in Africa, there are also many similarities in the procedure for working iron as well as the result. Two examples of iron working processes will be discussed here: Northern Yoruba (Nigeria) and Darfur (Sudan). Generally, iron working is a complex technological process that involves different stages of production: (1) prospecting and mining of iron ore, (2) smelting, and (3) smithing or forging.

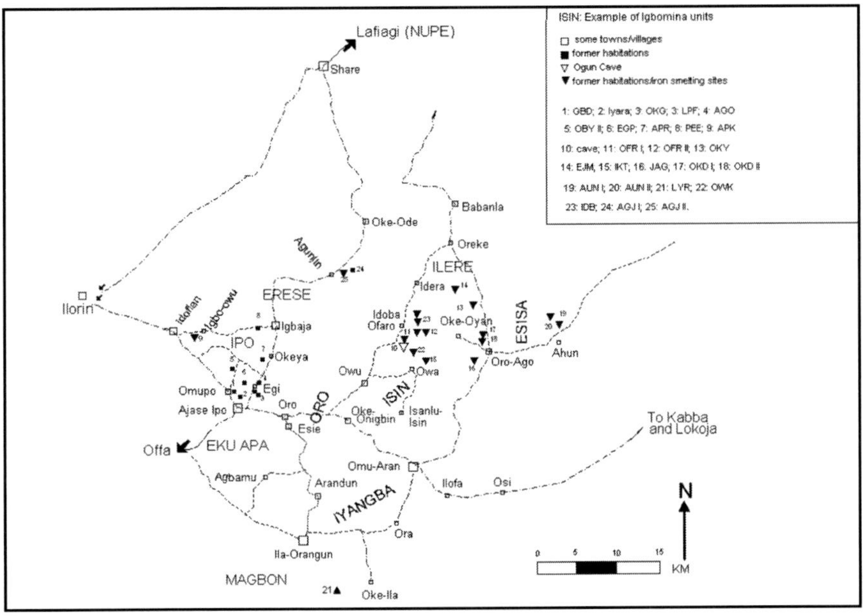

Figure 1.2: Map of Yorubaland showing Northern Yoruba area

Iron working (smelting and smithing) in Yorubaland was an extended family occupation with members congregated in areas with good quality iron ore. The type of sediment that yields the ore was a composition of laterite, quartz, and gravel of nodular particles of haematitic iron. Open-pit mining was conducted along riverbanks, especially during dry season. The ferruginous shale was mined from below the surface by means of a pick.[49] This ferruginous soil type in Yorubaland is called *egun*, and in Idofin in northern Yoruba, "the family praise name of a smelter includes *omo asawó l'órí egun*, meaning "the child of one who makes money from smelting ferruginous red soils."[50] It has also been claimed that since clay for pottery was obtained from the same source as iron stone, the origin of iron working might be connected to pottery making.[51] Early pottery furnace could reach temperatures similar to the minimum required to produce an iron bloom.[52]

The iron stone used by the Darfur smelters was ferricrete sandstone obtained from Jebel Tayu, a sandstone hill situated ca. 35 kms north west of Toumra.[53] Although located farther from the village, this was the preferred iron ore because of its high iron content. On the other hand, the ferricrete sandstone at Jebel Moro, and only about 13 kms to the village, has less iron content and only gave one third of iron.[54] The sandstones or iron stone are usually large lumps that must be crushed into smaller pieces before being poured into the furnace. In Darfur, the crushing of iron stone into smaller pieces was an activity reserved for women, and the only female task in the whole smelting process.[55]

In the Igbomina area of northern Yoruba where archaeological research has unearthed large evidence of iron working,[56] only little information on the process of iron smelting is preserved in the oral tradition (Fig. 1.2). In this area the practice of iron technology, particularly iron smelting, was attributed to non-indigene of the community. Oral traditions from Idera village suggest that ironworking was introduced and practiced at the old site of Oke-Oyan by a man called Ali who came to settle in the area. The smelting of iron was carried out in a shallow pit. *Adudu*, a circular clay structure, was built around the pit with firewood set into it, and tuyeres (clay pipe) were inserted to 'ewiri' (bellows) to blow air into the clay structure.[57] Some archaeological research in the area has reported evidence of iron smelting furnaces at some sites with walls standing up to 40 cm and between 1 m to about 2.5 m in diameter.[58] According to oral tradition, iron smelting usually began early in the morning till sunset, during which time no one was allowed to come near the smelting areas.

From the above, it is difficult to get a clear picture of the furnace used in northern Yoruba because of the scarcity of information. However, documented record of furnace types in the surrounding areas of northern Yoruba and elsewhere in Nigeria are available and might be useful in understanding the type of furnaces used in the region. The furnaces that have been found in various parts of Nigeria are mostly of the shaft and domed type.[59] Smaller cylindrical furnaces are found in northern Nigeria among Hausa blacksmiths from Tureta, Kurigi, Potiskum and the Pawa-Bakori area south of Katsina.[60] One distinct practice in the area was the use of clay pots in smelting.[61] Among communities in the hills of the Jos Plateau, the common form of furnace was a pit covered in a pot-like shape .with a hole in it.[62] Information from Afemai and Erhurun near Okene in the Niger-Benue Confluence shows that the furnace used was a clay-lined pit somewhat similar to that described from the Birom and also observed from the Dakakari.[63] Pole, describing Bellamy (1904) investigation of smelting processes at Ola-Igbi, 25 km north of Oyo in south-western Nigeria, states that "the furnace was mostly subterranean, with a conical shaft above ground domed on the outer surface. Below the pit under the shaft was a tunnel which provided access to tapping slag through a hole at the base of the pit."[64] Here, the greatest interior height of the chamber was about 5 feet (150 cm), and air pumped into the furnace through nine tuyere pipes while smelting lasting about 36 hours.[65] For northern Yoruba, it seems the furnace used was probably a clay-lined pit of the Niger-Benue Confluence or the conical shaft described for the Ola-Igbi, or a mixture of these two types.

In Darfur, Sudan, the furnace type employed by smelters is a shaft furnace.[66] Because of its nature, the slag from the smelt falls to the bottom of a pit with the bloom and then separated after the smelting. The construction of furnace was easy and quick, taken less than two hours.[67] The shaft furnace built in mud was less than 1 m high, with the walls about 4-5 centimeters thick.[68] A number of

Figure 1.3: Ironworking, Outcrop

holes (between 7 and 10) were created around the furnace to hold the tuyeres (cylindrical clay tubes) which measure about 30 cm. As Haaland indicated, the 10 holes with tuyeres were engaged by 10 men operating the bellows. The bellows which are made either of sheep or goat skin are family heirlooms and constitute one of the most important apparatus of iron working.

The Darfur smelters obtained the charcoal needed for smelting from two species of a tree called garad (Acacia nilotica and Tamarindus indicus).[69] These are hardwood that last longer in the furnace and could provide enough temperature for smelting. The trees used for charcoal are not available near the village. The wood had to be obtained from the forest and brought back to the village to be made into charcoal. Once the ore is crushed and the charcoal is ready, the furnace is prepared for the smelting to start. As described by Haaland,[70] at the bottom of the shaft furnace the smelters placed stalks of millet, wood, charcoal, and ore, and finally charcoal is again placed on top. The furnace is then lit through the tuyere and from the top of the furnace. The men gradually work the bellows by pumping it. After one hour additional ore and charcoal were added and this continued for another four hours when no more ore was added but the working of the bellows continued. The whole smelting exercise takes around 16 hours. The men working the bellows were relieved by other men to rest, eat, and drink, before they again took turns. When the smelt is completed, the furnace is destroyed and the smelters will bring out the day's efforts.[71]

The process used for producing iron throughout Africa during the precolonial times is known as bloomery. To put it simply, iron smelting was an act of producing 'bloom.' 'Bloom' is a rough mass containing nodules of metallic iron, slag waste, partially reacted ore, unburnt charcoal, and parts of lining of the furnace. The bloom was removed from the pit and allowed to cool off. In northern Yoruba, the bloom from the furnace was crushed on rock outcrop or bedrock located around the smelting area to remove the impurities and sorted to recover the metallic iron (Fig. 1.3). In some areas of Africa, bedrock hollows have been associated with ironworking, especially when found in the vicinities of iron working or at settlements with tradition of iron working. Ethno-archaeological study by Nicholas David at Sukur in northern Nigeria suggests some bedrock hollows as 'fining complexes' connected with iron working, especially for processing of blooms.[72] On the Jos Plateau in Nigeria, the core (or bloom) from the iron smelting furnace is pounded thoroughly with a stone hammer in a rock depression to remove the impurities (charcoal, slag) before they are handed over to the blacksmith.[73]

In Yorubaland, iron smelters rarely combined smithing with their own activities. At the end of the smelting the raw iron and the waste product were carried out, and the raw iron from the furnace was passed to the smiths who then make them into various objects. In Darfur, since iron smelting is also performed by the blacksmiths who also produce the finished tools and weapons, the separation of the bloom from the slag and other waste was carried out in a smithing forge.[74] The smith would heat the raw metal in a smithing furnace, until the metal was red hot and soft. It was then hammered on a rock or iron anvil and consolidate the metal fragments to a usable size. Generally in sub-Saharan Africa, smelting was a dry season activity. This is the time of the year when there are less agricultural works to interfere with the iron production. Also, dry season was preferable because rain and flooding did not interfere with the preparation of charcoal, the collection and transport of ore, the lightning of the furnace, and other activities related to iron smelting.

Archaeological research in northern Yorubaland has provided surface evidence of iron working in the form of slag and slag mounds, tuyere fragments (Fig. 1.4), blacksmith workshops with smithing apparatus (e.g., stone anvil) for iron-working (Fig. 1.5). Also, excavation at a number of sites in the area has yielded several pieces of iron slag, tuyere fragments, and metal objects. These materials are associated with a cultural level dated from the fifteenth century A.D.

Innovation in African Iron Technology

It has been recently suggested that iron produced from traditional African smelting furnace was of sufficient high carbon content to be considered steel.[75] According to this investigation, the medium to high level of carburization was achieved through the actual smelting process in the furnace.[76] These high carbon

Figure 1.4a and 1.4b: Ironworking, Tuyere

Figure 1.5: Smith Workshop

blooms were achieved in two ways in Africa. One is by inserting the tuyeres fairly deep inside the furnace structure, and the second way is having the other end of the tuyeres adjacent to the bellow's mouth, but not directly attached.[77] This innovation permits more air to enter naturally through the tuyere than was possible through the bellows alone. It has also been claimed that the length of insertion of the tuyere into the furnace allowed the preheating of the air in the tuyere up to 500-600°C before it passed through combustion chamber.[78] Earlier oral interviews and conversations with the Nyungwe smelters by Peter Schmidt indicated that Haya smelters in northwestern Tanzania positioned tuyeres far deep inside their furnaces.[79] Schmidt further conducted a complete ethnographic and technological study of the Haya ironworking process to test the preheating hypothesis. A series of six replicated Haya smelts were observed by Schmidt and Avery in 1976.[80] According to Schmidt and Avery, preheat temperatures of 500-600° were recorded for air inside the clay tuyeres in the furnace. Thus, the link between preheating and high furnace temperature was established, and that preheated air accounted for the operation of the blast zone immediately in front of the tuyeres at an excess of 1800°.[81]

Another important innovation in African iron working was the use of tall shaft furnace that relied on natural draft. In this case, air was naturally drawn into a number of holes around the base of the furnace and thereby excluded the need for people working the bellows.[82] Natural draft furnaces have been associated only with iron smelting. A recent study in Yatenga (Burkina Faso) has also documented the development of shaft furnace type to a larger size of 7 m high.[83] It has been suggested that once it is set up, it required no labor to operate it, and produce larger quantities of bloom that is carbon steel.[84]

Both of these processes are African innovations in iron metallurgy. The implications of this result are far reaching. Schmidt argued that the iron smelting technology in the BuHaya area was the same that had been employed to the excavated furnaces of more than 2000 years ago.[85] With preheating, the operation of preheated furnaces in Africa likely produced iron in a process altogether distinctive from Europe, and, more efficiently.[86]

Socio-Cultural Context of African Iron Working

The development of iron working in sub-Saharan Africa is best and most fully understood when placed in sociocultural context. African traditional iron working incorporates some elements of ritual and magic. Among the Luba, Ruund, Tshokwe, Luchazi, Rotse, Shona, Karanga and in Rwanda, furnaces are ornamented with breasts and regarded as a woman in labor.[87] Also, pregnancy is likened to cooking, the child undergoing an intra-uterine baking.[88] Ritual practices appear to be more common in iron smelting than in blacksmithing. On the other hand, rituals connected with political power are rarely found in smelting. In those ritual ceremonies associated with power, it is the forge (i.e., smithing) that has been very prominent.[89]

In most parts of sub-Saharan Africa, iron technology plays important role in origin myths. Common theme in oral traditions of some African groups is one that portrays the iron worker or smith as a divine hero or one that made 'civilization' possible. It has been suggested that the founder of the Kongo kingdom was a "master of smitheries."[90] The Tio king derived his prestige partly due to his association with smithing.[91] In Burundi, oral tradition also links the first Mwani, Ntare Rushati to the introduction of iron working.[92] The founder of the Luba state, Kalala Ilunga, is connected with iron and iron making.[93] It has been claimed

Figure 1.6: Ogun Cave

that the final 'rite' of a coronation, before the new ruler ascend his throne, is called "striking the anvil."[94] Similarly, in the oral tradition of the Kuba, an important episode known as "the hammers and the anvil" links iron technology with the establishment of kingship.[95] These types of traditions can be interpreted literally to mean that the political structures in some of these societies were founded by a smith. An alternative explanation was that iron workers, either smelters or smiths, had brought the major iron deposit in the area under their control. The example discussed below from the Yoruba area of Nigeria will provide further perspective on the ritual-religious connection of African iron working.

Ogun is one of the Yoruba deities or orisha popularly refer to as the god of hunting, iron, and warfare. The widely acknowledged opinion is that Ogun probably accompanied the spread of iron-making technology throughout sub-Saharan Africa as early as 2,000 years. This is linked with the beliefs that iron is sacred, that ironworkers are special members of the community whose profession of making iron made them to be feared or revered, and that workshops for ironworking are shrines or place of refuge for the dispossessed. It has been suggested that sacred ritual and its ideology may have been essential to ironworking as a way of remembering and continuing the steps and ingredients involved in the iron-working process.

The northern Yorubaland exhibited a strong mythology of Ogun. Oral traditions of Ogun and its association with the earth tradition have been pro-

Figure 1.7: A Cave Site

vided in the area. This fits well with the Yoruba myth of Ogun disappearance, by sinking into the earth, and with the close association of the Idoma Earth cult with hunting. The disappearance theory is often associated with mythical or legendary figures in Yoruba oral tradition as a way of expressing the transformation of an individual into orisa (deity). Such theory has been expressed for Sango, one of the early kings (Alaafin) of Oyo and the god of thunder and lightning in Yoruba pantheon. A certain cave located around Owode Ofaro and Owa Onire villages in Igbomina region has been used to explain Ogun disappearance (Fig. 1.6). According to the oral tradition, Ogun, accompanied by his dog, visited the two villages in the past. At the end of the visit that lasted several days, Ogun departed through a cave then known to the people as '*àjà ilẹ̀*' (underground tunnel). On his departure, Ogun informed the people that whenever they were faced with any difficulty, they should visit the cave and "call his name and he will answer them." The cave consequently became a shrine where sacrifice was made to Ogun from time to time. This practice continued until the seventeenth century when Ofaro and Owa villages were relocated. However, the ritual sacrifice to Ogun did not cease completely in the area as it was continued at the new settlements that were formed (e.g., Ofaro II, Owa-Orioke). In the nineteenth century, the worship of Ogun stopped as people were further displaced by wars and the conversion of the people to new religions –Christianity and Islam.

The Ogun cave's entrance (or opening) is about 1.6 m wide and 16 m long, while the cave's height varies between 1 m to 1.5 m, and the width about 40.3 m.

Figure 1.8: Iron Hill

The length of the cave and the deep interior could not be determined because of religious taboos. From the little examination conducted, the cave contained several small chambers, and on the surface of the cave floor were rock fragments and powdery sediments from the weathered rock (Fig. 1.7). Also, scattered pieces of pottery on the floor of the cave indicate evidence of past human activities inside the cave. A 1x1 m test excavation was carried out to examine sub-surface deposits. Materials recovered were mainly rocks, some pottery and small charcoal sample found below 0.3 m depth. More than 200 pottery sherds were collected from the cave (excavation and surface collection). Beside charcoal, no other organic remains were recovered in excavation. The excavated charcoal associated with pottery has produced a radiocarbon age of 410±60 BP, calibrated (2 sigma 95% probability) 1420 to 1640 C.E. (Beta 202615). This date is contemporaneous with the habitation sites of Ikotun (Owa Onire) and Ofaro I (Owode Ofaro) dated 405±100 BP (A13056) and 220±70 BP (Beta 88412) respectively.

The evidence linking the cave to iron working and probably to ritual use appears fairly strong. The cave was possibly a mine for iron ore or haematite rock, most of which littered the cave floor and extended to the outside. Extensive mining probably created the various chambers in the cave. The cave was an extension of large iron-bearing hill called '*Ipa Ogun*' (literally means 'path of Ogun') which is also dotted with pieces of haematite (Fig. 1.8). Iron ore-bearing haematites were mined in the cave and then transported elsewhere to be smelted in a furnace. The cave as a mine for iron ore later embraced the ceremonial or ritual function connected with the mythology of Ogun. This is similar to evidence from elsewhere in Yorubaland and in West Africa where forges are likened to earth shrines and the smith's status to that of the earth priest.

Spatial Relationship of Smelting Sites and Settlements

Unlike smithing workshop or forge, iron smelting sites are usually situated at the outskirts of the village. We observed during our research in Igbomina this discrepancy in the location of smelting and smithing forge in relation to habitation area. Idoba old site in Ilere is divided into two activity areas: Amu (iron smelting center), and the habitation proper known as Idoba. The distance between the two is about 1.5 kms. Other examples of locating iron smelting away from residential areas in Igbomina have been found at Owa Onire, Oke-Oyan, and old Oreke site. While alternative explanations for locating smelting sites off the residential area can be provided (e.g., safety, comfort, and proximity to source of fuel and other material), taboos and other rituals associated with iron smelting have been widely emphasized. This practice of locating iron smelting away from the habitation center may concur with local Yoruba belief that the furnace must not be too close to (nor too far removed from) human habitation. If it is too far away, according to local belief, the voices of the people will not penetrate the iron in the fire to give it the ringing quality that makes good iron.

In other parts of Africa, the location of smelting site at some distance from the habitation precinct follows other taboos. The Banjeli (Cameroon) emphasized sexual taboos in all stages of iron smelting, from mining to the construction of the furnace, and the preparation of charcoal to ensure that no contact was made with sexually active female. Also, excavation at Olifantspoort (South Africa) suggests that while forging was carried on within the village, smelting was not. Another reason for this practice is seeing in the parallel drawn between smelting and parturition. Liking it to the woman withdrawing to give birth to a child, smelting is conducted outside the habitation area of the village in order to keep the technical process secret as well as for symbolic discretion. However, probably not all African iron working follow this practice of situating iron smelting in the outskirt of residential areas. It has been documented that smelting at some Early Iron Age sites in Zambia and central Malawi were carried out within the village center. This raises the question of whether rituals and taboos as observed in African iron working was a late development, or was simply another example of the complexity of African iron working and the diversity and variation in African culture and practices. Generally smelting is usually carried out on the outskirt of village while the forge (smithing) in a relatively public place where villagers meet and chat and observe the smiths at work.

Iron Workers and Status

Blacksmithing is a specialized profession in most parts of Africa and it is an occupation universally performed only by men. In Darfur, the skills and knowledge required for the activity is transmitted within the family context and recruitment to the occupation is from father to son. In Yorubaland, and among the Nupe, any one could smelt iron, but the smiths because of their specialized knowledge, were 'a people apart.' In some areas of West Africa as well as in Sudan, the blacksmith occupation is organized like a caste structure. Being a blacksmith means having a certain inherited identity. Also, possessing this identity implies severe limitations or restrictions with other non-smithing inhabitants of the community. Thus intermarriage is prohibited outside the blacksmith group who now constitute an endogamous group. In Darfur, the segregation is manifested in beliefs about blacksmith's power to control thunder and lightning. Thus it is generally believed that people who have been exposed for crimes seek the service of the blacksmith to punish the offenders by a strike of lightning. This scenario is different among the Yoruba where blacksmiths are not clearly segregated or belong to caste system, and where the control of thunder and lightning and its wrath was not associated with Ogun (god of iron) but with Sango. In some other societies the blacksmiths are given high status. This has been reported among the Kpelle in Liberia and among the Suku of Congo. The stigmatization of blacksmith in societies that practiced occupational segregation was a major constraint both for members of blacksmith family and outsiders. For example, it prevented individuals from blacksmithing lineage to assume other identities, and skills in iron working acquired through socialization within the family units provided them with an occupational asset which they rigidly protected. In the same way, this stigmatization also reduced the interest of outsiders from non-blacksmithing lineage to embrace this occupation and thereby served to limit competition.

It seems this segregation and stigmatization of blacksmiths only affected their social status while their economic fortune appears to be protected. In the Sudan, the blacksmith-potters practiced agriculture and pastoralism on the same level as the rest of the local community. This also means that they are even better off economically than the rest of the population, since they have extra income from making and selling iron objects. This work puts the smith at the helms of most socio-economic endeavors, and sometimes even political. Smiths provide the tools used in cultivating, hunting and fishing, and prestige goods, regalia and bridewealth currencies used by the elites. As a Rundi blacksmith says: "We the blacksmiths we were rich [...] and respected as we were forging the hoes necessary to get food."

Whether iron working was a part-time or full-time profession depended on the nature of the African societies and the ability of the land to support the population. In most complex societies, iron production was practiced full time since the iron workers could depend on the other occupational groups of the large population for their other needs. In small-scale village societies, smiths combined iron working with other subsistence occupations through most of the year. Iron workers at Darfur in Sudan would usually do one smelt per year, normally at the end of the dry season. However, the frequency of demand for iron goods and market may determine whether iron workers devoted full-time or part-time to the profession. As a result, iron working may be conducted during the slow part of the year when less work is required on the farm in order to satisfy immediate and local demand. As societies grow in population and become more complex, individuals working as part-time smiths with full-time farming or herding can now support themselves as full-time iron workers producing iron goods for market purpose.

Decline of African Iron Working

By the time of the Atlantic era (ca. 1508 to 1850), African technology had been affected increasingly by the European presence. Increased coastal trade with Europeans and the gradual expansion of Europeans into the interior brought in large number of cheap European metal goods produced by the industrial revolution. However, the effect on indigenous African iron working may not be apparent before the eighteenth century, since European activities in most cases were restricted to the coastal areas. From the eighteenth century onward as European presence began to be felt in the interior of Africa, more European goods, including metals, were also finding their ways into the region. The European metal goods might have competed with indigenous iron manufactures that were already facing high labor costs in the wake of the slave trade and the scarcity of charcoal fuel in some areas after centuries of exploitation. Europeans on the coast also employed African iron workers and introduced new production methods such as coal fuel, European-style raised forges, and iron anvils. The Atlantic slave trade brought many African metallurgists to the Americas, and some of these craftsmen seem to have engaged in the production of metal in the Americas. In addition, with colonization of Africa by the Europeans and the introduction of cash crop economy in the nineteenth century, many Africans who were full-time or part-time iron smelters might have abandoned the profession completely and took to production of cash crops, now considered to be more lucrative and less laborious.

In northern Yorubaland, the decline of ironworking might have been associated to the frequent wars and raids of the early nineteenth century in the region. The history of northern Yoruba, for example, is closely linked with the dislocation of settlements. In response to war, or the threat of war, many people abandoned their settlements and relocated to more fortified sites in the area, while others retreated from the grassland into the neighboring forest belt to join existing settlements. We know from local traditions that the northern Yoruba communities (e.g., Ipo, Ila, Ilere, Yagba, Ijumu) were abandoned, sometimes on more than one occasion. It may be correct to say that the operation of crafts of the iron smelter, the brass caster, and the smiths became limited, but did not cease completely. The effect of war, and the concentration of population on hilltop, restricted ironworking activities to small areas, while the exchange of iron goods became possible only between communities located in close proximity of each other. The limited ironworking in the region during the wars was probably aimed at producing defensive, if not offensive, weapons. The iron smelting industry, already weakened by wars and raids, suffered further decline by the arrival of Europeans in Yorubaland from the late nineteenth century. The imported European iron materials replaced locally-smelted iron and became the main material used by blacksmiths in the region.

The deforestation of the environment also contributed to the decline of African iron working practices, particularly iron smelting, before the arrival of Europeans. Two important resources of the environment are vital to the operation and success of iron working. These are iron ore and the wood for charcoal fuel needed to transform the ore to iron. Charcoal needed for iron working was obtained from suitable tree-species. This may have several implications on the iron working industries. As Haaland states, "with an increasing iron production, increasing forest off-take is required and sooner or later a point will be reached where the exploitation exceeds regeneration." Where there is no restriction on the exploitation of forest trees and no regular re-planting of trees, there is likely to be a rapid increased in deforestation around iron production centers. However, the nature and extent of deforestation of an area depends on the species composition of the local flora and its suitability for charcoal production. Whatever the nature of the environment, whether forest or savanna, if it contains small amount of the essential tree species for charcoal production it is expected that this will be used quickly. On the other hand, if the suitable trees for charcoal are abundant in the environment the deforestation will be slower and the effect on iron smelting may not be immediately apparent.

Since suitable trees around the village depleted quickly, the iron workers must now travel long distances to procure suitable wood for charcoal. As reported by Haaland, for example, the blacksmiths of Toumra in Sudan walk for two days for wood needed for iron forging, while 25 or more years ago these trees were growing around the village. The local deforestation around the village has increased the problem of firewood collection and forcing people to relocate elsewhere. Also, this Sudanese village consisted 25 years ago of 25 huts for iron working and 150 families. In 1978, the village consisted of four huts for iron working and 25 families. However, beside difficult access to firewood as a major reason for migration, the village lies within the Sahel zone with less than 400 mm of yearly rain, which was not enough to support cultivation and pastoralism.

With the demise of bloomery process or iron smelting in Africa, smiths' workshops are now used to make tools from imported scrap iron. Today the blacksmiths produce mainly hoes, cutlasses, and axes used in cultivation, plus weapons such as knives and spears. Since the demand for their products is relatively limited, smiths are scattered among the local communities in Africa.

Conclusion

Not enough is yet known about when, with whom and how iron working began in sub-Saharan Africa. The archaeological dates available at present for the early iron working and the technical aspects of the industries could either reflect independent development at one or more centers in sub-Saharan Africa, or as diffusion through several sources or a single source. These early dates are in most cases too close in time that it is sometimes difficult to distinguish the donor centers from recipient centers. If the knowledge of iron working did come from Carthage or Meroe or both places to different parts of Africa, one might expect that the techniques used in the donor areas were transferred to recipient areas at times that were appreciably later. Also, dates cannot alone constitute evidence of spread, as other evidence are essential, such as use of similar methods in working iron as well as similarity in other socio-cultural features. But no matter the speculations of scholar, the fact remains that we still do not know how humans discovered that material such as rock could be turned into a metal.

The precolonial African iron working represents a unique process that required specialized knowledge for the selection and preparation of the iron ores, the selection of wood and preparation of the fuel, construction of the furnace and other furnace apparatus, the smelting and production of bloom, and the forging of bloom to an iron implements, and the question of appropriate rituals, social organization and economic base. Iron working in Africa is highly varied and complex. The variation in the technology and the procedures of iron working has generated a wide range in furnace types, amount of ore and fuel, time spent in smelting and the quality of bloom produced in the archaeological and ethnographic record. The study of African iron production has contributed to a better appreciation of the complexity of iron metallurgy, forcing scholars to move away from the issue of origins of iron working to more substantial topic like the technical aspect, and the socio-cultural, economic, and political context of African iron metallurgy. It is tempting to suggest that iron working in so vast a continent like Africa, was practiced by several groups with distinct material cultures. It is possible that some of the groups were distinguished from other related Iron Age groups by their use of different techniques for manufacturing iron, different pottery traditions, subsistence, and settlement systems.

Researchers of African iron working will continue to be challenged in different areas. There are still needs to finding and excavating undisturbed sites where iron smelting can be carefully and accurately dated. However, the need for reliable dates should not obscure other important issues that require attention. Some scholars have raised these issues before. First, more data is needed on the spatial relationship of smelting sites to habitation areas. Second, there should be rigorous research into the role of iron production in the development of socio-political complexity, particularly the emergence of complex polities (e.g., chiefdoms, states) in Africa. Thirdly, the environmental implication of iron production has only been considered by few archaeologists as a significant research focus. The disappearance of certain types of trees in areas of intense iron working may have contributed to the decline of iron smelting and need to be studied.

In many African oral traditions, it is often not known or remembered when exactly iron technology began. The people will claim that ever before any Europeans set foot on their soil they have been producing their own hoes, cutlasses, tools, and weapons made from iron. The Yoruba of Nigeria have a set of traditions of origins which locate the ancestral home of the people in areas outside their present abode. As suggested to us during one of our fieldworks in northern Yoruba, one of the reasons for migration of the groups to their present location was availability of resources for iron working. Iron ore is highly abundant and more recognizable. Iron ore deposit facilitated smelting and smithing and left their traces on the landscape. The presence of iron ore and other goods in Igbomina attracted people to the area. A migration story relating to iron working is told very succinctly in the tradition of Igbomina Ilere: a "man migrated from Ire in the present Ondo State of Nigeria to practice his profession of blacksmithing and to stay on a hill where he could find iron ores for his job."

Notes

1. A. J. Arkel, B. Fagan, and R. Summers, "The Iron Age in Sub-Saharan Africa," *Current Anthropology* 7 (1966): 451-484; R. Mauny, "Autour de l'introduction du fer en Afrique Occidentale," *Encyclopedie Mensuelle d'Outre-Mer* (1953): 109-110; P. L. Shinnie, ed., *The African Iron Age*. Oxford: Clarendon Press, 1971; R. E. Tylecote, "The Origin of Iron Smelting in Africa," *West African Journal of Archaeology* 5(1975):1-9; N. J. van der Merwe, "The Advent of Iron in Africa," in *The Coming of the Age of Iron*, edited by T. A. Wertime and J. D. Muhly, 463-506. New Haven, Connecticut, 1980.

2. J. G. D. Clark, *World Prehistory: A New Outline* (Cambridge: Cambridge University Press, 1969), 181.
3. F. J. Kense, "Archaeology in Anglophone West Africa," in *A History of African Archaeology,* edited by Peter Robershaw, 135-154. London: James Currey, 1990; P. V. Tobias, "History of physical anthropology in southern Africa," *American Journal of Physical Anthropology* 28(1985): 1-52.
4. Kense, "Archaeology," 140.
5. D. A. Welsby, *The Kingdom of Kush: The Napatan and Meroitic Empires* (Princeton, N.J.: M. Wiener Publishers, 1998).
6. A. F. Holl, "Metals and Precolonial African Society," in *Ancient African Metallurgy*, edited by M. S. Bisson, S. T. Childs, Philip de Barros, and A. F. C. Holl, 1-81 (Walnut Creek: Altamira Press, 2000), 6.
7. S. T. Childs and David Killick, "Indigenous African Metallurgy: Nature and Culture," *Annual Review of Anthropology* 22(1993): 317-337; L. M. Diop, "Traditional iron working and the Iron Age in Africa," *Bulletin de l'IFAN* 30 (1968):10-38; H. Lhote, "La connaissance du fer en Afrique occidentale," *Encyclopedie mensuelle d'Outre-Mer,* (1952): 269-272; D. W. Phillipson, *African Archaeology* (Cambridge: Cambridge University Press, 1985); P. R. Schmidt, "Cultural Meaning and History in African Myth," *International Journal of Oral History* 4(1983):183; P. R. Schmidt and S. T. Childs, "Innovation and Industry during the Early Iron Age in East Africa: The KM2 and KM3 Sites of Northern Tanzania," *African Archaeological Review* 3 (1985): 53-94.
8. Holl, "Metals and Precolonial," 7.
9. Kense, "The Initial Diffusion of Iron to Africa," in *African Iron Working: Ancient and Traditional,* edited by R. Haaland and P. Shinnie, 11-27 (Bergen: Norwegian University Press, 1985), 11-12.
10. Ibid.
11. C. L. Goucher, "Iron Is Iron 'til It Is Rust: Trade and Ecology in the Decline of West African Iron-Smelting," *Journal of African History* 22 (1981): 179-189; L. M. Pole, "Decline or survival? Iron production in West Africa from the seventeenth to twentieth centuries," *Journal of African History* 23(1982): 503-513.
12. Philip de Barros, "Changing Paradigms, Goals and Methods in the Archaeology of Francophone Africa," in *A History of African Archaeology*, edited by Peter Robertshaw, 155-172 (London: Heinemann, 1990); R. Haaland, "Iron Production, its Socio-Cultural Context and Ecological Implications," in *African Iron Working: ancient and traditional,* edited by R. Haaland and P. Shinnie (Bergen: Norwegian University Press, 1985) 50-72; P. T. Robertshaw, ed., *A History of African Archaeology* (London: James Currey, 1990); P. R. Schmidt, *Historical Archaeology: A Structural Approach in an African Culture* (Westport, Connecticut: Greenwood Press, 1978).
13. Holl, "Metals and Precolonial," 7.
14. Ibid.
15. Ibid.

16. Phillipson, *African Archaeology*, 149.
17. R. Mauny, "Datation au C14 des sites oust africains de l'age du fer," In *Actes du VI e Congres Panafricain de Prehistoire et d'Etudes du Quaternaire*, edited by H. J. Hugot, 533-539 (Chambery n.p.,1967), 533.
18. Mauny, "Essai sur l'histoire des metaux en Afrique occidentale," *Bulletin de l'Institut Francais d'Afrique Noire* 14(1952): 545-595.
19. B. G. Trigger, "The Myth of Meroe and the African Iron Age," *African Historical Studies* 2 (1969): 23-50; P. L. Shinnie, "Iron Working at Meroe," in *African Iron Working: ancient and traditional,* edited by Randi Haaland and Peter Shinnie, 28-35 (Bergen: Norwegian University Press, 1985).
20. A. F. Holl, "Transition from Late Stone Age to Iron Age in the Sudano-Sahelian Zone: A Case Study from the Perichadian Plain," in *The Archaeology of Africa: Food, Metals and Towns,* edited by T. Shaw, P. Sinclair, and A. Okpoko, 330-343 (London: Routledge, 1993); also by Holl, "Metallurgy, Iron Technology and Africa Late Holocene Societies," in *Traditionelles Eisenhandwerk in Africa,* edited by R. Klein-Arendt, 13-54 (Koln: Heinrich Barth Institut., 1997); Phillipson, *African Archaeology*, 152.
21. B. W. Andah, "Iron Age beginnings in West Africa: reflections and suggestions," *West African Journal of Archaeology* 9 (1983): 135-150; Diop, "Traditional iron working," 10-38; Lhote, "La connaissance," 269-272.
22. Lhote, Ibid.
23. Diop, "Traditional iron working," 37.
24. A. H. Sayce, "Second interim report on the excavation at Meroe in Ethiopia. Part II –the historical result," *Liverpool Annals of Archaeology and Anthropology* 4(1912): 53-65.
25. Trigger, "The Myth of Meroe," 47.
26. P. L. Shinnie and F. J. Kense, "Meroitic Iron Working," in *Meroitic Studies*, edited by N. B. Millet and A. L. Kelley, 17-28, *Meroitica* 6 (Berlin: Akademie Verlag, 1982).
27. Haaland, "Iron Production, its Socio-Cultural," 68.
28. D. Calvocoressi and Nicolas David, "A new survey of radiocarbon and thermoluminescence dates for West Africa," *Journal of African History* 20 (1979): 1-29.
29. Bernard Fagg, "Recent work in West Africa: new light on the Nok culture," *World Archaeology* 1, 1(1969): 41-50.
30. Calvocoressi and David, "A new survey," 10; Holl, "Metals," 14; E. E. Okafor, "Early Iron Smelting in Nsukka-Nigeria: Information from Slags and Residues" (PhD diss., University of Sheffield, 1992); also by the same author, Okafor, "New Evidence of Early Iron-Smelting from Southeastern Nigeria," in *The Archaeology of Africa: Food, Metals and Towns,* edited by T. Shaw, P. Sinclair, B. Andah, and A. Okpoko, 432-448 (London: Routledge, 1993).
31. Okafor, "Early Iron Smelting"; see also Okafor, "New Evidence," 437.
32. Kense, "The Initial Diffusion," 16.
33. Ibid, 16-17.

34. Holl, "Metals and Precolonial," 14.
35. E. Zangato, *Societes prehistoriques et megaliths dans le nord-ouest de la Republique-Centrafricaine*, BAR International Series 768 (Oxford, England: Archaeopress, 1999).
36. N. David, "Early Bantu expansion in the context of central African prehistory: 4000-1 B.C.," in *L'expasion bantoue* 3, edited by L. Bouquiaux, Societe d'Etudes Linguistiques et Anthropologiques de France 19 (1980): 609-647; Kense, "The Initial Diffusion," 17.
37. Holl, "Metals and Precolonial," 15; M. C. van Grunderbeek, "Essai de delimitation chronologique de l'Age du Fer Ancien au Burundi, au Rwanda et dans la region des grands lacs," *Azania* 27(1992): 53-80.
38. Schmidt, *Historical Archaeology*, 111, 235; P. S. Schmidt and D. H. Avery, "Complex iron smelting and prehistoric culture in Tanzania," *Science* 201, 4361 (1978):1085-1089.
39. Kense, "The Initial Diffusion," 16.
40. T. Maggs, "Some recent radiocarbon dates from Eastern and Southern Africa," *Journal of African History* 18(1977): 161-191.
41. M. Hall and J. C. Vogel, "Some recent radiocarbon dates from southern Africa," *Journal of African History* 21(1980): 431-455.
42. Kense, "The Initial Diffusion," 25.
43. C. C. Wrigley, "Speculations on the Economic Prehistory of Africa," *Journal of African History* 1, 2 (1960): 189-204.
44. Kense, "The Initial Diffusion," 18.
45. Calvocoressi and David, "A new survey," 11.
46. P. de Maret, "The Smith's Myth and the Origin of Leadership in Central Africa," in *African Iron Working,* edited by Randi Haaland and Peter Shinnie, 73-86 (Bergen: Norwegian University Press, 1985).
47. Philip Oyelaran, "Early Settlement and Archaeological Sequence of the Northeast Yorubaland," *African Archaeological Review* 15, 1(1998): 65-79.
48. David Aremu, "The Archaeology of Northeast Yorubaland with emphasis on early techniques of metal working," PhD dissertation, University of Ibadan, 1991.
49. Afolabi Ojo, *Yoruba Culture* (London: University of London Press, 1966), 97.
50. David Aremu, "Metallurgical Traditions: Origins, Technology and Social Implications," in *Precolonial Nigeria,* edited by Akin Ogundiran (Trenton, NJ: Africa World Press, 2005), 134.
51. Ibid.
52. S. T. Childs, "Style, technology, and iron smelting furnaces in Bantu-Speaking Africa," *Journal of Anthropological Archaeology* 10 (1991): 332-359; Eugenia Herbert, *Red Gold of Africa* (Wisconsin: University of Wisconsin Press, 1984), 4-5.
53. Haaland, "Iron Production, its Socio-Cultural," 53.
54. Ibid.

55. Ibid, 54.
56. Aribidesi Usman, *State-Periphery Relations and Sociopolitical Development in Igbominaland, North Central Yoruba*, Nigeria. British Archaeological Report Series 993 (Oxford: John and Erica Hedges Ltd. and Archaeopress, 2001), 125.
57. Ibid.
58. J. A. Aleru, "Archaeological Sites in Igbominaland, North Central Yorubaland, Kwara State, Nigeria: The Need for Systematic Archaeological Research," *West African Journal of Archaeology* 22 (1993): 37-50.
59. Aremu, "Metallurgical Traditions," 141.
60. Fagg, "Recent work," 46-47; P. J. Jaggar, "Kano City Blacksmiths: precolonial distribution, structure and organization." *Savanna* 2(1973): 11-25; J. E. Sutton, "Iron working around Zaria," *Zaria Archaeology Papers*, 8 and 12 (Zaria: Ahmadu Bello University, 1976).
61. J. Nicolaisen, "Afrikanske smede (African smiths)," *KULM* (1962): 71-105.
62. G. Berthoud, "Les Ganawuri du Nigeria," *Bulletin Annuel du Musée d'Ethnographie*, Geneva, 12 (1969): 24-28; R. M. Netting, *Hill Farmers of Nigeria* (Seattle: University of Washington Press, 1968), 56; C. L. Temple, *Notes on the Tribes, Provinces, Emirates and States of the Northern Provinces of Nigeria* (London: Frank Cass, 1963), 345.
63. D. Williams, *Icon and Image: a Study of Sacred and Secular Forms of African Classical Art* (New York: New York University Press, 1974), 64.
64. L. M. Pole, "Furnace Design and the Smelting Operation: A Survey of Written Reports of Iron Smelting in West Africa," in *African Iron Working*, edited by Randi Haaland and Peter Shinnie, 142-163 (Bergen: Norwegian University Press, 1985), 153.
65. Ibid.
66. Haaland, "Iron Production, its Socio-Cultural," 51.
67. Ibid.
68. Ibid., 52.
69. Ibid., 55.
70. Ibid., 56.
71. Ibid.
72. N. David, "The Ethnoarchaeology and Field Archaeology of Grinding at Sukur, Adamawa State, Nigeria," *African Archaeological Review* 15, 1 (1998): 13-63.
73. H. Sassoon, "Grinding grooves and pits in Northern Nigeria." Man 62, 233(1962):145.
74. Haaland, "Iron Production, its Socio-Cultural," 56.
75. D. H. Avery, and P. Schmidt, "A metallurgical study of the iron bloomery particularly as practiced in BaHaya," *Journal of Metals* 31 (1979): 14-20.
76. Kense, "The Initial Diffusion," 21.
77. Ibid.

78. N. J. van der Merwe, and D. H. Avery, "Pathways to Steel," *American Scientist* 70(1982): 146-155.
79. Peter Schmidt, *Historical Archaeology in Africa* (Lanham: AltaMira Press, 2006), 168.
80. Schmidt and Avery, "Complex iron smelting," 1087.
81. Ibid.
82. S. T. Childs, and E. W. Herbert, "Metallurgy and its Consequences," in *African Archaeology: A Critical Introduction,* edited by Ann Stahl, 276-300 (Oxford: Blackwell Publishing, 2005), 284; D. J. Killick, "A Little Known Extractive Process: Iron Smelting in Natural-Draft Furnaces," *Journal of Minerals, Metals and Materials* 43(1991): 62-64.
83. Childs and Herbert, ibid.
84. Ibid.; see also Bruno Martinelli, "On the threshold of intensive metallurgy: the choice of slow combustion in the Niger river bend (Burkina Faso and Mali," in *The Origins of iron metallurgy in Africa: New light on its antiquity, West and Central Africa,* edited by Hamady Bocoum, 165-188 (Paris: UNESCO, 2004).
85. Schmidt and Avery, "Complex iron smelting," 1085.
86. Ibid.
87. P. de Maret, "The Smith's Myth and the Origin of Leadership in Central Africa," in *African Iron Working,* edited by Randi Haaland and Peter Shinnie (Bergen: Norwegian University Press, 1985), 77.
88. Ibid.
89. Ibid.
90. Ibid., 73.
91. J. Vansina, *The Tio Kingdom of the Middle Congo 1880-1892* (London: Oxford University Press, 1973), 442.
92. G. Celis and E. Nzikobanyanka, *La Métallurgie Traditionnelle au Burundi: Techniques et Croyances* (Tervuren: Royal Museum for Central Africa, 1976), 11-12; de Maret, "The Smith's Myth," 74.
93. T. Reefe, *The Rainbow and the Kings* (Berkeley and Los Angeles: University of California Press, 1981), 81-82.
94. H. Womersley, *In the Glow of the Log Fire* (London: Peniel Press, 1975), 82-83.
95. J. Vansina, *The Children of Woot: A History of the Kuba People* (Madison: University of Wisconsin Press, 1978), 49-51.
96. de Maret, "The Smith's Myth," 73.
97. S. T. Barnes, "Many Faces of Ogun: Introduction to the First Edition," in *Africa's Ogun: Old World and New*, edited by Sandra Barnes (Bloomington: Indiana University Press, 1997), 4.
98. N. J. van der Merwe and D. H. Avery, "Science and Magic in African Technology: Traditional Iron Smelting in Malawi," *Africa* 57, 2 (1987): 143-172.

99. R. G. Armstrong, "The Etymology of the word 'Ogun'", in *Africa's Ogun: Old World and New*, edited by Sandra Barnes, 29-38 (Bloomington: Indiana University Press, 1997).
100. Usman, "State-Periphery," 118.
101. Okpoko, Alex I. "The Early Urban Centers and States in West Africa." *West African Journal of Archaeology* 17(1987): 243-265; Pole, "Decline or survival?," 503-513.
102. D. A. Adeniji, *Iron Mining and Smelting in Yorubaland*, (translated and edited by R.G. Armstrong), Occasional Publication No. 31 (Institute of African Studies, University of Ibadan, 1977).
103. Eugenia Herbert, *Iron, Gender, and Power: Rituals of Transformation in African Societies* (Bloomington: Indiana University Press, 1993).
104. R. J. Mason, "Background to the Transvaal Iron Age new discoveries at Olifantspoort and Broederstoom," *Journal of South African Institute of Mining and Metallurgy* 77(1974): 211-216.
105. de Maret, "The Smith's Myth," 77.
106. D. J. Killick, "Technology in its Social Setting: The Ironworkers of Kasungu, Malawi," PhD dissertation, Yale University, 1990.
107. Haaland, "Iron Production, its Socio-Cultural," 56.
108. S. F. Nadel, *A Black Byzantium: The Kingdom of Nupe in Nigeria* (London: Oxford University Press, 1942); Pole, "Decline or survival?," 511.
109. Haaland, "Iron Production, its Socio-Cultural," 57.
110. Ibid.
111. J. L. Gibbs, "The Kpelle and Liberia," in *Peoples of Africa*, edited by J. L. Gibbs, 197-240 (New York: Holt, Rinehart and Winston, 1965), 205.
112. de Maret, "The Smith's Myth," 76; I. Kopytoff, "The Suku of Southwestern Congo," in *Peoples of Africa*, edited by J. L. Gibbs (New York: Holt, Rinehart & Winston, 1965), 447.
113. Haaland, "Iron Production, its Socio-Cultural," 58.
114. Ibid.
115. Ibid., 60.
116. Celis and Nzikobanyanka, "La Metallurgie," 87; J. P. Chretien, "La production du fer au Burundi avant la mainmise coloniale, in Métallurgies africaines – nouvelles contributions, edited by Echard, N., Mémoires de la Société des Africanistes, 9(1983): 311-325. The quotation came from pages 320-21.
117. Holl, "Metals and Precolonial," 23.
118. Haaland, "Iron Production, its Socio-Cultural," 56.
119. Holl, "Metals and Precolonial," ibid.
120. A. Usman, "The Nineteenth-Century Black Atlantic," in *The Atlantic World 1450-2000,* edited by Toyin Falola and Kevin Roberts, (Bloomington: Indiana University Press, 2008) 114-134.
121. Haaland, "Iron Production, its Socio-Cultural," 61.

122. Ibid., 63.
123. Ibid.
124. Kense, "The initial Diffusion," 19.
125. Childs and Herbert, "Metallurgy."
126. W. H. Clarke, *Travels and Explorations in Yorubaland (1854-1858)* (Ibadan: Ibadan University Press, 1972), 152-53.
127. P. O. Dada, *A Brief History of Igbomina (Igboona)* (Ilorin, Nigeria: Matanmi Press, 1985), 21.

• Chapter 2 •

RETHINKING THE NOTION OF SCIENCE AND TECHNOLOGY IN AFRICA: A CASE OF HILL FURROWS, KILIMANJARO, TANZANIA

Elinami V. Swai

Introduction

It has always been a practice to identify science and technology as a unique epistemology that privileges western worldview, overshadowing certain science and technologies of other parts of the world such as Africa. One would hear, that is "science" and that is "what those people do," or that is "technology" and that is a "tradition." This chapter discusses one technology that for many years has been known as "traditional hill furrow." Hill furrow is an irrigation technology that has been used by various ethnic groups in East Africa, specifically in the regions along the Rift Valley like Engaruka and Sonjo in Arusha and Taita and Marakwet in Kenya. In Kilimanjaro for example, this technology sustained people for many years. People on the mountain managed and preserved their land and water sources through hill furrows and managed to achieve a very high and stable standard of living.

The main argument is that the discourse of hill furrows and other indigenous technologies within the earlier accounts by western travelers and colonialists were seen as just "traditional practices" without any scientific legitimacy or technological credibility. Large scale irrigation schemes, for example, were constantly referred to as 'irrigation technologies,' while hill furrows were referred to as "traditional irrigation" and such ideas have persisted to date. The early travelers and colonial encounter in East Africa and elsewhere in sub Saharan Africa were certainly pivotal in creating this binary division, indeed creating a model of modern irrigation scheme on the continent, making hill furrow technology look like inconsequential traditional activities that did not have any scientific basis.

In recent times, people have come to realize that traditional irrigation schemes are not really inconsequential or in conflict with Western or modern large irrigation schemes. In particular, there has been a growing interest in traditional irrigation systems and how they can be managed and sustained. This interest was exacerbated, in part, by a new emphasis on acknowledging local community initiatives to find alternative solutions to deteriorating economies with a goal to sustainable development. States and international organizations have seen a need for accepting and re-interpreting what are viable alternatives for economic growth,[1] local know how and indigenous practices being among them.

In the twentieth century, many traditional African agricultural irrigation systems were seen as archaic, without any trace of scientific method. Hill furrows are no exception to these conceptions, but of recent, these systems have been adapted to meet the challenge of new intensive agricultural practices, especially on the mountain slopes. This chapter will show that changes have not been wholesale; rather the local irrigation schemes have blended both traditional and modern irrigation techniques. The value of hill furrow technology to other economic sectors is also being recognized and exploited by both government and non-governmental groups.

Theoretical Framework

The common view of technology, as "applied science," constitutes a foundational element of the hegemony of advanced capitalism. In many ways, the concept of technology presupposes a unique epistemology (techno-epistemology) that focuses on the role of science in technology. However, understanding technology only in its technical side is deeply problematic; it overshadows the unique way people develop science and use its products to solve their problems. The model of science-technology relationship which has privileged a Western worldview is routinely and uncreative, without any innovative activity. In cultural studies, those applying (or, more often, simply presuming) western notions of science leave human culture and organization of society neatly beyond the purview of discussion.

Leaving aside the relationship between technology and science, it is important to recognize that the word technology has at least three different layers of meanings. First, technology is a form of knowledge.[2] Technological things are meaningless without the know-how to use them, repair them, design them and make them. That know-how cannot be captured in words. It is visual, even tactile, rather than simply verbal or mathematical. But it can also be systematized and taught.

Second, technology also refers to what people do as well as what they know. A hill furrow is a technology for irrigation, rather than arbitrary water canal, because it forms a part of a set of human activities. An irrigation system without human know-how is simply a useless system of things put together. An irrigation technology includes what irrigators do, as well as the utensils they use. So technology refers to human activities and practices. Third, there is a hardware definition of technology, in which it refers to sets of physical objects, for example the wood and soil.

In practice, hill furrows embody all three aspects, and often it is not useful to separate them. What is important is to recognize that science and technology develop in praxis. As such, technology is inherent to the continuously unfolding transaction between humans and the environment as they collectively modify with respect to their needs. Technology, as a result of human activity (in other words, science), therefore extends the human activity. Its unit of analysis is human activity, which takes into account both the collective nature of human beings and the cultural-historically determined transaction with the social and natural environment. Taken together this approach describes science and technology as a historical materialist dialectical human approach to life. This understanding helps us to recognize hill furrow as both a science and technology, bound up with culturally specific human practices.

It is important to note that science and technology in this chapter is understood as a microlevel formation that serves the survival of the collective. Hill furrow technology is thus conceived as a micro system of relations in which human needs are central. Its utility, then, is manifested in the social and cultural system; it is inferable from social actions, and actions leave traces both inside and outside human community. Such traces can be conceptualized as norms, taboos and social arrangement that surround the hill furrow technology which mediates human action in relation to the creation and maintenance of hill furrow. Accordingly, hill furrow can be seen as part of social practice (Vygotsky, 1978).

In this study, hill furrows in their practical sense, rather than their philosophical account are the central unit of analysis. As an example, I refer to hill furrows as both irrigation technology and cultural symbol with a particular role in Chagga society. As a technology, hill furrows are explicitly non-Eurocentric way of preserving water and irrigation. More so, the Chagga have attempted to preserve water and their environment in their own ways through the hill furrows. Drawing on Hai as a case study—and therewith articulating people's voice—allows us to overcome the colonial tendency that saw hill furrows as just 'simple irrigation technology.'

In what follows, then, we will first articulate in more detail in what respect current frameworks of science and technology fail to grasp the unique nature of hill furrow and how this limitation leads to its silencing. I will subsequently present an alternative cognitive framework developed to understand hill furrows. This framework allows us to articulate hill furrow as a technology emanating from human practice in its own right and to illustrate the many forms of social and cultural expressions by which it is constituted. Because of our methodological commitments to both an articulation of indigenous science and technology and decolonization of the minds (Mudimbe, 1994), we will illustrate this framework with the authentic case study of hill furrow that is both epistemologically and historically not rooted in western science. Finally, we will draw the implications of this study for the teaching and learning of science and technology in Africa.

As the most preferred irrigation system, the technology of hill furrow in the Chagga society has complex and dynamic history. Thus, it would be misleading to promote a picture of an unchanging Chagga society. The technology of hill furrows as irrigation infrastructure has been a traditional technology, practiced by the Chagga people for time immemorial and furrows have ensured sustainability of both social and physical environment of the Chagga societies for many years and was admired by early travelers and 'discoverers' who visited the eastern part of Africa. For example in early years of the 19th century, Rebmann, (1860, 242) a European traveler declared to have seen wonderful and sophisticated canals in Kilimanjaro, constructed by the use of 'small sticks.' According to Rebman, and later Stahl and H. M. Johnstone, the furrows were wonderfully built in every ridge on Mount Kilimanjaro.

While hill furrow technology appears to have been widespread in the pre-colonial period and some have persisted to the present day, both have undergone certain dynamics, including growth, transformation and decline. This fluctuation came about in the process of social transformation under colonial rule. The rise of colonial hegemony assumed that local people lacked the necessary science and expertise to manage the environment, including water. From the mid-1930's, the colonial government involved itself in water management using improvement projects,[3] while discouraging the pre-existing traditional hill furrows. During colonial period 'new' technology was introduced and the management of water was instituted to ensure that water was for the production of cash crops. The sophisticated system of agricultural water management, the ban of using water for other purposes save for cash crops and the lack of knowledge in maintaining the new irrigation system inhibited the motivation to maintain the new water sources and systems. For example, while the traditional hill furrow technology had stemmed from concern over food shortages in the region, the modern system of agricultural water management was geared to the production of high value cash crops (coffee) for export.

These changes drastically altered traditional hill furrows.[4] First, the management of water under colonial government was created with no awareness of local environment and local people's needs. This oversight affected the ways the local people managed water and irrigated their small holding *vihamba* (land). The resultant hunger, poverty and other complications on Kilimanjaro, which have persisted for many years, have not been related to the transformation that occurred from the alienation of hill furrow technology from the Chagga people.

While Chagga people's scientific knowledge and hill furrow technology may seem strategic and holistic alternative to the present environmental and poverty catastrophe, these have been eroded and eviscerated for hundreds of years. Chagga people have witnessed their traditional landscapes and flora and fauna radically altered as the widespread national adoption of machinery for water management ensued

However, there have been disagreements about the efficacy of hill furrows when compared to modern and large irrigation systems. Irrigation systems such as dams, piped water and other modern irrigation technologies have been seen to be more durable and efficient in conserving water. The central importance of irrigation schemes that this chapter argues for is their sustenance through systemic traditional norms, values and rituals. Hill furrows are not just system of irrigation, but were integrated within culture and were part of organization of society. They were meaningfully perceived and used as cultural expression.

In North Pare, for example, hill furrows were associated with gendered myths and symbols and only specific people were allowed to construct and maintain them. In many traditions in Kilimanjaro, women and children were prohibited from participating directly in matters of the furrow. Pregnant women specifically were banned from crossing the furrows.[5]

Hill furrow technology has emanated from people and has sustained them as they have regenerated ecosystem, thus becoming both the means and the ends of economic development. Unlike large scale planned irrigation projects, hill furrows are seen as a little more than haphazard and unplanned irrigation schemes. The goal of this chapter is to bridge this gap by showing that throughout history the Chagga have used hill furrows not only as their local science and technology to conserve water, soil and their physical environment, but also to create social cohesion. Through hill furrow technology, Chagga people have tapped into their local resources and have survived and strived for many years. Hill furrow technology has enabled people on the mountain to live communally, learnt from each other and from their own technology and have conserved water and soil, and have protect their physical environment since time immemorial.

During the colonial periods, coupled with capitalist philosophy of profit maximization and accumulation of property, this philosophy buttressed by legislation, in which land and water were the government property. Under the new legislation the colonial masters had powers manage and maintain water sources, while the natives were granted the right to access and use water. All these had the effect of vesting control over water in the Executive arm of Government.

This social and economic transformation shut off all viable alternative sources of water for the indigenous Chagga. Economic benefits accrued from this social transformation: water came to belong to the state and any individual wanting to access water from, whether on the mountains or from a river or stream, was supposed to have a water right. In Tanzania, the utilization of water is basically regulated by the 'Water Utilization (Control and Regulation) Act, 1974' (United Republic of Tanzania, 1974), with two later amendments: 'Water Utilization (Control And Regulation) Act (Amdt.) 1981 (No. 10)' (United Republic of Tanzania, 1981), mainly dealing with pollution issues, and the 'Water Utilization (General) (Amendment) Regulations, 1994' (United Republic of Tanzania, 1994a,b), which set out the fees for obtaining a water right and the annual fee for accessing water.

The construction of hill furrows as traditions and water dams and piped water as technology gave colonial masters control over land and water on the mountain. Thus, the series of measures over who should manage water and who should not, which technology was appropriate and which was not were consciously instituted to protect the colonial masters, while limiting the use of water among the 'natives.' More directly, colonial officials and environmentalists identified hill furrows as "unscientific" and "wasteful" (district annual reports) and therefore, a threat to soil fertility and harmful for agricultural irrigation. An important feature of the Colonial water law in Tanzania and in other African countries was what Okoth-Ogendo has called their 'essentially administrative character' (2000, 127).[6] The Colonial State failed to develop the content of both customary law and the imported English land law, with the result that much of what passed for water law was the law of water administration. This, together with the State's legislative interventions, strengthened its hand in water matters.

The presentation of hill furrow as harmful to the environment and to development at large was accompanied by the government decision to abandon them. Since 1980s, the government of Tanzania, with foreign aid, installed public standpipes. According to official statistics, by 1986 over 700,000 people or 60 per cent of the population of Kilimanjaro were being supplied with domestic water and the number of furrows on the Mountain for irrigation had halved from a thousand in 1960 to about five hundred in 1986.[7] Later on, the installed pipelines became a serious threat to the Chagga people's culture and social organization. While the banning and abandoning of hill furrows gained grounds, many Chagga cultural meanings and social expressions surrounding hill furrows dwindled. This transformation in Kilimanjaro explains two dynamics in Chagga society: the first, why the Chagga are suspicious of outsiders, and the second, lack of enthusiasm of any externally prompted 'innovations.' This has self-evidently been a problem, not only in the field of agriculture, but also in other fields such as education, health, and transport system (especially roads) services.

In recent decades there has been a considerable movement of people out of the mountain to apparently permanent settlements in other parts of the country, away from the mountain. Thus, the imposed irrigation schemes on the mountain shuttered the Chagga's historical and cultural sense of being. Their landscape has come to be not a home, but a place to visit only once in a year.

Most developmental and environmental theories assume that hill furrows are solely unscientific and inconsequential to people's livelihoods. If one takes time to study it, the knowledge used to build a furrow is as systematic as any scientific knowledge generated through a specific methodology. For example, Chagga people used a long time to observe their mountainous landscape and plants. They tasted the soil, observed the hillside and assessed the suitability of the soil to construct the furrows. They also maintained the furrows using seasons as their vision. The Chagga people knew when to start climbing the mountain to pave the way for water. The idea of observing the weather, the landscape, and the soil before constructing a furrow is as scientific as any. This long time observation and cautious management is what made furrows durable to date.

Although hill furrows are made of simple and "small" sticks," or simple equipments, rather than concrete cement and iron, the science and technology have worked. Science in this chapter is conceived as an organized knowledge, systematically generated through a specific methodology. Technology, on the other hand, is understood as applied science and its application (technology) to solve certain problems. Understood as scientific technology hill furrows can still be promoted to solve the problem of water on the mountain.

This chapter presents one viable alternative technology for economic growth, hill furrow. I aim at highlighting the intimate relationship between hill furrow and science and technology. Hill furrow is a mature technology that people in Kilimanjaro used for irrigation and other purposes prior to any contacts with the Europeans. The purpose is thus to deconstruct this relationship and to highlight the nature of technology in its own right as a human practice that may exist in relation to a multitude of understandings of the natural world.

Indeed, research in social studies of science and technology has shown that there are non-Western examples (Collins & Pinch, 1998), as in the case of hill furrow as a technology in its own right that is bound up with many more human practices. As well, sociocultural studies in science point to the validation of western notions of science and technology as the cause of silencing other science and technologies, especially those which articulate forms of human knowledge (Van Eijck & Roth, 2007). Such tendencies in modern society frustrate indigenous minorities to sustain the validity of their own worldviews and can be seen as forms of Eurocentrism, cultural imperialism, racism, and ethnocentrism.[8] Scholars have argued that traditional irrigation systems have equal value to modern and large irrigation schemes. I would argue that hill furrow technology has more value because it emanates from the people themselves and is maintained within traditional norms, values and rituals, and therefore, are more sustainable than modern ones.

There has been much discussion about sustainable development and the concept of sustainability in irrigation discourse. One basic principle of sustainability is that technology does "just happen." Instead, it is developed as a result of enabling conditions that are integrated into human cultures, social institutions and ideologies. Human traditional practices and their social arrangements are key to sustain any technology. It has been argued that a sustainable irrigation scheme should demonstrate longevity, a lack of long-term irreversible environmental degradation, and a degree of intra- and inter-generational equity (Rao and Rogers, 2006). Furthermore, sustainable development can be seen as being 'concerned with improving the overall quality of life as well as satisfying human needs.' Malley, *et al* point out that sustainable development 'appears to have great potential for reviving the idea of development as a multi-faceted, long-term, while taking care of intergenerational problem' (Malley *et al,* 2004). Clearly, the ability to develop irrigation schemes that secure a future in which people can transform economic opportunities and their lives into acceptable and environmentally consonant lifestyles is central to the concept of irrigation scheme. Few authors have attempted to examine extant irrigation systems that are sustainable and thus provide tangible examples of sustainability, although there have been a call to do so (Niemeijer, 1996).

Hill furrow irrigation technology demonstrates sustainability and therefore provides a tangible example of a valid science and technology that are needed to sustain the continent for future times. A full discussion about sustainability involves analysis of not only physical systems but also socioeconomic, cultural, and political systems (Crook, 1997). Given the constraints of space, this chapter concentrates on the cultural elements of hill furrow that has promoted sustainability in Kilimanjaro. (Descriptions and analysis of the other parts of hill furrow can be found in Fleuret, 1985; Adams and Anderson 1988). It may prove easier to analyze irrigation system from economic point of view than from cultural because of our focus on irrigation schemes as mechanism for modernization of our agricultural activities (Madulu, 2003).

In the 20th century, many traditional irrigation systems are seen as archaic and dependent on simple technology. Patrick Fleuret believes that hill furrow technology has not been given enough attention by scholars.[9] Also, an archaeologist from Kilimanjaro, F. T. Masao, was convinced that that the literature on hill furrows was too brief and often inaccurate.[10] He pointed out that technology of the *mfongo* (hill furrow) and the *nduwa* (collection ponds used for short-term water storage) is as old as the Chagga people. William Adams and David Anderson also lamented the ironic marginalization of a complex and dynamic history of hill furrow in Eastern Africa.[11] All the above writers point to a possible alternative irrigation technology that is sustainable and centered around human development.

Description of the Study Area: The Chagga People and the Geography of Hai District

Hai district is situated on the slopes of Mount Kilimanjaro, in the Kilimanjaro region, in northeast Tanzania. The district has an area of 13 000 km,[2] with multiple springs and rivers flowing from Mt. Kilimanjaro. Among scientists, Hai is known for its fauna, flora and notable as the cradle for human origins (Begun et al. 1997, Ward *et al.* 1999, Balter 2001). Interestingly, Tanzanian's hominoid fossils also include the earliest artifacts of learning, revealing knowledge in science and applied technologies (Leakey and Lewin 1992). Historically, the hill furrow technology in Hai district is not altogether dissimilar to the irrigation technology known from places along the Rift Valley, for example in Engaruka and Sonjo in Arusha.

In 1999, the Hai District had a population of about 154 000, with an average household size of 4.6. A household is defined as "people eating from the same house." The potential zone for agriculture on the mountain is very small, with only 46% of the 13 000 km,[2] being agriculturally productive terrain in Hai is scarcer than in the other parts of the region, with only one third of the land suitable for farming.[12] Each commune has a different ratio of land available at various altitudes and aspects, and this controls the cultivation options of farmers. The Hai residents have only 33% of productive surface but the 54% forest, and 13% pasture are also intensively cultivated.[13] Many rural communities in Hai, the locations where this study focused upon, are religious (Christian, 79% and Muslim 20%). Chagga language comprises a variety of dialects commonly spoken although Kiswahili is widely understood and spoken by the majority of the young generation. Economic activities include farming, dairy farming, commercial mining, and cottage industries. Wood is the main source of fuel in many homes. A few households use biogas (gas processed from waste products, such as cow dung and agricultural wastes).

The analysis of interviews with both men and women in Hai district on the irrigation canals indicates that hill furrow technology is well entrenched in people's sense of who they are in relation to their physical and spiritual environments. This cultural symbol has assured people in Hai and other places in Kilimanjaro of stable water for irrigation and preservation of soil from being carried away by the rains. For example, in traditional Chagga society, males were identified by the name of the furrow they lived. One would hear 'so and so is from, for example, Elmelek's furrow.' In Chagga, only men are allowed to own or to dig the furrow. As a chairman of Yuuri community recollected: "Digging furrows used to be men's jobs no woman was allowed to climb the hill because of water. Men's job was to bring water near the home and women's job was to bring it in the house. The two roles could not be switched."[14] This means that furrow was an important cultural symbol that could be used to identify people and adult males must be accepted in local furrow community before being considered and accepted as a respectable male. Sometimes one had to undergo initiation rites to be allowed in a furrow and join a group going to the mountain to bring water.

Initiation rites performed several purposes, such as marking of manhood and a man's strength of character. For example, those travelling away from their homes for trade or other business were assessed by the strength of their furrows to supply the needed water for the whole period of their absence. If that was not possible or unpredictable, a man was supposed to assign another man from the clan to watch over his furrow (which meant, his family) for him in the period of his absence. This had other repercussions, for better or for worse. Thus, an ideal Chagga man was identified by his furrow and women and children sang praises to a man with the best furrow. Daughters were advised to marry from a family with the best and strongest furrow. This strong cultural symbol made both men and women diligent in ensuring the safety of the family furrow. Women exercised zero grazing as a way to ensure animals were not grazed near the furrow. Everyone was aware of what the sanctions were and penalties for not doing their roles and these roles were not haphazardly assigned or assumed. For example, if a furrow was damaged as a result of an accident, one of the elders would sound a horn in the evening. This was known as 'Ighmba lya mfongo,' the call to the furrows. The next morning, everyone would leave their normal work and set about repairing the damage. Anyone who did not take part without good reason was fined. Thus, furrows were socio-cultural construction.

Furrow, as a technology emanating from science, has been used by the Chagga people to preserve water and irrigate their farms on the mountain. For many years, the Chagga have been using traditional science and technology to build extensive infrastructure of irrigation without depending on the government and water management experts. In the process, the Chagga have come to be identified, and have identified themselves as active participants in conserving their environment and for seeing to it that it continues to produce for their survival. The problem is that, from the time when the first European set his eye on this technology,[15] it has been ridiculed. Since extinction is forever a concern, the responsibility and role of scholars and researchers is to join the effort to preserve and promote it. This challenge is especially important because since time immemorial traditional science and technologies in Tanzania and the rest of Africa have been in conflict with modern technologies.

Thus, hill furrows among the Chagga people consist of social and cultural meanings and symbols, explicit and implicit, which people identify themselves with. Thus, hill furrows cease to be just a haphazard practice to cultural artifact and social practice. This definition is useful in the present context because it indexes cultural quality that is missing in national governments and international development agencies' debates, and it needs to be brought into clear view. Hill furrows as a technology enables us to understand economic development, and cultural meanings such as the source of fertility or manhood.[16] What differentiates hill furrows from other western technologies is the relative prominence of their social and cultural meanings and the kinds of emotion they instigate. While hill furrows are embedded with cultural meanings, western technologies, such as larger irrigation schemes, embody an abstract meaning of bringing about economic development. When a hill furrow is identified as a simple technology, its identification is purely physical, leaving aside its cultural symbol. The dual conceptual nature of hill furrows allows them to embody both material (applied science or a technology) and ideal (cultural norm and value). This view also asserts that hill furrows are there to stay.

To sum up, traditional technologies in Africa such as hill furrow cannot develop and strive in isolation from praxis. By uncritically accepting hill furrow as a simple practice of using sticks, we simultaneously accept a description of hill furrow technology within the western idea of African science and technology.[17] Hence, a view like this paves particular epistemological grounds for characterizing hill furrows, and other important indigenous practices as 'traditional practices of simple tools.' Such articulation creates an image of these practices that seems to be inferior to other human practices that have been acknowledged as scientific, such as scientifically planned water management technology. Such constructions have confined many African traditional practices into local and outdated mode, with nothing to offer modern science and technology. Thus, many traditional practices such as hill furrows are commonly absent in nationalist and official debates because of the dominant European thought that constructed them according to their terms and of their sovereign self-image.

As discussed in this chapter, hill furrows are appropriate technology to conserve water and environment centered on people's culture. They were designed and maintained to increase human quality of life first and foremost. The people in Hai and other places saw their furrows as the "end-use" that cemented relationships and created a sense of duty to both men and women without exploitation of gender. Similarly the improvement of irrigation projects do not take account of people's ownership and empowerment. In other words, the introduction of improved large irrigation schemes in Kilimanjaro was not to empower people, but rather to facilitate docile community acceptance of the conventional economic growth model. The focus of improved large irrigation scheme in Tanzania is on gross, macro-economic indicators as a sign of production and progress, thus rendering the model weak in acknowledging the output of hill furrow technology, particularly as they were more people oriented than economic.

Like many traditional practices, hill furrow technology in Kilimanjaro is not unique, static or unchanging. This is a reality that many scholars have not recognized. One only needs to point to the surviving furrows that continue to manage water and the environment among the Chagga people to acknowledge the importance embodied in their symbolism. One cannot, for example, explain sensibly how it is possible to see the thriving of plantains, trees and coffee in Kilimanjaro, a region that has the highest rural population densities in eastern Africa,[18] without some appreciation of the relevance of hill furrow technology. It is not easy to understand why the Chagga and other African groups have not been given credit for their scientific knowledge and skills in living harmoniously with their environment.

Practices of hill furrow were also characterized by specific rules. For example, each kind of furrow was identified by its furrow elder. When referring to the place, one had to be careful under whose furrow leadership was the place. As well, the furrow leaderships were considered political. They were celebrated in ceremonies by mentioning and singing each furrow by its leader's name. Hill furrow technology can be conceptualized as the way of knowing that was inherent to and intimately bound up with the specialized artifacts that mediated people's thinking and behavior. While the hill furrow technologies and cultural practices appear to have been widespread in the pre-colonial period and some have persisted to the present day, both have undergone certain dynamics, including growth, transformation and decline. Furthermore, the introduction of irrigation practices by outside 'developers' are by no means new, and there is already a history of innovation and attempted induced change. Knowledge of these twin dynamics of indigenous and induced change is fundamental to an understanding of the context within which the Chagga people exist.

While it will be too much of an exaggeration to suggest that the Chagga people had a universal perspective on the symbiotic relationship between their culture and physical environment, it is clear that they were certainly aware of the importance of communal collaboration and cultural cognition. Cultural cognition here refers to ways people's culture shape their ways of thinking and doing. Thus, while using their traditional science of water, soil and plants to create furrows and borders, this technology in turn ensured constant food production, water for irrigation and for other purposes as well as safe environment and medicine. Those who, in one way or another, went against the norms of the furrow were dealt with according to the norms agreed to by the whole community.

The framework for understanding hill furrows proposed in this study has several implications for discourse on science and technology. To begin with, it provides a means to overcome issues raised in studies that examine the definition and the nature of hill furrows in Hai Kilimanjaro, especially in the context of how people live their lives and arrange it around their furrows. More so, there is the issue that science, technology, and development studies are insufficiently informed by socio-cultural issues (Zeidler, Sadler, Simmons, & Howes, 2005).

As such, such studies fail to focus on local knowledges and practices that have been marginalized and misrepresented. As Escobar (1995:98) proposes, '[t]he remaking of development must start by examining local constructions, to the extent that they are the life and history of the people, that is, the conditions for and of change.' The framework proposed here provides the necessary epistemological ground for informing the ecologists and environmentalists in this respect. Ultimately, this may contribute to an acknowledgement of local people's knowledge and practices that support their larger developmental schemes.

I indicate the diversity and complexity of past experience with irrigation and land tenure in Kilimanjaro, and suggest that there is a pressing need for more research and for careful consideration before development projects be it in a small- or large-scale. This chapter demonstrates that the central dilemma in African countries and Tanzania in particular has been a lack of acknowledgement and support of traditional science and technology. The chapter historicized the technology of hill furrows and the process of its evisceration under colonial supervision that accompanied Tanzania's economic transformation.

I illuminate the role of traditional science and technology (hill furrows) in the sustainable environment (perseveration of water, soil and local plants), showing how people (not institutions, knowledge, practices, norms, values, roles, etc.) act, interact, reproduce, and transform the world (physical environment). The case I have made is that despite these obvious advantages, hill furrows in rural Tanzania have come to be viewed as unscientific and local practices. The question of traditional science and technology, which this chapter discusses, touches on a very broad range of cultural and environmental concerns.

Notes

1. J. Abbott, "Community participation and its relationship to community development,"*Community Development Journal*, 30, 2, (1995): 158–168; P. Freire, *Pedagogy of the oppressed*. Harmondsworth: Penguin, 1972; J. Friedman, *Empowerment: The politics of alternative development*. Oxford: Blackwell, 1992; A. Jones, "Community development training—A comparative approach," *Community Development Journal,* 27, 3, 1992, 199–210.
2. See Staudenmaier's (1985, pp. 103-20) outlining four characteristics of technological knowledge --scientific concepts, problematic data, engineering theory, and technological skill.
3. R. Burra and K. Van den Heuvel, *Traditional irrigation in Tanzania*, Vol. 1 (Dar es Salaam: SNV, 1987).
4. W. M. Adams, and D. M. Anderson, "Irrigation before Development: Indigenous and Induced Change in Agricultural Water Management in East Africa," *African Affairs*, Vol. 87, No. 349 (Oct., 1988):519-535.

5. See for example M. J. Sheridan, "An irrigation intake is like a uterus: Culture and agriculture in precolonial North Pare, Tanzania." *American Anthropologist* 104, 2002:1, 79-92.
6. H. Okoth-Ogendo, "Legislative Approaches to Customary Tenure and TenureReform in East Africa." *In Evolving Land Rights, Policy and Tenure in Africa*, eds C. Toulmin and J. Quan, 2000, 123–34. London: DFID/IIED/NRI.
7. Burra and Van den Heuvel, *Traditional irrigation in Tanzania.*
8. See V. Mudimbe, *The invention of Africa: gnosis, philosophy and the order of knowledge* (Bloomington: Indiana University Press, 1988).
9. P. Fleuret, "The Social Organization of Water Control in the Taita Hills, Kenya," *American Ethnologist*, Vol. 12, No. 1, 1985: 103-118.
10. See F. T. Masao, "The irrigation system in Uchagga: An ethno-historical approach," *Tanzania Notes and Records*, 75, 1974, 1-8.
11. See Adams and Anderson, "Irrigation before Development: Indigenous and Induced Change in Agricultural Water Management in East Africa," 2005, 519-535.
12. Interview with an agricultural officer in Hai, June 22, 2009.
13. Interview with an agricultural officer in Hai, June 22, 2009.
14. See for Example E. V. Swai, "Construction of Womanhood in Africa: The Case of Women in Rural Tanzania," PhD. thesis, Pennsylvania State University, 2006.
15. Rebman reached Kilimanjaro in 1860s; the technology of construction of canals along Mt. Kilimanjaro has been scorned as wasteful from the perspective of water management use of water. See for example L. R. Daluti, *Report on the Agro-social-economic Situation in Pangani River Catchment.* United Republic of Tanzania-Zonal Irrigation Office Moshi, 1994.
16. See M. J. Sheridan, "An irrigation intake is like a uterus: Culture and agriculture in pre-colonial North Pare, Tanzania," *American Anthropologist* 104:1, 2002, 79-92; and E. V. Swai, "Construction of Womanhood in Africa: The Case of Women in Rural Tanzania," PhD. Thesis, Pennsylvania State University, 2006.
17. Where technology is assumed to be a way of knowing, rather than an artifact produced from scientific investigation.
18. M. Tagseth, "Oral history and the development of indigenous irrigation: Methods and examples from Kilimanjaro, Tanzania," *Norsk Geografisk Tidsskrift_Norwegian Journal of Geography* Vol. 62, 9 2008, 22.

References

Abbott, J. 1995 "Community participation and its relationship to community development," *Community Development Journal*, 30, 2, 158–168.

Adams, W. M. and Anderson, D. M. 1988 "Irrigation before Development: Indigenous and Induced Change in Agricultural Water Management in East Africa," *African Affairs*, Vol. 87, No. 349, 519-535.

Amutabi, M. 2006 *The NGO Factor in Africa: The Case of Arrested Development in Kenya,* New York: Routledge.

Balter, M. 2001 "Paleontological rift in the Rift Valley," *Science*, 292, 198–201.

Begun, D., Ward, C. and Rose, M. 1997 *Function, Form and Fossils* (New York:Plenum Publishing).

Burra, R. & Van den Heuvel, K. 1987 *Traditional irrigation in Tanzania,* Vol. 1 SNV,Dar es Salaam.

Colding, J., and Folke, C. 1997 "The relations among threatened species, their protection, and taboos. Conservation Ecology [online]." Available from: http://www.ecologyandsociety.org/vol1/iss1/art6/manuscript.frames.html (Accessed March 6, 2009).

Escobar, A. 1995: *Encountering development. The making and unmaking of the Third-World.* Princeton, NJ: Princeton University Press.

Fleuret, P. 1985 "The Social Organization of Water Control in the Taita Hills, Kenya,"*American Ethnologist*, Vol. 12, No. 1, 103-118.

Freire, P. 1972 *Pedagogy of the oppressed.* Harmondsworth: Penguin.

Friedman, J. 1992 *Empowerment—The politics of alternative development.* Oxford: Blackwell.

Leakey, R. and Lewin, R. 1992 *Origins Reconsidered: In Search of What Makes Us Human* (New York: Doubleday).

Leont'ev, A. N. 1978. *Activity, consciousness, and personality.* Englewood Cliffs, NJ: Prentice-Hall.

Lewis, B. F., & Aikenhead, G. S. 2001 "Shifting perspectives from universalism to cross-culturalism," *Science Education*, 85, 3 – 5.

Madulu, N. F. 2003. "Linking poverty levels to water resource use and conflicts in rural-Tanzania," Phys Chem Earth Parts A/B/C. 28:911–917.

Malley, Z. J. U., B. Kayombo, T. J. Willcoks, P. W. Mtakwa, 2004. "Ngoro: an indigenous, sustainable and profitable soil, water and nutrient conservation system in Tanzania for sloping land," *Soil Tillage Res.* 77:47–58.

Masao, F. T. 1974 "The irrigation system in Uchagga: An ethno-historical approach," *Tanzania Notes and Records* 75, 1-8.

Mgumia, F. H. and Oba, G. 2003 "Potential role of sacred groves in biodiversity in Tanzania," Environmental conservation [Online]. [cited 2007 Feb 27] 30(3):259-266. Available from: journals.cambridge.org/article_S0376892903000250

Mudimbe, V. 1988 *The invention of Africa: gnosis, philosophy and the order of knowledge* (Bloomington: Indiana University Press).

Niemeijer, D. 1996. "The dynamics of African agricultural history: Is it time for a new development paradigm?," *Development and Change* 27:1, 87-110.

Oyewumi, O. 1997 *The Invention of Women: Making an African Sense of Western Gender Discourses.* Minneapolis: University of Minnesota Press.

Rao N. H, P. P. Rogers 2006. "Assessment of agricultural sustainability," *Curr. Sci.* 91:439–448.

Rebmann, J. 1860. Journey to Jagga. Krapf, D.J.L. (ed.) "Travels, Researches, and Missionary Labors During an Eighteen Years' Residence in Eastern." In *Norsk Geografisk Tidsskrift* 62 (2008).

Sheridan, M. J. 2002 "An irrigation intake is like a uterus: Culture and agriculture in pre-colonial North Pare, Tanzania," *American Anthropologist* 104, 1, 79-92.

Spivak, G. 1988. "Can the subaltern speak?" In C. Nelson & L. Grossberg (Eds.), *Marxism and the interpretation of culture* Urbana: University of Illinois Press. 271 – 313.

Swai, E. V. 2006 "Construction of Womanhood in Africa: The Case of Women in Rural Tanzania," PhD. thesis, Pennsylvania State University.

Tagseth, M. 2008 "Oral history and the development of indigenous irrigation. Methods and examples from Kilimanjaro, Tanzania," *Norsk Geografisk Tidsskrift_Norwegian Journal of Geography* Vol. 62, 9, 22.

United Republic of Tanzania Ministry of Water Energy and Minerals. 1994. *Report on the Meeting on Water Utilisation and Shortage in the Pangani River Basin.* United Republic of Tanzania , Ministry of Water Energy and Minerals, Moshi.

Van Eijck, M. W. and W. M. Roth, 2007. "Keeping the local local: Recalibrating the status of science and traditional ecological knowledge (TEK) in education," *Science Education,* 91, 926 – 947.

• Chapter 3 •

SCIENTIFIC AND TECHNOLOGICAL KNOWLEDGE AND PRACTICES AMONG THE YORUBA BEFORE COLONIALISM

Olukayode Segun Eesuola

Introduction

In all human societies, science and technology (S&T) often evolve due in part to human curiosity and desire to respond to needs and challenges of nature. This phenomenon is natural to all human existence. As all societies develop from one social stage and mode of production to another, new needs and challenges emerge, and in a causal relationship, the levels of scientific knowledge and technological engagement continue to advance. Except in the case of interruption, therefore, this is a pattern that all societies undergo, though the time they undergo it and the type they undergo often differ. Suffix to say, therefore, that the type of science and technology that develop in a particular society, the time it develops, as well as the speed at which it develops are all tied to the needs and challenges imposed on the society by the social stage and the dominant mode of production in the society.

This chapter posits that the Yoruba people of West Africa, like other peoples of the world, had developed and were still developing the science and technology relevant to the challenges of their environment before they were interrupted by colonialism. It draws evidence freely from Yoruba social structures, tools fabrication, medicine, pharmacy, communications, arts and equipment. The chapter insists that it was colonialism that halted the development of these scientific and technological processes, replacing them with other ones brought from Europe, and that this is the historical scenario that is responsible for the present claim that Africa is synonymous with technological backwardness. The chapter then posits in conclusion that for Africa to rank along the West in the current wind of globalization, it must revisit the significant science and technology it left behind in the face of colonialism.

The combined world of science and technology remains some sort of evidence of man's age long resolution to conquer and control his environment. This two-in-one term often emerges from man's dealings with his basic natural attributes such as feeding, shelter, security as well as self actualization. Science "merely provides principles"[1] upon which technology builds, and the pattern these attributes take in a particular environment often determines the pattern of technology that develops in that environment. It takes a snow prone society, for instance, for the members of the society to invent defrosting technological devices after a long period of scientific study and observation. It is like the old saying; necessity is the mother of inventions. Suffix to say then, that except in the case of intrusion or interference as was witnessed in Africa, every community of people naturally engages in scientific and technological realities of their time, and as necessitated by their daily survival and ambitions. Indeed, "in the field of science and technology, the history of Africa's self-development before foreign rule began has shown that these peoples achieved much in the past".[2] So, science and technology are one indivisible social venture that is driven by the nature of man's environmental challenges and ambitions.

What is Science and Technology?

Whatever we take as science and technology will definitely color our understanding of the twin concept, and will also determine our position on who had or did not have it in the past. It refers to knowledge as of facts, phenomena, laws and approximate causes gained and verified by observations, organized experiments and correct thinking.[3] We then take science as knowledge, and, for that matter, a kind of knowledge that is based on observations and empirical evidence that allow objective generalizations; and we further argue that to say a people do not have science is to derogatively refer to them as inferior species of homo sapiens who can neither partake in careful observation, nor think correctly at all. Then the twin word, technology: it is a combination of two Greek morphemes, techne, and logos, meaning skill and study respectively. Technology then implies practical application of scientific knowledge to issues that solve social problems and satisfy human needs and curiosities. "It is the technique of applying science for commercial and industrial purposes."[4]

From the foregoing clarification, it is clear and logical that like any other people in their societies, Africans must have possessed knowledge of, and engaged in a series of scientific and technological endeavors before colonialism. Their architects designed and built their houses. Their physicians healed their people. They fabricated tools and weapons used for the control of their environment, and they engaged in crop and animal farming for their immediate and remote survivals. All of these are scientific and technological knowledge among the African people before colonialism. It was when colonialism penetrated them that the environment as well as the mode of production changed and a new social structure evolved. This change discouraged and suppressed the advancement of indigenous scientific and technological knowledge and practices of the people. It is therefore strange to see modernization theorists such as Moore (1963) and Oludare (1999) who assert that modernization has equivalent to the transformation of African traditional societies from primitive to technological social formations. Such argument often implies that the traditional society is anti technology, and that technology only exists in, and began from the so-called modern/western societies. It also implies that the entire African continent, which includes the Yoruba people never had science and technology until they encountered western civilization through the scourge of colonialism.

Several scholars have pointed to evidence of S&T in pre-colonial Africa. Zeleza, for instance, confirms that the indigenous African society was an epitome of knowledge in all walks of life, and that the knowledge was so rich that some of them are toady still hidden to the world due to their lack of interest in making further research into them. Zeleza's opinion is that such abandonment occurred because of the action and inaction of western scholars and their African collaborators who discuss and appraise African knowledge with the tag of inferiority. In his words, "Knowledge permeated every aspect of African traditional society (including the areas of science and technology). Apart from a few anthropological essays on the aspects of this knowledge; African indigenous knowledge potentials remain largely untapped." Zeleza's opinion in the foregoing further implies that it was not exposure to western culture and ideas that brought science and technology to Africa, rather, people often think that Africa is technologically backward due to what he explains as the "myth of inferiority of African culture and African knowledge which has led to the condescending attitudes towards knowledge in Africa."[5]

In the foregoing, Zeleza feels strongly that currently the world neither recognizes that Africa had and still has scientific and technological knowledge, nor do most people know that their developments in these areas were thwarted by western penetration. As this appears to be an irreversible change, Zeleza then recommends that in order for Africa to check the ugly trend and take the frontline of scientific and technological knowledge for which history has already registered her, she must be more aggressive and pro active by liaising with other developed nations of the world, so that she can "facilitate transfer of capital (technological, financial, cultural and political) and help mediate in terms of knowledge production, the globalization of African scholarship and Africanization of global scholarship."[6]

But Oludare (1999) has an entirely different view. He sees science and technology as part of, if not the same, as culture. He argues that the imposition and establishment of European and American cultures created science and technology in Africa. He then concludes that "the relevant underdevelopment of African countries can be attributed in part, to absence of science and technological culture in most parts of Africa. Science and technology are mystified and pursued by a privileged few."[7]

This position obviously lags behind in two ways. It refuses to clarify whether the de mystification of science and technology is inherent in Africa, or the development arose after a period of history. Apart from being silent on the issue of when, Oludare does not also provide a systematic explanation on how science and technology became mystified in African societies. The silence on when and how of the so-called mystification of science creates and un-affordable suspense in his argument.

Other evidence abound that Africans had scientific knowledge and engaged in technological development before their contact with Europe. Color technology that is studied in Chemistry today began in ancient Egypt where Africans used animal products to achieve a wide array of colors. African blacksmiths had fabricated weapons and tools as far back as 1500 BC. It is now accepted worldwide that Egyptian Imhotep, AKA Prince of Peace, is the father of medicine, even modern medicine, and that amongst African people, "the appearance of metal and metal working technology is attested to in the archeological records of Africa by different kinds of evidences- artifacts made of metal tools used to work it."[8]

Resting on the thesis of social constructionism, this chapter argues that environmental demands and generational ambitions are the driving force of scientific and technological inventions, and when inventions are made, their knowledge advance with time, except in cases of calculated interruption that colonialism did to African and Yoruba civilization. Thus, the trend of technological backwardness in Africa was not inherent. Rodney (1972) contends that the foreigners that entered Africa in the name of colonialism deliberately and systematically worked for the underdevelopment of Africa. His view is that bringing Africa into the world economy at the time was to put them in a competitive disadvantage. Since Africa did not have the kind of science and technology present in Europe and America, which was the parameter for modernization, it became hard for Africans to compete with their foreign counterparts. The consequence of this was exploitation. Subsequently, Africa became a dependent entity that serves foreign interests. It became externally controlled and regulated by the metropolitan countries. Corroborating this view is the fact that the "backward regions assumed a dependency status (the last step before outright control) in relations to the metropolitan powers chiefly because the former were indebted to the latter."[9]

In a bid to assertively defend this position, this chapter interrogates, investigates, establishes and highlights the existence of science and technology among the Yoruba people of Africa before colonialism. The fact is that long before the advent of colonialism, the Yoruba people, like other peoples of the world, made tremendous advancements in scientific knowledge and its practical application, as evident in medicine, pharmacy, architecture, tools, musical and communications equipment.

Let us begin with medical and pharmaceutical sciences. In a comparative description of their medicine with that which developed in Europe, A. D. Buckley asserts that "Yoruba medicine is similar to European medicine in that the main thrust is to kill or expel from the body, the tiny invisible germs or insects (kokoro) and also worms, (aron) which inhabits small bags within the body."[10] He adds that "in summary, what the pre colonial Yoruba medicine strive to do, like European medicine, is to strive to destroy the agency that causes diseases." (ibid) Furthermore, the Yoruba physicians were neither quacks nor jacks-of-all-trades. Each concentrated on specific areas of various competence, with the exception of *adaunse,* the general practitioner, who had the capability of curing many ailments. Even in the absence of scanning machines and other contemporary equipment, the Yoruba people were very systematic and scientific in medical diagnosis. They resorted to Ifa divination where they were in doubt of a particular illness as corroborated here "that in those situations where there are many hidden factors... the gods clearly take over in Yoruba traditional thoughts."[11] Without suggesting that Ifa divination is absolutely accurate, one can argue that it was quite scientific because, unlike many other forms of divination which were based on magic, Ifa is a systematic knowledge that is so open that anyone can learn its principles and apply them with or without going through rituals. With this systematic and scientific practice, the medical prowess of the Yoruba people was remarkable in the treatment and management of even the so-called terminal diseases such as cancer and tuberculosis. Therefore, with systematic uses of leaves, herbs and other natural materials, coupled with the systematization of investigation, diagnosis and transfer of such knowledge, the Yoruba people had entrenched vast medico-scientific prowess before they were halted by the Europeans.

We can move to architecture. Shelter is one of the basic technological inventions that protect man against some environmental hazards that sometimes accompany nature: gnat, wild wind and even predators. It was in efforts to secure such protections that man advanced in modes and means of erecting houses in line with his immediate environmental challenges. The Yoruba people were no exceptions here. For some scientific reasons, their houses were usually made of mud and thatch. "The walls of the houses rising from 7-8 in height are built of mud, covered with tall grass, called bere or sege or ekan. In forestlands where these are not obtainable, a kind of broad leaf called gbodogi is used instead... the walls are plastered with black and sometimes red earth by the women."[12]

The reason that informed the choice of brick and grass is scientific, and it was used for such architectural technology that helped to keep the houses free from lizards, snakes and other reptiles. The red earth acts as temperature regulator, keeping the room cool all through the year. This marvelous brainchild of the pre-colonial Yoruba was a product of scientific knowledge worked-out by Yoruba technologists; and the knowledge kept advancing as time passed by. This is "the procedure of the unfolding of the imagination and the understanding of a fundamentally subjective relationship that is altogether appropriate to knowledge and which exhibits the reason for the pleasure of the object." [13] It is also scientifically interesting to note that the mud used for construction of houses is bullet proof and fire-resistant, reducing the fear of inferno that was then rampant in their environments. This is really the thought-product of scientists and technologists who had worked on several years of observations and applications.

Yoruba science and technology also extended to the field of agriculture. Food is fundamental to our needs, and we try to perfect the satisfaction of basic needs through the cultivation of crops and rearing of animals. The Yoruba gained ample knowledge of soil and animal fertility as well as crop propagation through scientific observation. Their hunters, while roving the forests in search of game, gained knowledge of, and documented the use of plants and animals. As Atanda observes, "through the knowledge so gained by the hunters, the people who wished to acquire new forest for plantation usually approached a hunter to direct them to a fertile soil."[14] Such scientific methods were also used in areas such as crop rotation, shifting cultivation, manuring, irrigation and erosion control. Technological inventions also advanced the fabrication of tools such as cutlass, hoe, and blacksmith blower as supported by the fact that the "single most important technological change underlying African agricultural development was the introduction of iron tools, notably the axe and the hoe, replacing wooden and stone tools." [15]

Aesthetic and general artistic impulses had a determining role in the genesis of Yoruba artistic productions at the very point where arts meet with science and technology: the area of sculpture, drums making, pottery, paintings, dyeing and weaving. These human activities, thought technically not called scientific, are primarily based on a scientific and technological presuppositions. The wood systematically carved to achieve sounds from dundun and bata talking drums that the Yoruba people had used since the 15th century are products of science and technology. The clay pot is a product under this category. It retains temperature and prevents the growth of micro-organisms on food, compared to the steel pot. Its manufacture was a product of technological thought, and its selective usage was a product of scientific observations.

Painting in the pre-colonial Yoruba land was conceived against the background of architectural settings, in such a manner that a building was incomplete without paintings. It is observable that the fundamental purpose of the paintings was, in addition to aesthetics, to strengthen the walls and depict every detail of human existence. Paints were made of specially processed cow stool called boto, and it is comparable to the state of communion with being in which man, according to Heidegger[16], finds himself as being-there (Da-sein), and as the shepherd of being.

For dyeing, women usually bought a quantity of yarn, bleached and dyed them in various aesthetic colors and sell them to weavers. Women, as well as men, engaged in weaving. Men weaved in small breadths and carried out their occupation in courtyards or secluded squares where they can stretch their warp. On the other hand, women fix their looms in the pizza of the house, close to the door of their apartments where they may be seen sitting on the ground, with their legs in a hole under the iron. They weaved cloths in broad pieces called ipele, two or three breadths forming a covering. The wool is wrapped round stout bamboo poles fixed to strong upright posts, top to bottom. There is a mechanism by which the breadths can be made to cross each other. Indeed, the process of cloth weaving in the pre-colonial Yoruba society is a product of scientific thought and technological invention as attested to by Bray, (1994) that "the Yoruba speaking area of Nigeria is one where two forms of weaving technology overlap, and both the upright single heddle loom used by women, and the narrow strings double heeded loom mostly used by men are found."[17]

These artistic works of science and technology constituted the crystallization of flux. They served as a resolution of the conflict between the repertoire of forms in the art seeking expression and appearances. It is the flux from the reproduction of sense data to the making of sense and a new creation out of them. It is in this regard Goethe seemed to have affirmed that nature and art appear to flee each other, yet before one can take a thought, they have found each other . This is exactly what science and technology is all about: the intercourse between nature and human art, and "the logic of art is the logic of integration and coordination, where the individual and the universe are fused together, while intuition and imagination transform the seasons and the intellectual experiences of the artist into an aesthetic continuum." [18]

In summary, we have so far established that contrary to Eurocentric scholarship that is sacrosanct in its claim of the European origin of science and technology and that Africans never had any knowledge of such until her contact with the west, Africans, like other peoples of the world had rich science and technology before they were infiltrated by colonialism. We use the pre colonial Yoruba people as a case study, and we extracted evidences from medicine, tools fabrication, mining, agriculture and other endeavors. We then feel that certain steps have to be taken by African scholars and policy makers in a bid to correct the situation. These steps will serve two purposes. First, it will take history out of pervasion and set it right. Second, and more importantly, it will lead to immediate re -discovery of several lost scientific and technological ingenuity of the African people.

Conclusion

Africa is a continent that has gone through many humiliating and unmitigated tragedies. The agonies of slavery, and the resultant psychological demoralization of Africans, the pervasive and outrageous swindling and pillaging of African resources, consequent upon colonialism and the new-age imperialism, the ceremonial but un concretized political autonomy granted the erstwhile colonies in Africa by the colonizing powers, and a host of other manifest woes, have cast Africans aside as an unfortunate conglomeration of ill-fated people inhabiting the continent. One of the areas of devastation is the suppression of science and technology of the African people. Although some scholars, especially the modernization theorists, posit that colonialism was a coin with two sides- exploitation of Africans on one hand and technological development on the other- others such as Rodney and Amin see it otherwise. In fact, Rodney argues rather bitterly that "it would be an act of the most brazen fraud to weigh the paltry social amenities provided during the colonial epoch against the exploitation and to arrive at the conclusion that the good outweighed the bad."[19]

There is no doubt that the 21st century dependency created a scientific and technological paradigm shift. The colonialists introduced a new set of technological ideas to Africa, took over political power to prepare a welcoming environment for their new science and technology and, as a result, changed the entire social order. These foreign ideals were then used as models for the measurement of civilization, science and technology. The African indigenous science and technology were termed "primitive" and "barbaric", and as such, were prevented from developing while the foreign ones were seen as being "progressive" and "modern". This led to cessation of efforts to advance the developments of the indigenous science and technology in Africa and by Africans.

There are two major propositions from our short discussion. One, the Yoruba people of West Africa developed, and were still developing science and technology necessitated by and relevant to their environment before the intrusion of colonialism. Second, the intrusion of colonialism suppressed, disjointed, altered and immobilized the scientific and technological advancements of the Yoruba people by creating a new social order which only supported the science and technology of the colonialists, thereby making them seem more relevant than the African ones. This condition developed the imposed technology and halted the growth of the indigenous one, and this is what the world now interprets as backwardness in African technology. From this background comes "the egregious error of the current sociology of development that has culminated in both blooming of the fundamental relationship between industrialized and non-industrialized societies, and in the development of theories that enhance western ethnocentrism."[20]

We recommend that if Africa must reclaim its technological competence after its deliberate suppression by Europe and America, it must first re-visit and rely on its languages, culture and worldviews. Fela Anikulapo-Kuti is one Pan Africanist singer who has been consistent in the issue of sociology of knowledge and its pervasion by the western world. He strongly believes that Europe and America did not teach African technology, matter-of-factly, it is in the reverse. In one of his songs, Perambulator, (1977) Anikulapo-Kuti insists:

Oyinbo no teach us notin	The Europeans taught us nothing
Teach us notin	Taught us nothing
Na we open dem eye	Rather, we civilized them

He then goes on to recommend rather figuratively in another song, Big Blind Country, that the African man needs to begin to look at issues from African perspectives, otherwise, European suppression of Africans will not stop. The lyrics:

We must to use African medicine	We need to use our African medicine
Make our eyes dey se well well o	So our eyes can see clearly
So we can se everything correct	So we can see the right way
Otherwise Kurukere no go stop o	Otherwise, evils of the west will not stop.

Unless and until this is done, the Yoruba people, and indeed all Africans, will continue to grope in the dark in search of identity, truth and meaning, especially when issues of science and technology are discussed around the world, and "until foreign institutions and cultures are removed from African land, that is when the African scientific and technological geniuses will be born."[21]

Notes

1. Remi Anifowose and Francis Eneumuo, eds. *Elements of Politics* (Lagos: Malthouse, 1999). 23.
2. Basil Davidson, *The Black Man's Burden* (New York: Times Books, 1992), 9.
3. *The International Webster's Dictionary* (USA: Typhoon International, 2004).
4. *The Oxford Paperback Encyclopedia* (London: Oxford University Press, 1998).
5. Paul Tiyambe Zeleza, "The Academic Diaspora and Knowledge Production in and on Africa" in *African Intellectuals: Rethinking Politics, Language and Development*, ed Thandila, M. (London: Zed Books,2005),208.
6. Ibid, 209.
7. Dare Oludare, *Why Africa is Poor* (Ikeja: Opinion Media, 1989), 26.
8. Joseph Vogel, ed, *Ancient African Metallurgy: the Socio Cultural Conflicts*, (Boston: Roman and Littlefield, 2000), 68.
9. James O'Connor, "The Meaning of Economic Imperialism," in K. T. Fann and Donald C. Hodges, ed, *Readings in U.S. Imperialism* (Boston: F. Porter Sergeant, 1971), 33.
10. Anthony D. Buckley, *Yoruba Medicine* (Oxford: Clarendon Press, 1985), 17.
11. Sodipo "Notes on Concepts of Cause and Chance in Yoruba Traditional Thought" in *Idea of African Philosophy*, eds, Olusegun Oladipo, (Ibadan: Hope Publications, 2000), 42.
12. Samuel Johnson, The *History of the Yorubas* (Lagos: CSS Ltd, 2001).
13. Hans-George Gadamer, *Truths and Methods* (New Heavens, Yale University Press, 1979), 40.
14. Toyin Falola, *Yoruba Gurus: The Indigenous Production of Knowledge in Africa*, (Trenton, NJ: Africa World Press , 1997), 42.
15. Walter Rodney, How *Europe Underdeveloped Africa* (Abuja, Panaf Publishing Incorporation, 1997 edition), 42.
16. Martin Heidegger, *Being and Time* (London: Blackwell, 1964).
17. Jennifer, M. Bray, *The Organization of Traditional Weaving in Iseyin*, 1968, in *www.articlebased.com*, November, 2009.
18. Chukwulozie, K. Anyawun, *The Nature of Black Culture Reality* (Washington DC: University of America Press, 1976), 384.
19. Walter Rodney, *How Europe Underdeveloped Africa*, 247.
20. D. Offiong, *Imperialism and Dependency* (Enugu: Fourth Dimenssion Publishing, 1980), 24.
21. Fela Anikulapo-Kuti, "I am PhD Common Sense," in Idowu Mabinuori, *Fela Why Black Man Carry Shit* (Ikeja: Opinion Media, 1986).

References

Amin, Sámi. *Unequal Development: an Essay of Social Formation in Capitalism,* 1976, in www.articlebased.com 2009.

Anifowose, Remi and Eneumuo, Francis, eds. *Elements of Politics.* Lagos: Malthouse, 1999.

Anikulapo-Kuti, Fela. *Perambulator.* Lagos , Africa 70, LPIO8897, 1977.

Anikulapo-Kuti, Fela. *Big Blind Country,* Africa 70, Lagos 1989, Unreleased.

Anikulapo-Kuti, Fela. "I am PhD Common Sense", in Idowu Mabinuori, Fela *Why Black Man Carry Shit.* Ikeja: Opinion Media, 1986..

Anyanwu, Chukwulozie, K. *The Nature of Black Culture Reality..* Washington DC: University of America Press, 1976..

Bray, Jennifer. M. *The Organization of Traditional Weaving in Iseyin,* 1968, in *www. articlebased.com* , November, 2009.

Buckley, Anthony D. *Yoruba Medicine.* Oxford: Clarendon Press, 1985.

Cordwell, J. *Naturalism and Stylization: Yoruba Art,* in Magazine of Art, 1953 , 0.220-225.

Goodman, Collins and Chant. *Pre Industrial Cities and Technology.* London: Rutledge and Open University, 1999.

Davidson, Basil. *The Black Man's Burden.* New York: Times Books, 1992.

Falola, Toyin. *Yoruba Gurus: The Indigenous Production of Knowledge in Africa,* Trenton, NJ: Africa World Press Inc, 1999.

Gadamer, Hans-George. *Truths and Methods.* New Heavens: Yale University Press, 1979.

Galtung, J. *The European Community: A Superpower in the Making,* London: George Allen and Unwin, 1973.

Heidegger, Martin. *Being and Time.* London: Blackwell, 1964.

Herodotus. *Histories II,* trans by George Rawlinson, New York: Tudor Publishing, 1956.

Johnson, Samuel. *History of the Yoruba.* Lagos: CSS Ltd, 2001.

Katznelson, I. et al, eds. *The Politics and Society Reader.* New York: David Mckay Co, 1974.

Moore, W. *Social Change,* New Jersey: Prentice Hall, 1963.

Ntuli, P Pitika. "Indigenous Knowledge Systems and the African Renaissance" In Hoppers, Catherine A. Odora, eds,. *Indigenous Knowledge and Creation of Knowledge System,* South Africa: New Education, 2002.

O'Connor, James. "The Meaning of Economic Imperialism" , in K. T. Fann and Donald C. Hodges, eds, *Readings in U.S. Imperialism.* Boston: F.Porter Sergeant, 1971.

Offiong, D. *Imperialism and Dependency,* Enugu: Fourth Dimension Publishing , 1980.

Oludare, Dare. *Why Africa is Poor.* Ikeja: Opinion Media, 1989.

Oluwole, Sophie, eds. *The Essentials of African Studies,* Lagos: Mayodele Ajayi Enterprises, 1997.

Oxford Paperback Encyclopedia, London: Oxford University Press, 1998.

Rawlinson, George, trans II, *History of Herodotus,* New York: Tudor Publishing, 1956.

Rodney, Walter. *How Europe Underdeveloped Africa*, Abuja: Panaf Publishing Incorporation, 1997.

Sodipo. "Notes on Concepts of Cause and Chance in Yoruba Traditional Thought" in *Idea of African Philosophy*, eds, Olusegun Oladipo, Ibadan: Hope Publications, 2000.

Tausch, A. *Towards a Socio Liberal Theory of World Development*, New York: St Martins, 1993.

The International Webster's Dictionary, USA: Typhoon International, 2004.

Vogel, Joseph, ed. Ancient *African Metallurgy: the Socio Cultural Conflicts.* Boston: Roman and Littlefield, 2000.

Zeleza, Paul Tiyambe. "The Academic Diaspora and Knowledge Production in and on Africa" in *African Intellectuals: Rethinking Politics, Language and Development*, ed, Thandika Mkandawire, M. London: Zed Books, 2005.

• Chapter 4 •

YORUBA NUMERAL SYSTEM, EPISTEMIC ROOT IN IFA ORACULAR CORPUS AND THE CHALLENGES OF SCIENCE AND TECHNOLOGY

Nelson O. Fashina

Introduction

The Yoruba people and their language are regarded as one of the earliest in world history, based on current research. They are primarily located in Western Nigeria, but their geographical spread extends to Togo, Ghana, Benin Republic, Sierra Leone and Liberia in West Africa. The Yoruba Diasporic continuities cover United Kingdom, USA, Brazil, Cuba and the Caribbean. With about 40 million Yoruba speaking people all over the world, it is one of the major classical linguistic cultures to develop a "written grammar and vocabulary" in black Africa (Campbell, 1995: 588). Thus, it is one of the classical African languages with high literacy and hence the potential capability to annex the resources of global initiative of improved knowledge and civilization through science and technology. While major works have been done to improve the Yoruba language communicative efficiency through linguistic resources, up till this early beginning of the 21st century, only a little work has been done in the field of research into the origins, nature and characteristic features of classical Yoruba numerals, apart from Fakinlede's proto-scientific attempt at categorizing and developing some scientific formula for already existing Yoruba numeral words (Fakinlede, 2005). The researcher expresses deep concern that Yoruba language study has delayed so much to develop a modern science and technology compliant numeral system. Fakinlede's worry is understandable because as a chemistry Research Fellow based in the United States, he is driven by the innovation conscious desire to make scientific discoveries compliant to, and teachable in, his Yoruba mother tongue. Thus, the researcher argues that world civilization and technological development is fast dynamic and that this civilization is dictated by the pace and advancement of knowledge in the sciences and technology which can only be appropriated and applied to acquisition of knowledge in teaching and education via the indigenous language of every race. He laments in the old fashion that comparative level of development as those of the Western and Eastern parts of the globe may be unattainable by the Yoruba nation if, in the 21st century, the Yoruba language still depends on what he calls 'heavily wordy' numeric system of counting which has persisted in Yoruba pristine education and indigenous technology to date. Fakinlede (2003: 226) asserts that the benefits of scientific knowledge through Yoruba language and numerals are not to be forfeited:

> ... in order for these benefits to be realizable, the current capabilities of the Yoruba language have to be evaluated. While Yoruba is fully developed as a medium of communication, art, literature, religion, and commerce, and can be used as a medium for teaching elementary mathematics, its current scope as a language of teaching science above the rudimentary stage is limited. This limitation directly impedes the academic community's ability to make Yoruba textbooks available even for primary schools.

Fakinlede's concern is informed by a sense of nationalism and genuine commitment to the emancipation of Nigeria and Africa from the standpoint of his Yoruba language, culture and technological development. No wonder Bola Ige, the Yoruba patriarch, described Fakinlede in the "Foreword" to the latter's titanic *Yoruba – English Dictionary* (2003:6): "He is an avowed and proud Yoruba nationalist" who "wrote with great knowledge and understanding of Yoruba history, culture, and place in the world." Bola Ige, himself a qualified linguist and writer, opines that new world developments indicate that "a language is no longer a mere means of conversation between two people of the same ethnic group, but a language is a suitable tool of communication with the world, whatever the mother tongue."(p. 6). This view embraces the globalization drive of modern world civilization, and thus implies the need to make our languages, indeed the Yoruba language, accessible, literate, and compliant to modern world dynamics as a means of contributing the knowledge it carries to the ocean of global knowledge. Evidently, as we shall show in this chapter, the bulk of knowledge that the Yoruba language carries is sufficiently derived from Ifa oracular divination corpus. This apparently is one of the missing gaps in Fakinlede's seminal work. An understanding of the source of Yoruba numeral would assist indigenous language education planners like A. B. Fafunwa, J. I. Macauley and J.A.F. Sokoya (1989), Fakinlede (2003, 2005) to achieve more success in the endeavor.

They would be able to route for the rationalism, thought system, philosophy of knowledge and environment, and the logic of indigenous calculus.

In the light of the above, and scholars' worries that the Yoruba may not have an existing science/technology compliant numeral system, this chapter examines the Yoruba language against the background of claims and counter claims about its numeric inefficiency. The chapter practically proves that the Yoruba use words as counting system just as the English language. And this shows that even the early Yoruba had grasp of numeric knowledge such as in concepts of correspondence, classification, and hierarchical order. Truth is that the Yoruba counting system is not wordier than that of English which is a global international language of education, arts, commerce, politics, science and technology. In that wise, Yoruba counting system is adequate for research and practical work in modern science and technology. Many other scholars have made similar hasty calls and justifications for the Yoruba to change from using the relatively global Arabic numeral (1, 2, 3, etc.). In spite of this critical concern about creating a science compliant numeral system in Yoruba, there is the need to understand that both 'globalization' and 'localization' in their extreme left and right parallels may have historical, cultural and cosmological consequences. Succumbing to pressures of 'change' may spell regrettable catastrophe for both hybrid cultures and civilization and sacred indigenous traditions which would be desecrated through swift modern adaptations to 'new' values. Ever since the Arabic numerals were evolved for counting by the Western, Eastern and most colonized nations and technologies of the world, it has been sustained for technological development, education, commerce, politics and other international transactions. And the practice has been without negative impact on the original cultures of the people. Therefore, the fact that numeric system exists in Yoruba language makes it easier to dialogue with other global language cultures for the purpose of evolving composite world knowledge of inter-language counting, measuring, weighting etc. for the purpose of commerce, education, politics, culture, and religion in application to world development plans.

Therefore, the mission of this chapter is first to find out if Yoruba language has actual numeral figures or symbols other than words expressing figures; to find out if the Yoruba words expressing figures are in any way wordier than the English, Roman and Arabic alternatives. In order to present the pragmatic reasons that validate our research findings, this chapter discovers as it conceptualizes the (i) formulaic alternation of additive and subtractive counting system, which invariably underscores early Yoruba knowledge and existence of a proto-arithmetic and mathematical science; (ii) the sequential use of two types of numeric counting codes, namely the five digit count numeric node and the ten digit count numeric node; (iii) the Yoruba distance and weight measurement system; (iv) the Ifá numerical figures as integrated in modern Yoruba counting system; and this manifests in two ways, namely the Yoruba cosmic symbols/ signs of the Odù Ifá, and the proto numeric bonded pairs as seen in Ọ̀yẹ̀kú méjì verse of Ifá divination. The chapter sets to present the picture, nature, properties and characteristic features of Yoruba counting system and its origin in Ifá Divination wisdom, and how these 'numeric' resources provide template of structure for frameworks of science and technology.

Writing in his book, *From One to Zero* (1985: 6), Georges Ifra observes that "Early in this century there were still peoples in Africa, Oceania, and America who could not clearly perceive or precisely express numbers greater than 4. To them numbers beyond that point were vague, general notions related to physical plurality." This claim appears abysmally incorrect in the sense that historical, anthropological and archaeological accounts proved the existence of early civilization among the African peoples, including the Yoruba, the Egyptians and others, even scores of centuries before the early twentieth century mentioned in Ifra's book. In Yoruba, for example, the Ifá divination corpus which dates back to the era before Christ had evolved and later provided a source of numeric counting system to certain hundreds of limits as integrated into and expounded upon in the modern Yoruba language of today. If Ifra's claims were true then, men in every society, including the Egyptians, the Greek and the Romans whose early civilization is incontestable did not even count figures or "express numbers greater than 4" until the early twentieth century. How then did the Egyptians design their hieroglyphs and constructed their titanic pyramids? Were men not able to even count their fingers beyond 4? Were women not counting their lunar menstrual cycles? Answers to these questions are proofs of inaccuracies in Ifra's conjectural claims. Man is indeed a creature with inborn consciousness of numeric values. And the African Yoruba is no exception. George Campbell in his book, *Concise Compendium of World Languages* (1995: 588) confirms that "Yoruba was one of the earliest ... African languages to be codified in the shape of a written grammar and a vocabulary ... and the credit goes to Samuel Crowther, a Yoruba who was sold as slave in 1821, freed by the British, baptized and ordained to serve as a missionary in the Yoruba country."

To both ancient and modern practice of knowledge, the value of numeric understanding, measurement, geometry, arithmetic and mathematics was placed in a second to none order of relevance. Thus, the practice of abstract thinking and liberal value-loaded body of knowledge and of philosophy was regarded with suspicion except it had proof of relative empirical evidence whereby the facts are observable. Scientists did not 'believe in magic'. Rather, they thought that 'reason and proofs define science' (Joy Hakim, 2005:298). The same is applicable to pristine Yoruba 'classical science' and philosophy, which is expressed in the works of verbal arts like poetry, story-telling, story-singing, chanting, drumming, as well as plastic arts like pot making, sculpture, drawing, painting, engraving, designing, iron casting and smelting, goldsmith, blacksmith, weaving of cloths, dyeing, laceration of human cheeks for inscription of ethnic marks, embalmment and other types of traditional medical practice and divination therapy. All these vocations involved calculations of gravitational weight and force attraction between objects in relation. Numeric, arithmetic and geometric measurements were inextricable to these pristine Yoruba technologies.

Relative quantitative and qualitative analyses and similar knowledge to the likes of Pythagoras rules and parallelograms may have unconsciously influenced the measurement and crafting of semi-circles, horizontal and diagonal lines and shapes of ethnic marks made on human cheeks. Here, it is not being proved that pre-colonial Yoruba people invented the likes of the scientific terms of Isaac Newton's formula for the force (F) of gravitation:

$$F = G\frac{m1m2}{R2}$$

m= mass

r = distance between the masses

G = the universal gravitational constant

But, there is obvious proof of application of proto-Newtonian universal rules to basic weight and distance measurement methods. These rules are similar to the use of <u>kilogram</u> for weighing the mass of substance. The Yoruba term for weight measurement is **ìdiwọ̀n or ìwọ̀n.** The <u>meter</u> is for measuring length or distance, and the Yoruba words measuring it are **ìwọ̀n ọwọ, ìwọ̀n ẹsẹ́ or ìlàsẹ̀,** and **saarè** for measuring the radius of land space. The <u>second</u>, <u>minute</u>, <u>hour</u> is for measuring time. The Yoruba sexegesimal words for measuring time are **ìsẹ̀jú** (for minute), **ìsẹ́jú ààyá** (for second), **wákàtí** (for hour)**.** The fact that the Yoruba had words for measure in the three measurement categories is a proof that the Yoruba also had some knowledge of 'the universal gravitational constant' (UGC) which is expressed in Newton's Meters Square per Kilogram Squared (Nm2/kg2). According to Paul Fleisher (2001:18), "a newton is how much push or pull is needed to accelerate a mass of 1 kilogram (1 kg) 1 meter per second per second (1 m/s2). G = 0.0000000000667 Nm2/kg2. That's 10 zeroes after the decimal point----gravity is extremely weak". Although it was not unlikely that Yoruba language numeric measures were based on informed observatory guesses rather than precision and accuracies, yet, there is evidence that the pre-colonial Yoruba knew the density of water, soil, various rocks, metals, etc. And they used this relative knowledge of quantitative measurement of materials in creating arts, crafts, architecture, building and cosmic medicine.

In order to do this, they used a counting system which derived from the Ifa divination knowledge about the principle of 'bonded pairs,' whereby each **Ogbè** or **Odù** exists in pairs of binaries. The near equivalent Yoruba word for pair is **méjì** or **èjì** (two or twin). From this, we derive the inextricable pairs - that is the characteristic numeric value of each of the sixteen great or major **Odù** or **Ogbè** with their embedded or integrated coefficients of Additions, Subtractions and Multiplications. This requires explicit illustration and analysis. Thus we have

Yoruba Counting System	Arabic Numerals	Sixteen Major Odùs of Ifá
Eení	(1.)	Èjì Ogbè
Eèjì	(2.)	Ọ̀yẹ̀kú Méjì
Ẹ́ẹ̀ta	(3.)	Ìwòrì Méjì
Ẹ́ẹ̀rin	(4.)	Òdí Méjì
Aárún	(5.)	Ìrósùn Méjì
Ẹ́ẹ̀fà	(6.)	Ọ̀wọ́nrín Méjì
Eéje	(7)	Ọ̀bàrà Méjì
Ẹ́ẹ̀jọ	(8.)	Ọ̀kànràn Méjì
Ẹ́ẹ̀sán	(9.)	Ògúndá Méjì
Ẹ́ẹ̀wá	(10.)	Ọ̀sá Méji
Mọ́kànláá	(11.)	Ika Mejì
Méjìláá	(12.)	Òtúurupọ̀n Méjì
Mẹ́tàláá	(13.)	Òtúá Méjì
Mẹ́rìnláá	(14.)	Ìrẹtẹ̀ Méjì
Márùndílógún	(15.)	Ọ̀sẹ́ Méjì
Mẹ́rìndínlógún	(16.)	Òfún Méjì

Each Odù or Ogbè is represented by actual Yoruba numerals, which appear like Roman numerals but, evidently, are not sourced from the Roman for two reasons: One, the Yoruba people have been using this numeral for centuries when men could barely count their fingers and toes. Two, this method of counting derives from the Yoruba peasant agriculture when pegs and stakes of yam tendrils were counted on farm. Among the Yorùbá, we can identify three counting systems, namely, the vertical stroke or Roman figure-like type, the word-count system which is most widely used, and the distance and weight counting/measurement system. We shall explain, illustrate and derive their rules of usage.

The Roman Figure-like Type

In this category, the number of strokes corresponds to their numeric equivalence in Arabic numerals: I = 1, II = 2, III = 3, IIII = 4, IIIII = 5, IIIIII

= 6, IIIIII = 7, IIIIIII = 8, IIIIIIII = 9, IIIIIIIII = 10. Now, from this, the Yoruba can calculate in tens, hundreds and thousands using the addition, subtraction, and multiplication device. Thus, to count One Hundred by using this category of numeral system, the Yoruba would make ten strokes ten times as follows: IIIIIIIII x IIIIIIIII (**ìdì mẹ́wá lọ́nà mẹ́wá** or **ìlé mẹ́wá lọ́nà mẹ́wá** : that is ten modules in ten places) = 100. In that continuum, Ten Thousand will be IIIIIIIII x IIIIIIIII x II (ìdì Ọgọ́rún lọ́nà ọgọ́rún: that is hundred modules in hundred places) = 10,000.

This counting category has limits. It may not count in millions. The reason is that at the period of its invention, the quantity of things to be counted was limited in number and quantity, including cowries used as money, houses, horses, slaves, kings, chiefs, serfs, soldiers, farm produce, and even human population. And this is applicable to other ancient historical societies of the world. In modern times, however, quality and values of money and materials are low, but quantities are astronomically multiplying. The challenge that this trend poses to modern Yoruba language has been well attended in the advancement and use of modern Yoruba word counting system for quantity evaluation. Examples are presented and evaluated below:

Yoruba Word Counting System

This is the second category. It uses words to count, just like in English. Again it would be misleading to assume that this form of word-counting derives from Yoruba contact with the British through colonialism. The Yoruba had established some advanced monarchies and cultural civilizations of their empires before the British colonized Africa. This form of counting is more technical than the former. It uses additive and subtractive formula. In the table below, the first column presents the Yoruba word numerals, the second names the corresponding meaning or word figure in, and the third enumerates the similarity and differences between the two.

Yoruba Word Counting System	English version	Area of Conflation (Yorùbá/ English)
Eení/Ọ̀kan (**numeric node**)	One	3- alphabets, same num value
Eèjì	Two	,, ,,
Ẹ́ẹ́ta	Three	,, ,,
Ẹ́ẹ́rin	Four	4-alphabets
Aárún	Five	,, ,,
Ẹ́ẹ́fà	Six	3- alphabets
Eéje	Seven	,, ,,
Ẹ́ẹ́jo	Eight	,, ,,
Ẹ́ẹ́sán	Nine	4-alphabets
Ẹ́ẹ́wá	Ten	3-alphabets

Yoruba Additive and Subtractive Formula for Counting

As from the numbers below, that is ọ̀kànlá to infinity, the Yoruba and English word-numerals share the same number value for all their initial numeric morphemes. Thus, to get the next higher number from **eéwá** (ten) to eérìnlá (fourteen), you add ení (one), èjì(two), ẹ́ẹ̀ta(three), erin(four) as follows: ẹ́ẹ̀wá + oókan = oókànlá (eleven), ẹ́ẹ̀wá + eéjì = eéjìlá (twelve), ẹ́ẹ̀wá + ẹ́ẹ̀ta = ẹẹ̀tàlá (thirteen), ẹẹwa + ẹ́ẹ̀rin = eérìnlá (fourteen). This is the Yoruba additive formula for counting. Thus, as from oókànlá to ẹ́ẹ̀rìnlá, you begin the addition of the òkàn to èrìn in order to get higher number. This additive principle ranges from oókànlá to ẹ́rìnlá, then the subtractive takes over from àrùndínl**<u>ógún</u>** (fifteen) to òkàndínlógún (nineteen) and stops there. Next higher number is Ogún (twenty). And the next counting formula after Ogún is again additive. **'lé'** is the Yoruba additive morpheme, while **'dín'** is the subtractive morpheme. And they are used one after the other in consecutive alternation after approximately five figures each in a continuum of counting from okanla (eleven) to infinity. See the emboldened syllables in the numeral table below. However, a variant of the additive 'le' as in the example in the numeral table from Oókàn**lá** occurs with the slight change from 'le' to 'la'. Within this context of Yoruba additive and subtraction formula for counting, the change from vowel */é/ to /á/* does not alter or change the meaning. They both mean 'to add' or 'in excess of' or 'higher than'. In some cases, where the additive 'lé' or 'lá' stands as a word on its own and is not used as suffix or bound morpheme within a word, the additive morphemes 'lé' and 'lá', each takes initial vowel */o/* i.e. 'ó lé' as in : 'ogún **ó lé** meta' (literally meaning 'twenty in excess of three: 23). Given the same context, the subtractive morpheme 'din' ('less of') will also take an initial vowel */o/* i.e. 'ó dín' as in 'ogún **ó dín** meta' (meaning 'twenty less of three': 17). Below, Ọ̀kànlá is a combination of **numeric counting node (òkàn: one) + an additive suffix (lá: excess of).** Within the context of the usage of **'lá'** as an additive, there is inherent deletion. **'lá'** encodes the following: 'ó **l**é méw**á**'. But, in order to avoid wordiness, 'ó **l**é méw**á**' as a clause undergoes syntactic deletion transformation, and the remnant is **'lá'**. This makes for easier counting as in saying **ẹ́ẹ̀rìn lá** instead of saying **ẹ́ẹ̀rin ó lé mẹ́wá.**

Thus written together with its nodal number, it will be 'oókan lé mẹ́wá' (meaning, one in excess of ten: 10 +1) = oókànlá. The syllabe**'lá'** retains this function throughout, whenever it co-occurs with a numeric node as suffix. This form is found in oókànlá, eéjìlá, eétàlá, and ẹ́ẹ̀rìnlá below.

<u>Oókàn</u>lá (node + additive)	Eleven
<u>Eéjì</u>lá	**<u>Tw</u>**elve
<u>Ẹẹtà</u>lá	**<u>Th</u>**irteen
<u>Ẹ́ẹ̀rìn</u>lá	**<u>Four</u>**teen

All four categories above display the features described as **node + additive** in this chapter. The next five categories below have th**e node + subtractive** formula. Fuller explanations are provided below the following Yoruba and English word-figures:

Aárùndínlógún (node+subtractive)	**Fift**een
Ẹ̀ẹ̀rìndínlógún	**Six**teen
Ẹ́ẹ́tàdínlógún	**Seven**teen
Eéjìdínlógún	**Eight**een
Oókàndínlógún	**Nine**teen

In the five categories above, Aárún, Ẹ́ẹ́rin, Eẹ́ta, Eéjì and Oókan, the emboldened initial position disyllabic morphemes are figure **nodes** (one, two, three, four, and five). '**dín**' is the subtractive formula. The last disyllables 'lógún' in each case is additive formula. Less of deletion transformation, it is written as 'ó lé ní ogún' (meaning 'an x number in excess of twenty', where x may be any number as the case may be).

The next category of five begins with another node figure, 'ogún'(twenty). The additive '**lé**' is used to add oókan, eéjì. ẹ̀ẹ́ta, ẹ̀ẹ́rin (one, two, three, four) to the nodal figure in order to get correct counting up to twenty four – 'ẹ̀ẹ̀rínlélógún' – 'four in excess of twenty'.

Ogún (node)	Twenty
Oókànlélógún (additive + node)	Twenty one

Explanation

It is necessary to give further explanation here for the purpose of clarity. The concatenation of node and additive here also functions the same way as we have in **oókànlá** in the table above.

Analysis

Oókan = one
lé = additive + adjectival (in excess of or in addition to).
lógún = l + ogún(twenty).

The alphabet 'l' as a letter is a lateral, liquid sound which is an adverbial particle whose English near equivalence is 'by' or 'than' or 'more than'. Thus, taken together, the word figure, oókànlélógún, literally means 'one more than twenty' and that is 21. In that counting continuum, you progressively add ookan, eéjì, eta, ẹ̀ẹ̀rin (one, two, three, four) to ogún (twenty) in order to get higher numbers, just the same method as in English word numerals viz: twenty one = 20 +1, twenty two = 20 + 2, twenty three = 20+3, tweny four = 20 + 4. It will not be logical to claim that this structural similarity between English and Yoruba word numeral system is an evidence of colonial influence. This is because the Yorubahad started using their own word numeral system many centuries before the British colonized Nigeria.

Eéjìlélógún	20 +2	Twenty two
Ẹ́ẹ̀tàlélógún	20 + 3	Twenty three
Ẹ̀ẹ̀rìnlélógún	20 + 4	Twenty four

The next category of five word-numbers is the **node + subtractive** type. The node is 'ọgbọ̀n' (thirty). To get aárùndínlógbón (meaning 'five less of thirty'), that is twenty five, you subtract five from thirty i.e. 25 = 30 -5). This counting equation applies till you get to 'oókàndínlógbón' (twenty nine), as shown below.

Aárùndínlógbòn	Twenty five
Ẹ́ẹ̀rìndínlọ́gbọ̀n	Twenty six
Ẹ́ẹ̀tàdínlọ́gbọ]n	Twenty seven
Eéjìdínlọ́gbọ̀n	Twenty eight
Oókàndínlọ́gbọ̀n	Twenty nine
Ọgbọ̀n (Node)	Thirty
Oókànlélọ́gbọ̀n	Thirty one

The Five Digit Additive and Subtractive Alternation Formula

This rule is not prescriptive, but descriptive of the internal laws that we observe as governing the progressive and regressive counting system in Yorùbá. Starting from oókànlá (eleven), we observe that the rule of counting, as earlier pointed out in this chapter, alternates from additive (lé or lá) to subtractive (din). Thus, oókànlá to ẹ́ẹ̀rìnlá (11 to 14), uses additive formula by adding I, 2, 3, 4 to ẹ́ẹ̀wá (10) in order to get the next higher number. But, starting from aárùndínlógún (15), the subtractive formula operates. You regressively subtract aárún, ẹ́ẹ̀rin, ẹ́ẹ̀ta, eeji] oo]kan (5, 4, 3, 2, 1) in order to count from 15 to 19. Then you get to **ogún** (twenty) which is another **numeric node.**

The Ten Digit Count Numeric Nodes

The numeric nodes in Yoruba are **ẹ́ẹ̀wá (10), ogún(20), ọgbọ̀n (30), ogójì** literally 'ogún méjì': 20 x 2 **(40), àádọ́ta (50), ọgọ́ta** literally 'ogún mẹ́ta': 20x 3 **(60),** àádórin **(70), ọgọ́rin** literally ' ogún mẹ́rin': 20 x 4**(80), àádọ́run (90),**

ọgọ́rún literally 'ogún márún': 20 X 5(**100**) . The observed governing rule here is that you get a numeric node to which you add or subtract after the count of 10 digits each.

The Yoruba distance and weight counting / measurement system

Yoruba use the following to measure distance, as in foot, yards, meters, kilometers, etc: ẹsẹ̀ bàtà kan, ẹsẹ̀ bàtà méjì, ẹsẹ̀ bàtà mẹ́ta, etc measuring one, foot, two feet, three feet, etc. They also use ìlawọ́ or ìlasẹ̀ kan, ìlasẹ̀ méjì, ìlasẹ̀ mẹ́ta, etc for one step, two steps, three steps, etc. Similarly, they use saarè for measuring land radius, while weight of substance is weighed, using ìwọ̀n, ìdiwọ̀n, ìgbònwọ́, etc.

The Yoruba numeral figures integrated in Ifá corpus

Earlier, we have taken a pragmatic step to analyze the forms and functions of Yoruba numeral words. Like the English, we have seen that the Yoruba have words that they use as figures to count. And every one seems to know. But, this chapter has gone beyond mere knowing that the Yoruba has words they use in the same way the entire world uses Arabic numerals to count. The chapter has studied the numeral system and developed a grammar of it. Some scholars in African study are wont to accept that the Yoruba has words for naming figures, but would also complain that Yoruba language has no actual numeral figures for counting that is similar to the Arabic and Roman forms. The latter claim is wrong. This chapter confirms that the Yoruba had figures for counting. But, this was swallowed up by the introduction of the Arabic numerals brought about by British colonial education. As we shall soon find out in this chapter, the Yoruba counting figures are still being used till date in Ifá divination science, wisdom and philosophy. Ifá embodies the entire gamut of Yoruba pristine knowledge, and it encodes the past, present, future and the 'fourth stage' of "the unborn" which interlocks between the other three spaces (Soyinka, 1976).

Quite unlike the Roman version, the Yoruba numeral is a simple finger or stick count whereby one stroke I = 1, two strokes II = two, III = three, IIII = four, IIIII = five, etc. Here, the number of strokes corresponds to the actual value of the numeral in say Arabic form. Therefore, 10 = ten strokes, 20 = twenty strokes, and 100 = One hundred strokes up to infinity. In this category, the similarity between the Yoruba and the Roman numeral terminates so prematurely at III (three: 3). But the Roman numeral begins to vary from four which is written as IV. Thus, while the Yoruba form their numerals from one vertical stroke I, "the Romans formed all their numbers from 7 numerals. For numbers consisting of more than 1 numeral, the usual rule was that you added the numerals together to get the value of the number" (Zelda King, *The Story of our Numbers: The History of Arabic Numerals.* New York: The Rosen Publishing Group, 2004). The seven Roman numbers are I = 1, V = 5, X = 10, L = 50, C =100, D = 500, M = 1,000. Thus, in Arabic, III = 1+1+1+1 = 3; XV = 10+5 = 15; CC = 100 +100 = 200, etc. In order to get a lower number, you have to subtract the lower number on the left from the higher number on the right, i.e. (4): IV = 5 – 1= 4, IX = 10 -1 = 9, XC = 100 – 10 = 90. And to get higher number, you have to add the first number to the number on the right, i.e. 6, 7, and 8 are VI= 5+1, VII = 5+ 2, and VIII = 5+3, respectively.

But in Yoruba quasi-numeral system, the numeral counting is in form of modules or bunches. Each module (called ìlé- kíní = one module) may consist of either a set of 3, 6, and 9 in case of certain categories of counting. Thus, 3 sets of 3 (ìlé- mẹ́ta:) = 3 x 3 = 9; 3 sets of 4 (ìlé –mẹ́rin : 3 each set of 3 in four places) = 3 X 4 = 12, and so on. 6 sets of 6 (ìlé mẹ́fà lọ́nà mẹ́fà: each set of six in six places) = 6 x 6 = 36 (mẹ́rìndínlógójì); 3 sets of 36 (ìlé mẹ́ta lọ́nà mẹ́rìndínlógójì) = 3x 36 = 108 (Ọgọ́rún ólé mẹ́jọ).

Yoruba numeric cosmic symbols/signs of the Odù Ifá

1) Èjì Ogbè		2) Ọ̀yẹ́kú Méjì		3) Ìwórí Méjì		4) Òdí Méjì	
I	I	II	II	II	II	I	I
I	I	II	II	I	I	II	II
I	I	II	II	I	I	II	II
I	I	II	II	II	II	I	I
5) ìrósùn Méjì		6) Ọ̀wọ́nrín Méjì		7) Ọ̀bàrà Méjì		8) Ọ̀kànràn Méjì	
I	I	II	II	I	I	II	II
I	I	II	II	II	II	II	II
II	II	I	I	II	II	II	II
II	II	II	II	II	I	I	I
9) Ògúndá Méjì		10) Ọ̀sá Méjì		11) Ìká Méjì		12) Òtúúrupọ̀n Méjì	
I	I	II	II	II	II	II	II
I	I	II	II	I	I	I	I
I	I	I	I	II	II	I	I
II	II	II	II	II	II	I	I
13) Òtúá Méjì		14) Ìrẹtẹ̀ Méjì		15) Ọ̀sẹ́ Méjì		16) Ofun Meji	
I	I	I	I	I	I	II	II
II	II	I	I	II	II	I	I
I	I	II	II	I	I	II	II
I	I	I	I	II	II	I	I

The working of this numeric system for Ifá Divination Corpus is metaphysical and scientific. This knowledge is universal, and there is reason to believe that since this formula predates the age of Western scientific revolution and discoveries, the latter might have had a lot to gain from the former through migration, dispersion and colonialism. A reading of the above data table would reveal that the numerals symbolically occur either in single- bonded pairs or in double-bonded pairs. Each bonded pair, whether single or double, exists in relation or alternation of a Negative co-occurring with a Positive – one on the left, and the other on the right. This notion of bonded pairs is the basis of chemical engineering, physics and biology. Before the discovery of science, Ifa divination wisdom and philosophy of the Yoruba people discovered that the principle of universal laws is based on contrariety – coexistence of contrasts such as good and bad, light and darkness, white and black, day and night, heaven and earth, length and breadth, male and female, water and fire, air and earth, solid and liquids, back and front, etc. Ifá divination recognized that whenever these binaries interact, they emit energies of checks and balances that make things to work and produce effects. This explains the reason why phenomena in all the compartments of Ifá wisdom and philosophy exist in bonded pairs, which enables substances, plants, animals and man to connect to certain powerful universal radiation whose effect can activate or repel, as the case may be. No wonder we have Èjì Ogbè, Ọ̀yẹ̀kú méjì, ìwórí méjì, Òdí méjì, ìrósùn méjì, Ọ̀wọ́nrín méjì, Ọ̀bàrà méjì, Ọ̀kànràn méjì, Ògúndá méjì, Ọ̀sá méjì, ìká méjì, Òtúúrúpọ̀n méjì, Òtúá méjì, ìrẹtẹ̀ méjì, Òfún méjì. The words eéjì and méjì are the same, meaning 'pairs'. They are numeric adjectives that postmodify their nouns. In this sphere, Ifá wisdom and philosophy had long before the present age posed a big challenge to science and technology. And the latter had responded by converting this knowledge into compact uses for development and civilization. Thus, for example, electricity is generated from a combination of bonded pairs of anode + cathode, light energy conductor + insulator, ovule + spermatozoa, etc. Although, Ifá as Yoruba science and philosophy did not actually produce electricity for human consumption, it however provided observatory discovery and knowledge about it. For example, Fashina (2000: 4) has revealed that the knowledge of 'electricity' energy is symbolized in the marriage and acts of Ṣàngó the god of thunder, lightning and electricity, and Ọya the goddess of water and rain in the pantheon of Yoruba gods. Ṣàngó and Ọya are husband and wife. Ṣàngó conducts electrons while Ọya generates neurons that repel. Ṣàngó is fire while Ọya is water. Water repels fire. This is an integrative science of nature which had been known to the corpus of Yoruba wisdom thousands of years before Isaac Newton lived. According to the Italian statesman, Niccolo Machiavelli (1469 – 1527), "It must be considered that nothing is more difficult to carry out, nor more doubtful of success, nor more dangerous to handle, than to initiate a new order of things" (cf. Hakim, 2005: 11). The knowledge of pairs in chemistry whereby the formula for water is H2O was not known to Western scientists until after 1860. John Dalton, six decades before, did not know that two atoms that are alike cannot attract light energy but can only repel each other. In fact, it was not until the 20th century that Western science was able to understand the resistance of diatomic molecules at the Karisruhe conference. (Hakim, 2005: 301). But, in Ifa divination knowledge, both Òdí méjì and Ọ̀kànràn méjì talks of the electron-like and neuron-like magnetic flow in male and female, which makes them to attract each other as opposite sex. But, if you place male and male together or female and female together as can be figured in two-like atoms, they will repel each other since the same type of energy substance flows in them. This knowledge is cast in descriptive anecdotes and allegories of copulation in the great Ifa corpus.

One facet of Ifá is a natural philosophy that understands the laws and principles of the universe, using categories of numeric and word counting system for measuring distance, evaluating quality, reckoning weights, and counting quantity. It is easy to fault some of its methods as being not empirical and scientific. But, the truth is that the early thinkers of the nature of the universe like Pythagoras, Aristotle, Plato and so on did not call their endeavor as 'science'. The word was an invention of William Whewell, an English classics scholar in his book, *The Philosophy of Inductive Sciences* (1840). He said, "We need very much a name to a cultivator" of natural philosophy. "I should like to call him a scientist". Science started with observation and experiment. The precursors of science – Isaac Newton, Francis Bacon (a contemporary of Shakespeare and Queen Elizabeth I), Robert Hooke, Christian Huygens (a Dutchman), never called it "science". But, the scientific methods used in modern times began in the works of Bacon and the French philosopher Rene Descartes.

Although English may have no symbols other than the Arabic form for representing figures, the Yoruba had proto-figure symbols for counting, specifically in Ifa proto-scientific knowledge. But they have language specific words for representing numerals. And the Yoruba language words for representing numeric figures is not in any way more wordy or untidy than that of English which is the most powerful language of science and technology in the world. If the entire Western world is still using the Arabic numerals for counting, why do we need a different numeral for counting in Yorùbá? There is a universal uniformity in the numeral system of the world, which helps to advance the claim that human beings originated from the same source of creation. Therefore, all other forms of relative numeric variation is just a matter of difference in symbols and words ('logos'), and the standard measurement techniques are relative to a people's econometric perception of quantity. The core universal natural numbers that count from zero (0) is a template from which other kinds of world numeric systems take their form. Thus, we propose that there is the integration of universal measurement technique in the pristine Yoruba numeric counting system.

This requires little or no modern compliant change in most world languages, and Yoruba should be no exception.

Now, to develop modern compliant Yoruba numerals for mere counting and arithmetic figures will be quite an effort. But, it's possible. In that case, certain delicate logographic signs would have to be intricately assigned to the word numeric forms we are presently using. But, may I ask, why not retain the present Arabic form, as it is being used in many international languages?

A Source/Origin of Yoruba Counting System from Ifa Divination Corpus

There is a difference between 'numeral system' and 'numerology'. The latter is a grand esoteric factor behind the delineation of some category of Yoruba numerals. Each numeral name under the numerological category is descriptive. It has sacred mytho-historical background which is rooted in certain verses of Ifá literary corpus. Thus, the geometry and arithmetical progression of the Yoruba numbers has implication for certain cryptic meanings in esoteric drive towards 'efficacy'. For example, the Yoruba word-numerals - 'eení, eèjì, ẹ̀ẹ̀ta, ẹ̀ẹ́rin - up to eẹ́rìndìnlógún (that is 1, 2, 3, to 16), are human 'life path' soul numbers. The mystical application of figures to interpretation and prediction or foretelling of human life fortune is called numerology - "study of the supposed magical power of numbers" (Catherine Soanes, et. al, 2005:616). In Ifá divination for example, there are sixteen major 'Odù'. Each major Odù contains sixteen lesser ones. Therefore, from time immemorial, Ifa priests usually counted the obvious total number of Odù by finding the value of 16 in 16 places: 16 X 16 = 256. Hence, by their knowledge of proto-type multiplication by addition and subtraction, the pre-colonial Yoruba could count from 1 to 256. Counting by the sixteen major Odú- Ifa, the Yoruba talks of "Àwọn àgba agba mẹ̀rìndínlógún olufè" – meaning the sixteen arch-deities of Ilé Ifẹ̀, the mythic-historical cradle of Yoruba land. They talk of Odù mẹ́rìndínlógún (16 Odùs). When a fellow Ifá initiate greets a Babaláwo (Ifá Diviner), the latter may respond using numeric count idiom as follows: "Awo ń jẹ́ gbè'je, gbẹ̀jọ, gbẹ̀rìndìnlógún olufẹ̀" – meaning "the priest answers seven, eight and sixteen times, which is the number of Ifẹ̀ arch-deities!". This is a cultic code of greeting. Here again, we see the immanence of numeric style of mystic greetings. The Yoruba also talk of "Àwọn igba irúnmolẹ̀", that is, some "two hundred deities" in the pantheon of Yoruba gods. The following esoteric poetic chant is taken from one of the verses of Ọ̀yẹ̀kú Méjì, the second of the sixteen great Odùs of Ifá divination corpus, as chanted by Awo Àpáta Adéníji, the Erinmi Ò'n pìtàn Ifá of Ìwó town in Ọ̀sun State, Nigeria.

Extract from Oyekú Méjì

Èjèè'jì ni mo'gbè n' ògbe **ò`kan**	We homage the double (**two**) exponent figure, not the single **one**.
A dífá fún **èjèe'jì** òrẹ́	It divined for **two** twin-like friends
Tí wọ́n n'tì ikọ̀lé ọ̀run bọ̀ wá sí ìkọ̀ lé ayé	Who jointly descend from heaven to earth.
A dífá fún Igbó, Igbóléti	It divined for **one** named Igbóléti
A dífá fún Òkè Òkéníyì	It divined for **two** named Òkè Òkéníyì
Òkè ò jẹ	Òkè had no food
Òkè ò mu	Òkè had no drink
Òkè ò simi	Òkè had no rest
Òkè ò bàyè	Òkè had no respite
Wọ́n waá ní o kára n' lẹ̀ ó jàre	He was therefore asked to acquit himself
Ẹbọ ni kó wá ṣe	And come to offer sacrifices
Óní "kílẹbọ? kílẹbọ?"	Òkè quickly inquired: "What is the sacrifice"? (2ce)
Wọ́n ní kó rú Aṣẹ́	He was asked to sacrifice Aṣẹ́
Ótún ní "kílẹbọ? Kílẹbọ?"	Again, he asked: "What next sacrifice?" (2ce)
Wọ́n ní kó ru Agbọ̀n.	He was asked to sacrifice Agbọ̀n
Ó gbọ rírú ẹbọ, ó rú	He complied to carry the sacrifice
Ó gbọ́ àtùkànsè ó tù	He complied to appease his primal destiny
Kò pé, kò jìnà,	In no distant time,
Ó n'ṣẹ́gun ikú lọ́tún	Òkè was conquering death on the right.
Ón'gbọn wèrèpè àrùn dànù lósì	Òkè was wasping off a hornet of diseases on the left
Òkè ni yoó ṣẹ́gun **irínwó** ọkọ́	Òkè (mountain) is bound to conquer **400** (**irínwó**) hoes
Òkè ni yóo ṣẹ́gun **ọ̀tàlélégbèje** àdá	Òkè is destined to overcome**1,400(Ọ̀tàlélégbèje)** machetes.

* (As chanted in Yoruba by Awo Àpáta Adéńijì, the Erinmi Onpìtàn Ifá of Ìwó town, Òsun State, Nigeria (translation mine)

Use of numeric allusions in Ifá divination

What we want to illustrate here is the immanent use of numeric figures as a metaphysical theme in the above extract from Ọ̀yẹ̀kú Méjì verse in Ifá divination corpus. And this points to the fact that pristine Yoruba had knowledge of counting system when Ifá wisdom corpus possibly became the center and circumference of Yoruba philosophy and epistemology. The figures or word figures in this chant are put in bold caps. They are seen in:

a.) **Word figures in Ifa divination** i.e., line 1 = ẹ̀jì (two), and ọ̀kàn (one):

Here the word figures are used to count number of persons. But, the verse uses numeric allusion as a means of presenting the cosmic and metaphysical origin of figures. Thus, by origin, the number **one** encodes the possibility and existence of **two**. In this mythography, the figures **one** and **two** are twin-like friends who both descended from heaven to earth. The idea of duality of existence, or "bonded pairs" of molecules and atoms in chemistry and physics is comparable to this numeric metaphysics. This strain of duality becomes the organic principle in this Ifá verse (ẹsẹ Ifá) as we see the consistent alternation of binary opposites in the proto numeric pairs exemplified below:

b.) **The proto numeric/ bonded pairs** in that Ifá verse:

Line 1	ẹ̀jì/ ọ̀kàn
Line 3	ìkọ̀lé ò`run/ ìkọ̀lé ayé
Line 4	Igbó / Igbólétí
Line 5	Òkè/ Òkéníyì,
Lines 6 & 7	Òkè ò jẹ/ Òkè ò mu
Lines 8 & 9	Òkè ò simi/ Òkè ò bàyè
Line 12	kílẹbọ/ kílẹbọ
Lines 13 & 15	Aṣẹ́/ Agbọ̀n
Lines 16 & 17	Ógbọ́/ógbọ́
Lines 18 & 19	ó rú/ ó tù
Line 20	kòpẹ́/ kò jìnà
Lines 21 & 22	ikú lọ́tún/ àrùn lósì
Lines 23 & 24	sẹ́gun irínwó ọkọ́/ ṣẹ́gun ọ̀tàlélégbèje àdá

Now, the above bonded pairs exemplify the proof that Yoruba mythology has an integral knowledge of abstract mathematical and statistical reasoning. These pairs are metonymic of word-figures and allusive of proto numerals. The second party in each pair is also a semantic extension which, by every stretch of imagination, has symbolic allusion to a better state of affairs. For instance, the divination wisdom valorizes **two** above **one,** and the **dual** above the **singular.** So, the second embodies the first, and it is the additive conjugation of the second to the first that makes for duality or binaries which has better vigor than the singular. This is a complex Yoruba metaphysical view of the universe. And, of course, in nature there are dualities such as man/woman, water/fire, light/darkness, land/sky, good/bad, night/day, left/right, north/south, east/west, God/Satan, etc. In existential philosophy, it is proven that the existence of one element in binary relation calls to the existence of the other. There is man because there is woman. We know that a thing is good because there is the existence of its binary opposite called 'bad'.

Lines 23 and 24, "ṣẹ́gun irínwó ọkọ́" (four hundred/ 400 hoes)/ ṣẹ́gun ọ̀tàlélégbèje àdá (one thousand, four fifty/ 1,450 machets), point out the pristine knowledge of the Yoruba in the arithmetic of advanced figure counting, at least up to thousands. Literally, Òkè means 'mountain'. So, in a Yoruba adage that derives from life experience through exposure to primal technology of the Stone Age and the Iron Age, the Yoruba say proverbially:

Igba ọkọ	Two hundred hoes,
Igba àdá	two hundred machetes,
Wọ́n n b'ókè e ṣ'ọ̀tẹ̀	they ally to surmount the mountain
Ara wọn ni wọ́n n ṣe o jàre!	They all are acting in vain

In the above allegorical adage, it is further confirmed that the Yoruba have a culture of speaking, thinking, reasoning and calculating in actual figures or word-figures. It is possible, perhaps, that greater percentage of Yoruba folk-tales, proverbs, anecdotes, allegories, adages and other cultural practices derive from Ifá divination wisdom, philosophy and metaphysics. Thus, we may argue the possibility that aspects of the origin and source of Yoruba numerate system derives from Ifá divination corpus.

Conclusion

In conclusion, this chapter makes use of library and oral archival resources as well as formal and informal interviews of relevant informants in the field of African studies, science and technology. The ideas are synchronized and applied to the investigation of our research hypothesis or research questions. Invariably, we endeavor to define the origins of the Yorùbá—who they are, the population of Yoruba-speaking peoples, their geographical coverage, and their locations in West Africa and dispersions in Diaspora around the world. We also explore the oral and written significance of Yoruba language for economic, educational, cultural, artistic, religious, esoteric, political and administrative communication purposes. Above all, we attempt to redefine Yoruba numerals and investigate their characteristic features and functions as language of counting, measurement, and for quantitative, qualitative, arithmetic and geometric evaluation, some of which are rooted in and sourced from Ifá divination wisdom. This chapter discovers as it conceptualizes the (i) formulaic alternation of additive and subtractive counting system, which invariably underscores early Yoruba knowledge and existence of a proto-arithmetic and mathematical science, (ii) the sequential use of two types of numeric counting codes, namely the five digit count numeric node and the ten digit count numeric node. And this manifests in two ways, namely the Yoruba cosmic symbols/ signs of the Odu Ifa, and the proto numeric bonded pairs as seen in Ọ̀yẹ̀kú méjì in Ifá divination verse. These are evidences of the existence of systematic knowledge and of proto-scientific philosophy by the pre-colonial Yoruba people. The chapter sets to present the picture, nature, properties and characteristic features of Yoruba counting system and its possible origin in Ifá divination wisdom. It also explains how these 'numeric' resources provide template of structure for frameworks of science and technology in Yoruba language.

Notes

1. Available resources from archaeological, historical and linguistic evidences link the Yoruba to ancient cultures and civilizations of the world such as East Asia, the Jews, etc. Also see Robert Smith, *Kingdoms of the Yoruba*. Madison, University of Wisconsin Press, 1988; George L. Campbell, *Concise Compendium of the World's Language*. New York/London, Routledge, 1995.
2. Chief Bola Ige, legal luminary, social critic and former Governor, Oyo State fondly called the Cicero from Esa Oke, his native home, was a most respected leader of the Afenifere, a pan-Yoruba political group. His assassination in active duty as Nigeria's Federal Minister of Justice in 2002, under the Obasanjo Presidency, still keeps Nigerian politics restive.
3. Ifa oracular divination is a corpus of Yoruba pristine wisdom, knowledge, philosophy and metaphysics which has been with and has been used by the Yoruba from the earliest times dating around 10,000 and 8,000 BC. This account is based on the archaeological excavation of one the oldest human remains, skeleton, at Iwo Eleru near Akure, present day capital of Ondo State, Nigeria. See *Nigeria Magazine* 1972; A.O. Olubunmi, *The Rise and fall of the Yoruba Race 10,000 BC – 1960 AD*, Ile Ife: 199 publishers, 2007; F. Willet, *Ife in the History of West African Art*, London, 1967.
4. This table, showing the features of major sixteen forms of Ifa divination, is adapted from Wande Abimbola's *The Sixteen Great Poems of Ifa*. UNESCO, 1966.

Bibliography

Aaboe, Asger. 1964. *Episodes from the Early History of Mathematics*. New York.

Abimbola, Wande 1960. *The Sixteen Great Poems of Ifa. UNESCO.*

Acharya, Subrata Kumar.1993. "The Transition from the Numerical from the Decimal System in the Inscriptions of Orissa". *Journal of the Epigraphical Society of India*, 19: 52-62.

Aiyejina, Funso. 2010. "Esu Elegbara: A source of an African Alter/Native Theory of African Literature and Criticism". CBAAC Occasional Monographs, No. 15.

Ajayi, Adegboyega and S. Idowu Fadarebo 2009. *Oral Traditions in Black and African Culture*. Lagos: Concept Publications Ltd for Centre for Black and African Arts and civilization(CBAAC).

Anderson, W. French. 1956. "Arithmetical Computations in Roman Numerals," *Classical Philology*, 51(3): 145- 150.

Armstrong, Robert G.1962. *Yoruba Numerals*. London: Oxford University Press.

Babawale, Tunde, Akin Alao and Tony Onwumah 2010. *New Frontiers in the Teaching of African and diaspora History and Culture*. Lagos: Concept Publications Ltd for Centre for Black and African Arts and Civilization (CBAAC).

Bag, A.K. 1990. "Ritual Geometry in India and its Parallelism in other Cultural Areas," *Indian Journal of History of Science* 25(1-4): 4-19.

Blocksma, Mary 2002. *Necessary Numbers: an Everyday Guide to Sizes, Measures, and More*. California: Portable Press.

Fafunwa, A.B., J. I. Macauley and J. A. F. Sokoya 1989. *Education in Mother Tongue: The Ife Primary Research Project*, 1970-1978, Ibadan: Ibadan University Press.

Fagg, William Buller 1980. *Yoruba Beadwork: art of Nigeria*. New York: Rizzol.

Fakinlede, Kayode J. "Modification of Yoruba Numeral System for Use in Science, Technology and Mathematics", in *Yoruba Creativity: Fiction, Language, Life and Songs* Toyin Falola and Ann Genova (eds.) Trenton: New Jersey: Africa World Press, 2005.

______. 2003. *Yoruba Modern Practical Dictionary*. New York: Hippocrene Books.

Fashina, Nelson 2001. "Iron and Metallurgy, World Technology and the Modern Supernatural," in Commentaries. http://www.nigeriaworld.com.

Fleisher, Paul 2001. *Relativity and Quantum Mechanics: Principles of Modern Physics*. New York: Lerner Publications, 2001.

Hakim, Joy 2005. *The Story of Science. Newton at the Center.* Washington D.C.: Smithsonian Books

______. 2007. *The Story of Science. Einstein adds a New Dimension*, Washington: Smithsonian Books.

Healki, Thomas 1996. *Creative Ritual: Combining Yoruba, Santeria, and Western MagicTraditions*, Minnesota: S. Weiser.

Hetfield, Jamie 1996. *The Yoruba of West Africa*. New York: Rosen Publishing Group.

Ifra, Georges 1985. *From One to Zero: a Universal History of Numbers.* New York: Viking Penguin, Inc.

Layiwola, Dele (ed.) 1999. *A Handbook of Methodology in African Studies*. Ibadan: John Archers Ltd.

Smith, Robert Sydney 1988. *Kingdoms of the Yoruba*. Wisconsin: University of Wisconsin Press.

Urdang, Laurence 1986. *Dictionary of Numerical Allusions*. New York: Facts on File Publications.

Wilson, H.W. 2001. *Facts About World Languages: an Encyclopedia of the World's Major Languages, Past and Present*. New York.

• Chapter 5 •

EUROCENTRICIZATION OF KNOWLEDGE SYSTEM IN AFRICA

Elinami V. Swai

Introduction

In many rural communities in Africa, people believe that human illnesses originate in two ways: those caused by natural phenomena, and those which are man-made— through evil eye or by witchcraft. The illnesses that originate from natural phenomenon, it is believed, can be cured in modern hospitals. Those from evil eye or witchcraft, it is believed, have no cure in any modern hospital. The unnatural ones can be cured through traditional medicine, sorcery or through miracles. This knowledge system, which in this chapter is termed as indigenous epistemology or African knowledge system, is not confined to illness alone, but can be found in other social, economic, political and cultural phenomena. Indigenous epistemology in modern Africa provides a window into a deepening understanding of competing knowledge ideologies in Africa. While empirical knowledge system developed from the west has come to dominate, African indigenous epistemology or knowledge system has resisted this marginalization and many people continue to tap into it whenever possible. The main argument in this chapter is that throughout history African indigenous knowledge systems have survived and their potentials are far important than mainstream pundits would have us know.

This chapter attempts to analyze the contingent historical processes that explain the following. First, the new-found prominence and socio-political authority of the "knowledge system" that became almost exclusively Western with some sort of certificates at the expense of indigenous knowledge system, traditional wisdom and insights. Second, the role of modern education establishment and its authority in reconstituting "modern Africa"; and lastly, how these two transformations left in their wake a popular discourse that rationalized the domicile of western knowledge in non-western world, thus the 'eurocentricization' of science and technology. At the heart of this study, then, is the question of how the people of Africa, women and men, came to be perceived as research subjects (e.g., what they know and should learn), and as teaching objects (e.g., unintelligent and ignorant). These perceptions were implicated by the intersecting historical processes of colonization and modernization, the intersection that also explains the transformation of the culture[1] in Africa.

The Context

Within colonial and postcolonial Africa, the discourse of education and knowledge was certainly ubiquitous propagated through competing ideologies, that portrayed Africans as ignorant and devoid of any intelligence, elevating colonial powers and consistently downgrading the intelligence of African people. Colonial texts provide particularly transparent examples of the power/knowledge relationship because many of the representations contained therein are based on primitive/modern, black/white, female/male binaries of the colonial *episteme* that appear anachronistic to the present-day reader. The concept of the episteme that was developed by Foucault (1977) helps explain these different readings of education and knowledge because it suggests that the principles and rules by which education and knowledge are conceptualized remain elusive. For this reason, I begin my analysis of education and knowledge by looking at the colonial episteme and "the distances, the oppositions, [and] the differences" that defined education and knowledge in colonial texts and which continue to define education and knowledge in the post-colonial texts.

In detailing African knowledge systems, since the invasion of the continent by aliens, Africans have been forced to accommodate other ways of knowing and have taken western epistemology as superior than their own. Particularly during colonial times, followed by the political independence that started in the 1950s in many African countries, modern science and formal education came to hold an added level of legitimacy and credibility, while indigenous systems of knowledge production became illegitimate and obsolete. While modern science and formal education came to be seen as potential to promote economic growth and development, indigenous knowledge systems came to be at the receiving end, seen as a cause for all social and economic maladies in Africa. Both the colonial and postcolonial officials were certainly pivotal in creating this binary division, indeed forging a new template of education and knowledge system in modern Africa. During this period, science and technology taught in schools appeared to be the most promising domain for investment in the people, as means of liberating them from myths brought about by indigenous knowledge. While indigenous knowledge, such as rainmaking, indigenous healing and the like seemed as holding people behind the shackles of darkness, Western education seemed to liberate people and bring equality in society. Educated individuals were seen mainly as a means of advancing the collective good of society,[2] indigenous healers or rainmakers were seen mainly as operating outside society, and who should be fought without any further justification because they act against "us."

Language, social skills and belief systems that saw African people successfully manipulate the harsh environment of Africa for a long time before Europeans arrived were interrupted. The tight social control of African people through indigenous knowledge systems and activities was contaminated by colonial powers and missionary activities. Generally up until the 1970s, no positive images and concepts of African indigenous knowledge systems were presented through the many public institutions. African knowledge systems were invisible and no inclusion of indigenous perspectives was given within the formal institutions of education.

On the topic of neglecting knowledge systems, Africa has certainly been a longstanding, privileged field of study, especially when related to science and technology.[3] To discuss colonial and postcolonial Africa in terms of the emerging cultural and social practices brought by the colonial encounter is not a novel idea; most noteworthy is Walter Rodney's *How Europe Underdeveloped Africa*.[4] Indigenous metaphysical beliefs came to be challenged through religious education[5] and African epistemological beliefs came under attack in secular education.[6] In the process of modernization formal education replaced African knowledge systems. Gender historians have gone further to show how African culture and gender relations more specifically were affected by social and political transformation in Africa.[7] As well, African scholars[8] are calling for the acknowledgement of African knowledge systems. In fact, the neglect of knowledge systems and traditions in Africa as a field of study has led the way in scholarly discussions about the inextricable ties between Africa and the West under colonial rule, from which researchers of other fields have taken their cues in writing their own historical narratives.

The study of how indigenous knowledge systems in Africa became ingredients of colonial and post colonial texts, and the way in which they came to be articulated and sustained is missing from this scholarly wisdom. Put simply, historical narratives have traditionally paid much attention to the various ways in which the term indigenous knowledge system is fragmented, but consistently theorizing it as monolithic and creating in it a sense of irrelevance. The myth of obsoleteness of African knowledge system still exerts its influence on people and researchers, leaving African knowledge systems in the ambiguous position to which they are so understood: if, on the one hand, African knowledge systems are unscientific with no potency for development on the other hand, and on the other, representing backwardness and primitivity.

Little has been written about the ways in which the imperial spread of the idea of formal education and knowledge, which in this chapter is termed as eurocentricization of knowledge system in Africa, facilitated this sense of community. Indeed, the historical process of modernization of knowledge under colonial and postcolonial supervision that accompanied Africa's economic transformation ultimately yielded a different rationale for African culture and knowledge system, and thus African social organization. It is this powerful cultural and social re-organization, or rather re-interpretation of education and knowledge justified by the acquisition of discrete and fragmented knowledge, which underpins the expression, the "eurocentricization" of knowledge system. This chapter, then, attempts to uncover the rationale for formal education and modern knowledge system as a way of understanding how African knowledge system and social relations were impacted by social and cultural transformations, most visibly characterized by shifts in work patterns, a growing modernity, and by implication, notions of literate, educated, but also and concomitantly, by the notions of ignorance and illiteracy propounded in modern schools.

Isolating African knowledge system as the primary analytical construct to explain historical changes in Africa is not entirely new scholarship. Increasingly, studies on the intersection of imperialist politics, modern education and social transformation have been on the upswing in Africa. This scholarship converges in one critical argument: the role of modern education in legitimating and being legitimated by state policies or put simply, the marginalization of African knowledge system in state policies. In sum, during colonial, and part of post-colonial period, increasingly in Tanzania as elsewhere in Africa, modern education became the official rationale of knowledge attached to human value. By adding human value to this matrix of marginalization of African knowledge system, we see more clearly how Africans, their knowledge as well as their culture became not only transformed in modern terms, but also engendered as well.

When colonialists took over Africa, they had an agenda to dictate what knowledge was. This agenda aimed at marginalizing local knowledge, without acknowledging its efficacy in curing diseases, such as diabetes, that had defied Western medicine. The agenda was also to claim superiority, without acknowledging one's potential in solving social problems. This has been the case for many traditional healers' knowledge of healing and herbs such as the *mwarubaini* (neem tree) whose medicinal properties were discovered by African medical practitioners, and yet today the tree parts are sold globally without the Africans benefiting from patent, trademark, and property rights like their Western pharmaceutical counterparts. Africans are denied opportunities to gain from the results of their knowledge and ideas. If any collaboration occurred between Africans and the Westerners, attempts were made to inculcate into them Western ways of thinking instead of promoting their ideas.

Eurocentricization

Before discussing the "eurocentricization" of science, technology and knowledge in Africa, however, this chapter introduces some theoretical strands touching on ways in which African education and knowledge systems have been undermined or marginalized. Eurocentricization here refers to an institutional process of eurocentrizing knowledge brought by colonial reforms, the most critical of which African education and knowledge systems were banned in modern schools and colleges. Definitions of Eurocentricization come from Eurocentric, which simply means a grand narrative of Euro/American rationality work that draws on other cultural traditions and represents different racial perspectives. Eurocentric is biased master narratives that subsume 'Other' cultures under universalistic theories of evolution portraying them as minority cultures. Eurocentricization does not suggest a cultural transformation of Africans by engendering a Eurocentric outlook toward people and the world or development of human resources.[7]

Zeleza (2007) points out that the main challenge for Africa is to transcend the alterity of negative difference embedded in Eurocentric epistemology, taking note that autonomy, supremacy, sovereignty, and hegemony are possible for Africa. Zeleza is optimistic that it is possible for Africa to transcend negative difference. This is because Africa is already and has always been at the center of human history, notwithstanding the fantasies of Eurocentric historiography that has inferiorized African difference by turning it into a difference in time and a difference in space.

Two more points need to be emphasized regarding eurocentricism. First, temporal differentiation is articulated in evolutionary terms and historical stages in which the West is always ahead in a linear development, initiating progressive change that others are fated to imitate. Second, Eurocentric critics have gone further and provided theoretical insights into how such discourses either actively construct the other or are implicated in their construction. From this perspective, cultural differentiation posits central and marginal places, territorializes social development—civilization and barbarism—so that levels of scientific innovations become measures of human worth and historical agency in which the West, once again, is not only placed at the center but its global expansion and interventions become imperative humanistic acts of magnanimity.

The concept of eurocentricism offers a powerful entry point, not only because eurocentricism conjoined African and European knowledge systems, but also because we have over the last fifty years developed the critical tools to undress the emperor. This is to argue that the recentering of Afrocentric in Africa only makes sense if it is accompanied by the recentering of eurocentricism in European knowledge systems. But the project of recentering Africa goes beyond that. It requires constructing a new global education and knowledge system on this fragile planet. Even as we deconstruct eurocentricism in Africa, African science and technology need to be rearticulated more.

In detailing African cosmology since the invasion of the continent by aliens, the people have been forced to accommodate other ways of knowing which they take as facts. African ways of knowing, which could be termed as knowledge system, in its own rights, came to be considered an epiphenomenon of modern society. However, in at least three cases in the history of modern Tanzania, indigenous knowledge system has assumed more stable forms of formal education and knowledge. At these important moments, indigenous knowledge system came to represent valid and possible alternatives to sustainable development. Further, it is precisely when indigenous knowledge system assumed that credible space for sustainable development that African knowledge system acquired the status of global recognition. We are going to analyze here those moments, in which competent people with knowledge system to organize their society came to be identified as ignorant and illiterates. These representations were not necessarily based on what these people did or did not do, but a reflection of the notions of what educated and knowledgeable people thought and acted upon.

Indigenous knowledge systems in Africa are a reflection of concrete events, life patterns and other social phenomena that are culturally shared. They are also acutely spiritual, observed through cosmic and natural occurrences, such as rhythmic patterning of seasons and predictability of land formations and other earthly events. All these realities became the roots of African knowledge systems, which were reflected in everyday activities. Ngwabi Bhebe (2000) makes this point:

> Before the European conquest of Africa, Africans had built up a pool of knowledge and technology which they used to sustain agriculture, human and animal health, industrial production involving food processing, metallurgy, leather tanning, timber seasoning, fermentation of beverages, making of dyes, mining and architectural engineering. But political subjugation by Europe so traumatized Africans that many of them lost confidence in and looked down upon their own culture, forcing some of them to view and embrace Christianity and Islam as a progressive move but without totally losing their old cosmology or basic beliefs (pp. 7 - 8).

There are foundational characteristics of indigenous education and indigenous knowledge systems, which are not in conflict with formal education, but could be taken as complimentary. A few of these characteristics are included here to provide handmarks for the reader.

A sacred view of nature permeates its foundational process of teaching and learning:

- Its processes adhere to the principle of mutual reciprocity between humans and all other things.
- It presents something for everyone to learn, at every stage of life.
- It recognizes the levels of maturity and readiness to learn in the developmental processes of human brain.
- It recognizes language as a sacred expression of breath and incorporates this orientation in all its foundations.
- Its purpose is to teach a way of life that sustains both the individual and the community.
- It provides the rule of conduct and social ethic.
- It recognized the human being as a social being who learn through others.[9]

What was different about the reinvigorated importance of formal education and modern knowledge system in Africa is not related to any weakness of indigenous knowledge system in solving the problems of Africa. Rather, African knowledge systems came to hold marginalized position by relating it with formal education in modern world. The act of associating and disassociating of African knowledge systems and formal education and modern knowledge was not accidental, but part of colonial tradition. Colonial tradition always put the boundary between them and the 'other' – the division between indigenous and tradition on the one side, and formal education and modern knowledge on the other. The two sides of this binary pair, were not equal. African cosmology was positioned on the negative side of the binary.

There are several reasons given for the negation of African knowledge systems. One is the differences of opinion among colonialists and missionaries regarding the relationship between African knowledge systems and modernization reforms tied to capitalist mode of production. What this chapter argues is that once colonial officials set their feet in Africa and the African nationalists followed them, there was an exercise of a socio-political authority that constructed what should be termed as knowledge and what should not.

Indigenous Knowledge System Extinction

Concern for the place of indigenous knowledge in African education has existed as a persistent problem for Africans,[10] and it has been particularly a challenge in the provision of formal education in African societies.[11] What is new is that postcolonial elites and scholars have become active in dialoguing ways to resuscitate African knowledge systems (Wass 1999, Cox 2000).[12] Their concern is that, the extinction of African knowledge systems extends beyond the African continent and affecting the continent on a global scale.. For example, nearly one-half of the African's 6000–7000 languages are facing rapid extinction,[13] 2400 are endangered and 50% of these are considered dying, with the possibility that up to 90% will disappear in the 21st century.[14] These languages express indigenous knowledge systems. Africa alone has at least 1000 distinct indigenous languages (Bamgbose 1991) and at least 200 of these are facing language death.[15] In their passing, several thousand years of science knowledge may become extinct. Since extinction is forever a concern, responsibility and role for African scholars is to join the effort to preserve and promote indigenous knowledge in Africa. This challenge is especially important for those scholars who recognize that education and knowledge systems are best constructed through people's lived experiences. And, especially, given there is international agreement that indigenous knowledge systems are the best alternative to natural resource management and sustainable livelihoods,[16] it is high time that post colonial scholars acknowledge and support it for socio-economic or even political development.

To date, the official knowledge in most sub-Saharan countries is that obtained from the books written by the Westerners and education policies have excluded the use of indigenous knowledge and languages. This has resulted into having educational liability where we see poor performance, high school drop-out rates, inappropriate and culturally irrelevant materials, and exclusion of students from participation in their own learning (see Bunyi 1999, Bamgbose 2000, Zerner 2000).

In indigenous belief system, physical landmarks within the African environment provided places of spiritual identification. These are places where the spirits dwell and were the resting places of the ancestral spirits. African intrusion on the landscape came by way of ceremonies, storytelling and signposting. These activities are complex and reaffirmed obligations of Africans to law, teachings and culture. The interaction with the land presented Africans with the ability to change the natural environment to suit their social, spiritual and economic well-being. The voices that come from the forests were not simply those of the spirits, but also those of African ancestors. The belief is that these beings still talk and sing to African people from their location; and that the ancestral spirits are wise: through their work and through infinite time they sculptured the landscape and taught how the country should be run.

At the time of political independence in 1961, the education sector was in "utter negligence...the country was in tatters and had great deficits in many sectors, the worst in education."[17] The number of college graduates did not exceed ten. With the population of about 10 million people scattered all over the country, could ten 'educated' people solve all the problems facing the country, let alone know them? While this was not enough, post independence Tanzania continued to depend on modern education and graduants from formal education to solve all the problems facing the country. Indigenous knowledge systems were put into the back bench. In short, because modern education was able to attract a steady flow of students after political independence in Tanzania, while people continue to shun away from indigenous knowledge systems and wisdom.

No less significant in creating the ignorance and illiterate Africans were the more focused eurcentricization reforms during the signaling of the end of indigenous leadership authority such as that of Mangi (Chagga), Mtemi (Nyamwezi), and Mtwa (Hehe). Such indigenous leadership authority was illegitimized in the discourse of illiteracy and its place District Commissioners (DCs) were instituted. The new leadership was considered as educated and knowledgeable in solving modern problems in modern ways. As political ideology, the discourse of ignorance and illiteracy focused on individuals who had not acquired formal education, regardless of how much he or she could contribute to society. Once it was accepted that those with formal education were the only ones who had capacity to reason, then it became possible to render indigenous knowledge obsolete.

In this new grid, the educators, followed by religious leaders, were the supreme knowledge authorities, since only they could academically isolate the source of ignorance and illiteracy or identify the nature of ignorance and how to remove it. Moreover, precisely because of the infused importance of modern education and religion, educators and religious leaders were given added administrative authority by the government. They alone were in a position to judge on critical educational and academic matters, such as who should be accepted in school, who should go to another level and who should not. By contrast, community leaders and indigenous experts such as healers and special teachers were considered as ignorant and uneducated.

It is important to remember, however, that colonialism was not just about separating people from their knowledge system or suppressing their urge to attain modern education. It was also about arranging, representing and redistributing knowledge. For example, Africans were taken to Europe as scientific samples where they were put into show rooms and scientific laboratory to be observed as samples. These individuals were catalogued, studied and scientific knowledge was generated about them.[18] Equally, plants and animals were not spared in this scientific discovery. Many plants and animals were taken from Africa for the same purposes, to be observed and knowledge about them became the European discovery. Britain became a global laboratory for scientific research and source of knowledge, while Africa became the source of material for scientific discovery.

Since the post-WWII period, Africans had been active participants in curriculum reforms with the help of imported British scientific texts. Many conversations have taken place regarding the need to apply scientific measures to eradicate ignorance and illiteracy. The efforts of African leaders to appropriate this imported scientific knowledge and entrench it in the curriculum became a tradition in African recent history. Thus, just as colonial officials did earlier, when many African countries became independent they initiated educational reforms. In postcolonial reform documents, the African political leaders and education policymakers proclaimed commitment to scientific knowledge from the primary to higher levels. In summary, eurocetricization of knowledge in Africa had the indelible effect of rendering Western education the privilege of "the knowledge." Specifically, curriculum reforms that more closely adapted western notions of scientific knowledge were particularly effective in transforming African knowledge system into western scientific materials; these two areas reconfigured what could be considered as *knowledge* and what was not, shaping the template of the modern African knowledge system.

The African optimistic outlook of the 1960s' and 1970s' education based on scientific knowledge coincided with a renewed and unprecedented popularity of notions of knowledge in western countries (Schultz 1961). This contributed to creating a general atmosphere characterized by a euphoria, faith, and confidence in science and technology, a means for development and an effective instrument for achieving national economic development.

To recapitulate, scientific knowledge was not new to Africa; western notions of science, however, are. Western scientific knowledge dealt mainly with introducing new or modified academic subjects, such as history, geography and literature. This fragmented subjects rendered the otherwise continuous knowledge system in Africa obsolete. The authority of these fragmented subjects in the name of knowledge was contingent on the state's funding and students were awarded certificates and diplomas at the end of the course. This gave them legitimacy to gain access to the world of work in government institutions..In addition to institutional authority, government employment guaranteed and deepened academic investment in the fate and survival of the state apparatus. Indeed, it is this growing symbiotic relationship between western education and the state that helps explain how the discourse of ignorant and illiterate Africans became ensconced in nationalism.

Those who attained certificates and diplomas were further made exclusive by their modern, specialized knowledge, which solidified their authority, not simply as experts, but arguably as socio-political agents of the state. They, after all, spread state-sponsored modernization through their government employment, looking down upon their counterparts as ignorant and illiterates.

Thanks to the rising prominence of adult education in general, and the authority of indigenous knowledge system in Africa, a recent initiative of many governments to move towards a policy of revitalization of indigenous knowledge, we now see communities taking over wide-ranging responsibilities from developing work and income-generation initiatives, land repossession and infrastructure preparation. Such initiatives have given new impetus to those who had been identified as ignorant and illiterate as they support the view that people have power and are capable of attaining sustainable change not only when they are schooled, but only when conditions exist for real community involvement in setting goals within a supportive governmental environment.

Since the mid-1970s, the debates on resuscitating indigenous knowledge in Africa have taken center stage, not only at the level of individual nation-state in Africa, but as an international concern, particularly at the United Nations and its various specialized agencies. In Africa, while the standard of living of a few individuals has soared from bare subsistence to affluence, the majority of people are subjected to destitution. Moreover, as was pointed out by Gottleib (1996), the continuous growth of technological and social power—and the attendant religious and political ideologies that support them—have strengthened the belief of humankind's fundamental difference attaining knowledge and acting upon her or his environment and this has created more gap between the haves and the have not's. It is in this womb that the embryo of contemporary urge to return to indigenous knowledge began to fertilize.

Many scholars in Africa have noted with concern, to varying degrees, that Africa's educational system has suffered chronic under funding, a lack of resources, low teacher morale resulting from poor working conditions, ineffective teacher training and professional development and relaxed supervision of trained personnel.[19] These problems have been a major concern for some local educationists and policy makers who have supported school reforms in many African countries.

Tanzania undertook a series of educational reforms under the auspices of the World Bank/International Monetary Fund (IMF) Structural Adjustment Programs in the 1980s. Part of these reforms brought structural changes to the delivery of education in Tanzania.[20] Tanzania experimented on community school programs which encouraged schools to be involved in community activities. This new structure replaced the old system of schools as elite places where local communities were banned from trespassing. There have also been changes aimed at promoting indigenous languages, culture and arts in the school curriculum. While the changes have been well intentioned, the implementation process has been riddled with difficulties that have thrown the entire educational system off its stride. With these issues in mind, it is significant that we question the objective of the World Bank and IMF's involvement in the implementation of educational reforms that appear to serve the external interests of global labor and capital.

Although the reforms sought to promote indigenous knowledge systems through their incorporation in educational system, the education system suffered from the fact that the current education policy in Tanzania, as Brock-Utne and Holmarsdottir (2004) describe it, is "confusing, contradictory and ambiguous."[21] The reforms did not consider the socio-political factors that could implicate the implementation of successful reforms, for example, working with the local knowledge of educators and providing incentives for field implementers to have a stake in the reform, nor did the reforms respond to the socio-cultural values that impinge on how education is understood in local contexts. While it is common knowledge that active participation in classroom promotes learning, one common Tanzania narrative concerns the separation between the students, their knowledge and language in classroom interactions. In their study of language policy in Tanzania, Brock-Utne and Holmarsdottir for example, cited Puja's study in the following manner:

> ...study by Grace Puja (2001) comprising 73 female students at the University of Dar es Salaam only two of the 73 (3%) identified themselves as Swahili, yet over 63% of the participants stated that they speak Kiswahili most of the time in their homes. Those who speak Kiswahili most of the time at home do so irrespective of the parental level of education, type of occupation, racial, ethnic and religious backgrounds. Another fact mentioned in this study is that Puja had worked out her interview guide in English (for a PhD taken in Canada) and initially began interviewing in English, yet only eight of the 34 Tanzanian female undergraduates that she interviewed, chose to be interviewed in English, the rest preferred Kiswahili.[22]

It is important to note that throughout Tanzania's history, educational reforms have not powerfully targeted questions of language and indigenous knowledge and their implications for schooling. Research in the region indicates that current teaching mostly encourages passive sitting and listening.[23] Education reforms in Africa have failed to critically address questions of local knowledge and specifically of language. The works of Brock-Utne have been insightful and relevant to the discussion of local language in education in Africa. Brock-Utne has demonstrated how language is one of the most important factors militating against the dissemination of knowledge and skills in Africa, arguing that the imposed medium of communication is the root cause of failure in rapid social and economic well-being of the majority of people in Africa. According to Brock-Utne, there is a strong correlation between underdevelopment and the use of a foreign language as the official language of a given country in Africa (e.g., English, French or Portuguese). She poses some questions: Why have African countries failed to help people improve their social, economic, and political activities via their mother tongues? Why insist on their learning English or French first before modern technology could be introduced to them?[24]

To sum up, the African knowledge system, whether efficient in solving African problems or not, was rendered obsolete and subordinated by the reconstituted view of "scientific knowledge" that had been brought about by colonial officials and adapted by postcolonial elites. Both the externally imposed eurocentricization of knowledge system in Africa and the hitherto unrecognized internally active participation of education reforms were activities that gave a brand to African knowledge system. This chapter has attempted to analyze the contingent historical processes, the prominence and socio-political authority of the "knowledge system" in Africa. It assessed the role that modern education establishment and the role it played in reconstituting the "modern Africans" in colonial and postcolonial contexts. It also demonstrated how both colonial and post colonial officials left in their wake a popular discourse that rationalized the domicile of western knowledge in non-western world, thus 'eurocentricization' of science and technology in Africa.

Through a variety of helpful theoretical interventions from postcolonial cultural studies, this chapter has attempted to argue on three interrelated points. First, Africa's colonial encounter deepened the symbiotic relationship between modern education and state hegemony building. The nature of colonial education, however, demanded an eurocentricization of African knowledge, yet not altogether at the expense of the majority of Africans. Consequently, colonial education proved hegemonic not only for the colonial masters, but also for some privileged Africans. Second, eurocentricization of knowledge, as a historical process, inscribed and encoded few privileges within the very inter-workings of education institutions and government so that access to power was already circumscribed, gendered and unequal. That is, eurocentricization of knowledge redefined social roles and remodelled social and economic structures to ultimately reify the most powerful ontological definition of personhood in modern terms, namely, that of educated/illiterate; smart/ignorant. This ontological category of African citizen during colonial period was in many ways a modernized version of a traditional role of the majority of Africans. Eurocentricization of knowledge system in Africa undermined a prior and more active role that Africans did to develop their continent, in relation to colonial version of development that ushered people in new and hidden forms of development embedded with social control, often going unnoticed due to the presumed and necessary benevolent association of formal education, modern knowledge and development.

This study has attempted to show how the black race was also implicated by Africa's larger socio-economic changes, whose reverberations went beyond the process of eurocentricization of knowledge system to include new patterns of consumption, a growing social disparity and consequently changes to (or reifications of) social space in knowledge production. This chapter has shown how formal education and modern knowledge provided the authority that solidified capitalism's relationship to nationalism within a certain popular discourse of ignorant and illiterate precisely by eurocentricizing knowledge in the name of modern education.

Implications Towards Knowledge Democracy

Research is needed on the role of African knowledge system as practiced in traditional African societies and in modern schools, including empirical evidence of how different worldviews affect learning and construction of knowledge. African knowledge system was mainly operational in traditional societies and may be quite relevant in the current situation. Indigenous knowledge as a worldview seeks to ensure that people and communities were actively involved in learning and solving their perceived problems. Knowledge was created by the people and for the people. It ensured that knowledge, skills, and attitudes were integrated in the learning process and promoted the survival of the community.

Genuine democratic knowledge then should not only be for the realization of economic development but also for the body of knowledge that is to be produced.[25] In the periods of social transition, education becomes central to our future well-being. Only if African knowledge systems and their ways of attaining it are placed at the center of our education can Africans continue to develop their capacities and transform their educational institutions and respond openly and imaginatively to the current period of change. By altering who controls knowledge, the knowledge produced, and indeed, the very definition of what constitutes knowledge may also change. This implies that more research on the contribution of indigenous knowledge system is required in order to understand more the potential that indigenous knowledge systems holds, and their contributions to the global economy.

Notes

1. Culture here is understood as life pattern and behavior acquired and transmitted through traditional (i.e., historically derived and selected) ideas and actions.
2. Derek Mulenga, "Mwalimu Julius Nyerere: a critical review of his contributions to adult education and Postcolonialism," *International Journal of Lifelong Education*, Vol. 20, No. 6, 2001, 446–470.
3. See C. Peters, *New Light on Dark Africa* (London: Ward, Lock, and Co., 1891); J. Thomson, *Through Masaii Land* (London: Frank Cass and Co., 1968); P. Gerdes, *Ethnomathematics and Education in Africa* (Stockholm: Stockholm University, Institute of International Education, 1995); and O. Tsuma, *Science Education in the African Context* (Nairobi: The Jomo Kenyatta Foundation, 1998).
4. See Walter Rodney, *How Europe Underdeveloped Africa*. Nairobi: Heinemann, 1980.
5. See J. Mbiti, *African Religions and Philosophy* (Oxford: Heinemann Educational Publishers, 1969).
6. Ngugi wa Thiongo, *Decolonising the Mind: The Politics of Language in African Literature* (Portsmouth, NH: Heinemann, 1986); W. Cobern, "Science and a social constructivist view of science education," in. W. Cobern, ed., *Socio-cultural Perspectives*

on Science Education: An International Dialogue (Dordrecht: Kluwer Academic Publishers, 1998), 7–23; O. Tsuma, *Science Education in the African Context* (Nairobi: The Jomo Kenyatta Foundation, 1998); G. Bunyi, "Rethinking the place of African indigenous languages in African education," *International Journal of Educational Development*, 19, 337–350, 1999; and R. Mosha, *The Heartbeat of Indigenous Africa: A Study of Chagga Educational System* (New York: Falmer Press, 2000).

7. See Frances Vavrus, "Uncoupling the Articulation Between Girls' Education and Tradition in Tanzania," *Gender and Education*, Vol. 14, No. 4, 2002, pp. 367–389; Oyeronke Oyewumi, *The Invention of Women: Making an African Sense of Western Gender Discourse* (Minneapolis: University of Minnesota Press, 1997).
8. See for example M. B. Ramose, "In search of an African philosophy of education," in *South African, Journal of Higher Education*, Volume 18, No. 3, 2004; D. M. Mengara, (ed), *Images of Africa: Stereotypes and Realities.* Trenton and Asmara: Africa World Press, 2001; and P. Crossman, "Perceptions of Africanization or Endogenization at African Universities: Issues and Recommendations" article in Zeleza P. T. and Olukoshi, A. (eds), *African Universities in the Twentieth Century*, Volume 11. CODESRIA: Dakar, 2004.
9. A. K. Bangura, "Ubuntugogy: An African educational paradigm that transcends pedagogy, andragogy, ergonagy and heutagogy," *Journal of Third World Studies*, 22(2), 2005, 13-53.
10. See A. Bamgbose, *Language and Exclusion: The Consequences of Language Policies in Africa* (Munster: LIT Verlag, 2000); and G. Bunyi, "Rethinking the place of African indigenous languages in African education," *International Journal of Educational Development*, 19, 1999, 337–350.
11. A. Cleghorn, "Primary level science in Kenya: constructing meaning through English and indigenous languages," *Qualitative Studies in Education*, 5, 311–323; and M. Rollnick, "Current issues and perspectives on secondary language learning of science," *Studies in Science Education*, 35, 2000, 93–122.
12. M. Wass, "Taking note of language extinction," 1999. Available online: http:/ www.colorado.edu/lec/alis/articles/langext.html (accessed, 02.22.2010); R. Cox, "Will tribal knowledge survive the millennium?" *Science*, 287, 2000, 44–45.
13. M. Brenzinger, *Language Death: Factual and Theoretical Explorations with Special Reference to East Africa* (Berlin: Mouton de Gruyter, 1992); and M. Krauss, "The World's Languages in Crisis," *Language*, 68, 1992, 4–10.
14. B. Wuethrich, "Learning the world's languages – before they vanish," *Science*, 288, 2000, 1156–1159.
15. G. Sommer, "A survey of language death in Africa," in M. Brenzinger (ed.), *Language Death: Factual and theoretical explorations with special reference to East Africa* (Berlin: Mouton de Gruyter, 1992), 301–417.
16. See D. Brokensha, D. Warren and O. Werner Indigenous knowledge systems and development (Lanham: University Press of America, 1980); P. Richards *Indigenous agricultural revolution: ecology and food production in west Africa* (London: Hutchinson, 1985); B. DeWalt, "Using indigenous knowledge to improve agriculture and natural resource management," *Human Organisation* 53, 1994 123-31;

and M. D. Warren, L. J. Slikkerveer, and D. Brokensha (eds) *The cultural dimension of development: indigenous knowledge systems* (London: Intermediate Technology Publications, 1995).

17. See Editorial Comment, "Education in Tanzania since Independence," *Papers in Education and Development,* Number 22, 2002, v-xiii.
18. Reference can be made to Khoisan Sarah Baartmann who was taken from the Cape Colony in South Africa and exhibited at the Picadilly Circus in London because of the purported abnormality of her sexual organs. Knowledge extracted from her was that she and her women folks in general were suffering from steato-pygia (an enlargement of the buttocks) and elongated labia.
19. See P. W. Achola, *Implementing educational policies in Zambia* (Washington, D.C.: The World Bank, 1990); M. Carnoy and J. Samoff, *Education and social transition in the Third World* (Princeton, NJ: Princeton University Press, 1990); K. Banya, "Illiteracy, colonial legacy and education: the case of modern Sierra Leone," *Comparative Education*, 29(2), 1993, 159–70; and K. Banya and J. Elu, "Implementing basic education: an African experience," *International Review of Education*, 43(5/6), 1997, 481–496.
20. K. Osaki, "Environmental Education Research in Eastern and Southern Africa Region: Emerging Issues and Concerns," *Papers in Education and Development*, 20, 2004, 73-82.
21. B. Brock-Utne and H. Holmarsdottir, "Language policies and practices in Tanzania and South Africa: problems and challenges," *International Journal of Educational Development* 24, 2004, 68.
22. Ibid., 68-9.
23. Osaki, "Environmental Education Research in Eastern & Southern Africa Region," 73-82.
24. B. A. Fafunwa, "Using national languages in education: a challenge to African educators." In: Unesco–Unicef: *African Thoughts on the Prospects of Education for All.* Selections from papers commissioned for the Regional Consultation on Education for All, Dakar, 27–30 November, 1989, 103.
25. O. Fals-Borda and M. Bahman, *Action and Knowledge: Breaking the Monopoly with Participatory Action-Research* (New York: The Apex Press, 1991).

References

Achola, P. W. 1990 *Implementing educational policies in Zambia* (Washington, D.C., The World Bank.

Bamgbose, A. 2000 *Language and Exclusion: The Consequences of Language Policies in Africa* (Munster: LIT Verlag).

Banya, K. 1993 "Illiteracy, colonial legacy and education: the case of modern Sierra Leone," *Comparative Education*, 29(2), 159–70.

Banya K. and J. Elu, 1997 "Implementing basic education: an African experience," *International Review of Education*, 43(5/6), 481–496.

Bhebe, N. 2000 *The Historical Dimensions of Democracy and Human Rights in Zimbabwe: Pre-Colonial and Colonial Legacies* v. 1.

Brenzinger, M. 1992 *Language Death: Factual and Theoretical Explorations with Special Reference to East Africa* Berlin: Mouton de Gruyter.

Brock-Utne, B. and Holmarsdottir, H. 2004 "Language policies and practices in Tanzania and South Africa: problems and challenges," *International Journal of Educational Development* 24, 68.

Bunyi, G. 1999 "Rethinking the place of African indigenous languages in African education," *International Journal of Educational Development*, 19, 337–350.

Carnoy, M. & Samoff, J. 1990 *Education and social transition in the Third World* Princeton, NJ: Princeton University Press.

Cleghorn, A. "Primary level science in Kenya: constructing meaning through English and indigenous languages," *Qualitative Studies in Education*, 5, 311–323.

Crossman, P. 2004 "Perceptions of Africanization or Endogenization at African Universities: Issues and Recommendations" in P. T. Zeleza and A. Olukoshi, (eds), *African Universities in the Twentieth Century*, Volume II. CODESRIA: Dakar.

Cobern, W. 1998 "Science and a social constructivist view of science education," in. W. Cobern, ed., *Socio-cultural Perspectives on Science Education: An International Dialogue* (Dordrecht: Kluwer Academic Publishers, 7–23.

Fafunwa, B. A. 1989 "Using national languages in education: a challenge to African educators." In Unesco–Unicef, *African Thoughts on the Prospects of Education for All.* Selections from papers commissioned for the Regional Consultation on Education for All, Dakar, 27–30 November, 103.

Fals-Borda, O. and Bahman, M. 1991 *Action and Knowledge: Breaking the Monopoly with Participatory Action-Research* New York: The Apex Press. Gerdes, P. 1995 *Ethnomathematics and Education in Africa* Stockholm: Stockholm University, Institute of International Education.

Foucault, M. 1977 History of systems of thought, in: D. F. Bouchard (Ed.), *Language, Counter-Memory, Practice:Selected Essays and Interviews by Michel Foucault* (Ithaca, Cornell University Press).

Krauss, M. 1992 "The World's Languages in Crisis," *Language*, 68, 4–10.

Mbiti, J. 1969 *African Religions and Philosophy* Oxford: Heinemann Educational Publishers,

Mengara, D. M. (ed). *Images of Africa: Stereotypes and Realities.* Trenton and Asmara:Africa World Press.

Mosha, R. 2000 *The Heartbeat of Indigenous Africa: A Study of Chagga Educational System* New York: Falmer Press.

Mulenga, D. 2001 Mwalimu Julius Nyerere: a critical review of his contributions to adult education and Postcolonialism, *International Journal of Lifelong Education*, Vol. 20, No. 6, 446–470.

Ngwabi, B. 2000 *Christianity and Traditional Religion in Western Zimbabwe, 1859-1923.* Gaberone, Botswana: Mambo Press.

Osaki, K. 2004 "Environmental Education Research in Eastern and Southern Africa Region: Emerging Issues and Concerns," *Papers in Education and Development*, 20, 73-82.

Oyewumi, O. 1997 *The Invention of Women: Making an African Sense of Western Gender Discourse*, Minneapolis: University of Minnesota Press.

Papers in Education and Development (PED), 2002 Number 22, Editorial Comments, v-xiii.

Peters, C. 1891 *New Light on Dark Africa* London: Ward, Lock, and Co.

Ramose, M.B. 2004 "In search of an African philosophy of education", in *South African-Journal of Higher Education*, Volume 18, No. 3.

Rollnick, M. 2000 "Current issues and perspectives on secondary language learning of science" *Studies in Science Education*, 35, 93–122.

Rodney, W. 1980 *How Europe Underdeveloped Africa*. Nairobi: Heinemann.

Sommer, G. 1992 "A survey of language death in Africa." In M. Brenzinger (ed.), *Language Death: Factual and theoretical explorations with special reference to East Africa* Berlin: Mouton de Gruyter.

Thiongo, N. 1986 *Decolonising the Mind: The Politics of Language in African Literature* Portsmouth, NH: Heinemann.

Thomson, J. 1968 *Through Masaii Land*, London: Frank Cass and Co.

Tsuma, O. 1998 *Science Education in the African Context*, Nairobi: The Jomo Kenyatta Foundation.

Vavrus, F. 2002 "Uncoupling the Articulation Between Girls' Education and Tradition in Tanzania," *Gender and Education*, Vol. 14, No. 4, 367–389.

Wass, M. 2000 "Taking note of language extinction," 1999. Available online: http:/www.colorado.edu/lec/alis/articles/langext.html (accessed, 02.22.2010); R. Cox, Will tribal knowledge survive the millennium? Science, 287, 44–45.

Wuethrich, B. 2000 "Learning the world's languages – before they vanish." *Science*, 288, 1156–1159.

Zeleza, P. T. 1997 *Manufacturing African Studies and Crises*, Dakar: Codesria Book Series.

Part II

CROSS-SECTION OF ENVIRONMENTAL ISSUES AND THREATS IN AFRICA

• Chapter 6 •

FROM JESSE TO IJEGUN: POVERTY, ENVIRONMENTAL DEGRADATION AND THE QUEST FOR THE REDISTRIBUTION OF NIGERIA'S OIL WEALTH

Monsuru Muritala

Introduction

From Jesse to Ijegun, the carnage occasioned by the pipeline explosions no doubt remains monumental. Prior to the discovery of oil and gas in Nigeria, there were other resources capable of transforming the society from traditional to a developed economy, but events of over four decades have shown that mismanagement, corruption, environmental degradation and poor leadership seem to have turned what ordinarily should be a blessing to a "curse".

This chapter highlights recurring calamities associated with oil and gas in Nigeria. The chapter raises some questions such as why people keep risking their lives in spite of the obvious dangers of destroying and scooping petrol from pipelines. Is environmental degradation caused by sabotage or negligence on the part of the multinational oil companies? And can we attribute prevalent poverty to pipeline vandalization?

Oil exploration and Nigeria's underdevelopment are inseperable. The fact remains that Nigeria had generated over $400 billion from over forty years of exploration, and yet it has no remarkable effects on the infrastructure or the standard of living of the majority of its citizens. Despite the huge foreign exchange earnings, the country is still ravaged by poverty, hunger and impoverished educational system required to produce the manpower needed for a competitive economy.

The inability of the federal government to stop smuggling, bunkering and other oil-related crimes compounds the problems. This has resulted in social unrest and several industrial crises in the recent past.[1] Steve Azaiki argues that injustice never breeds harmony; peace, according to him, will continue to elude the nation if nothing concrete is done to remedy inequity in income and distribution.[2] One major threat to the operation of the oil industry is the politics of oil, the mainstay of the economy and a determinant of international diplomacy and national security. Put differently, the industry is challenged by the politics of revenue allocation and resource control, which has changed the orientation of an average Nigerian who is forever scheming to get into a position from which he will most benefit from oil money.[3] Colonial legacy partly accounts for Nigeria's undue reliance on proceeds from oil. As it would be recalled, colonial capitalism did not factor the development of Nigeria into its agenda; thus, the extraction of raw materials and their shipment to the metropolitan factories remain the major focus. It is evident that the colonial authorities consolidated its stronghold on the natural resources by first establishing political control over them. Daniel Omoweh argues that the British upturned the pre-colonial land system in the colony and imposed its land policy on the colonized.[4] Thus, the systematic overthrow of powers of the elders and traditional rulers to hold land in trust for the people. Besides, the British consolidated their economic hegemony with many acts and laws, prominent among which was the Mineral Act of 1912. This vested sole ownership of not just the minerals but also the land where they were mined in the hands of the government.

By 1914 however, the Mineral Act of 1912 was amended as the Mineral Act of 1914; thus, the colonial state granted the Royal Dutch/Shell group of companies and the British petroleum license and monopoly to prospect for, and exploit petroleum in the colony in 1938.[5] Consequently, the stage was set for the unending theatre of environmental destruction, unemployment, mass displacement of inhabitants, exploitation of oil and mineral resources without concern for the environment. This is because the acts, as argued by Omoweh Daniel, not only defined the content of the ownership of minerals and land, but also turned the people who lived in the host mining communities into squatters in their ancestral homes. By 1960, when Nigeria became politically independent, one would have expected a diversification of the economy by those who were entrusted with the management of the post-colonial state, but they continued the inherited exploitative policies and structures of the colonial masters in both the mineral sector and the larger economy.[6]

It has been observed that the government at independence, pre-occupied with the motives of earning more revenues, ventured into the country's mineral sector without a technological base, and an energy policy; the argument here is that the weak technological base of Nigeria at this period put her at disadvantage as far as the control and protection of the environment are concerned. This is because the established state agencies such as the Nigerian Mining Corporation and Nigerian Coal Corporation were charged with the duty of collecting rent and royalties for the state from foreign operators before they even ventured into production.[7] In fact, the liberalization of the oil exploration in Nigeria, following the amendment of the Mineral Act of 1958 and 1960, impacted greater negative consequences that were later to unfold.

Leadership, Corruption and Mismanagement of Resources

Nigeria, with more than two million barrels of oil produced per day is, no doubt, Africa's leading oil exporter.[8] The country has the potential to build a prosperous economy but attendant corruption, inconsistencies and gross mismanagement prevent it from doing well. As the economic life blood, oil fetches over 90% of foreign exchange revenue. The industry brings in about 85% of the government's total revenue.[9] Azaiki argues that during the boom period of between 1988 and 1993, the government earned an average of $10 billion annually. But despite the wealth generated by Nigeria's natural resources, poverty is widespread and social indicators place it among the five poorest countries in the world.

The economic backwardness has been attributed to the recklessness of the military rulers in Nigeria by appreciable number of the citizenry, but events have proved otherwise that democratically elected government is equally guilty of this cankerworm that has eaten deep into the fabrics of our society. In fact, the present democratic dispensation is still faced with the major socio-economic challenges of persistent high unemployment rate (29%), poverty rate (70%) illiteracy rate of (35%) and large wealth disparities.[10] Arguably however, the corruption perpetrated in the various government offices and establishments culminated in poor standard of living of the present generation, the organization of civil society and the prospects of the next generation as well as having direct bearing on the challenges of environmental sustainability. Consequently, it brought about unequal access to socio-economic benefits and equal treatment before the law.

It is against this background that one would analyze why hundreds of deprived people were roasted to death in Jesse, Isikwato, Abule-Egba and Ijegun to mention a few places in an attempt to scoop oil from pipeline burst as well as the vandalization of petroleum pipeline for survival. The poser here is how could a rational person's sense of judgment be clouded as to risk his or her life in an atmosphere of highly inflammable object? The answer is not far fetched: social injustices brought about by extreme poverty that resulted from social-economic exclusion. Besides, the suffering in the midst of plenty remains the consequences of poor leadership associated with corruption and gross mismanagement of resources. Corruption, as conceived by Obiageli Ezekwesili, has the capacity to deplete our values, morals, and also eat deep into the fabrics of our social, economic and political life.[11] Thus, John Ikubaje corroborated this when he opines that government crises, economic woes and development predicaments in Nigeria are attributable to corruption.

A conceptual explanation of corruption has been attempted by various scholars, while the modernist school, as represented by Huntington, argued that corruption may be more prevalent in some cultures than others but in most cases it is more prevalent during the most intense phases of modernization. But some others opined that colonialism brought different values and administrative practices that promote bureaucratic corruption in Africa, Asia and Latin America.[12] Hyden corroborated this view by referring to corruption as "colonial heritage".[13] Still on the colonial legacy, Kwame Nkrumah long ago argued that the colonial officers were corrupt in their administration as they exploited their colonies for the development of their home countries through diverse means.[14] The exploration of these colonies by the colonial officials for personal enrichment and colonial governments' development, according to Ikubaje, made Africans to also see the new state as an exploitation entity for private realm development.[15]

Although, the "Africans' position" sounds convincing, some scholars have nonetheless debunked the position by arguing that corruption is prevalent in countries like Liberia, Nepal and Thailand to mention but a few places where colonial governments were never established.[16] But for the purpose of this chapter, the resource curse theory of corruption and what some scholars' term "devil excrement" and "paradox of plenty" remains central to the analysis. The proponents of this theory opined that the extraction of resources lowers the wealth of a country unless the funds generated are invested in other forms.[17] The resource curse does not refer to the mere possession of petroleum or other minerals, but rather to countries that are overwhelmingly dependent on resource/oil revenues. This dependence is generally measured by the extent to which natural resources dominate total exports or by the ratio of natural resources export to Gross Domestic Product, a figure that can range from as low as 4.9% to 85%.[18] This theory links corruption to underdevelopment in countries like Cameroon, Nigeria and Equatorial Guinea with regards to mismanagement of natural resources.

Using the revenue generation approach, the petroleum sector remains the bedrock upon which the grand corruption that sustains other forms of corrupt practices in the country is perpetrated. The sector accounts for over 70% of the country's revenue at all levels of government (federal, state and local government); 40% of the GDP, and more than 85% of foreign exchange earnings.[19] According to one estimate, over $400 billion of oil money has been stolen by Nigerian present and past leaders from independence in 1960 to date.[20] He further stressed that the records of corrupt practices on agricultural resources (cocoa, groundnut and palm oil) as major source of export earnings (1960-early 1970) was less compared to what it is under oil exportation period.[21] A careful study of the nature of corruption in the Nigerian oil sector indicates several problems: underpayment of taxation, royalties and gas flaring penalties by oil companies, bribery from oil companies to the Nigerian government officials; tax evasion by oil companies, personalization of oil rigs contract awards by minister of petroleum, oil bunkering and under declaration of oil revenue by the Nigerian National Petroleum Corporation(NNPC) and the Central Bank of Nigeria (CBN) have contributed significantly to oil resource curse in Nigeria.

Poor leadership has been identified as the major problem confronting the economic breakthrough of Nigeria. From independence, the country past leaders (military and civilian) successfully squandered and are still squandering the nations wealth without remorse. Prominent amongst this was the $12.4 billion oil windfall from the first Gulf war allegedly mismanaged by former military Head of state Gen. Babangida.[22] The late Gen. Sanni Abacha personalized the Nigerian Central Bank by issuing cheques at will for billions of naira for personal use.[23] He was listed among the ten most corrupt world leaders with $2.5 billion "loot".[24] Gen.Yakubu Gowon, during his administration, spoke publicly that money was not a problem to Nigeria which could therefore afford to embark on any project whatever its cost. Thus, he squandered wealth on white elephant projects: FESTAC 77, Cement Armada, Universal Free Primary Education, the Sports jamboree, etc.[25] The democratic regimes, both past and present are equally guilty of corruption and squandering of the nation's resources. During the second republic Alhaji Umaru Dikko, the transport and aviation minister of the period, was alleged to have amassed a fortune of over $1 billion in three years.[26] The Obasanjo administration, with his anti-corruption campaign, was an epitome of corruption. The trial by the Economic and Financial Crime Commission (EFCC) of Governors James Ibori, Orji Uzor Kalu, Boni Haruna, and Jolly Nyames for misappropriating millions of dollars under their custody says it all.

Be that as it may, the money laundering allegations against Governors Diepreye Alamieyeseigha and Joshua Dariye of Bayelsa and Plateau States respectively revealed the corruptible nature of our leaders to the outside world. Besides, corruption has become so entrenched that virtually all the strata of the Nigerian society are afflicted with the phenomenal. It is therefore not surprising that in 2007, Nigeria was ranked 148 in the list of the most corrupt countries in the world. But kudos to the various anti-corruption campaigns, the recent survey ranked Nigeria 22 out of the 47 nations surveyed in Africa.[27] Consequently, the selfish and greedy attitude of Nigerians in the positions of authority in both private and public sectors have retarded the growth of the economy in the past and present. Nigeria is blessed with enviable mineral resources that over the years have been grossly mismanaged by our successive visionless leaders thereby translating to economic wants, joblessness, marginal employment, poverty, social disharmony, criminality and insecurity.

Environmental Degradation, Poverty and Agitation for Resource Control

Economic development devoid of environmental sustainability is not only retrogressive but injurious to the well being of human existence. Globally, the issue of environmental protection, conservation and sustainability has been a continuous matter of concern. This is because human beings have benefited and still benefiting from the environment with little returns to the environment. The earth is being threatened by climate change and global warming resulting from careless human activities in the course of technological advancement, industrial processes, environmental pollution and economic activities. In a bid to rescue the earth from environmental disaster, several earth summits had been held from Stockholm to Japan and conferences are being convened from time to time by scholars, policy makers and world leaders.

Environmental discourse is central to the analysis of this chapter because the purpose is to establish the socio-economic consequences of neglecting the environment as well as the mismanagement of natural resources. As a matter of fact in Nigeria, every known law on environmental safety has been violated.[28] For instance, if the operating oil companies had carried out their duties as specified in the Oil Pipelines Act Cap145 of 1948, Oil Pipelines Act 1965, No. 24 and the Oil Pipelines Act Chapter 338, Laws of the Federation of Nigeria 1990 which provide, among other things, for the payment of compensation for economic crops and property damaged, payment of survey fees and other miscellaneous matters,[29] environmental degradation would have been minimized and community relations strengthened. Environmental degradation is everywhere, from desertification in the north to gully erosion in the south-east and gas flaring and oil spillage in the Niger-Delta. The most serious cases associated with oil exploration occur in the Niger Delta.

The Niger Delta is located in the South-South geographical region of Nigeria and has a population of about 20 million people of different nationalities.[30] Divided into nine states of the federation, Abia, Akwa-Ibom, Bayelsa, Cross-River, Delta, Edo, Imo, Ondo and Rivers, these states are otherwise known as the oil producing states. With special reference to Bayelsa, Delta, Rivers, Cross-Rivers and Akwa-Ibom their environmental problems require serious and urgent attention. The operations of the petroleum industry, poor urban development and gross misuse of natural resources have translated to poverty, illiteracy, rural-urban migration, overcrowding of cities and socio-economic problems. By their topography, the primary occupations of the people include fishing, farming, forest products gathering, and crafts which are done at subsistence levels, basically to keep body and soul together.[31] Oil extraction, according to Steve Azaiki, has negatively impacted the socio-physical environment, threatening the fragile peasant economy, bio-diversity, social livelihood and the very survival of the region. A 1995 World Bank assessment of industrial wastes in Rivers State shows that petroleum industry is the largest source of pollution, generating most of the 5500 tons of hazardous wastes produced annually.[32] It further notes that oil spills and leak occur hundreds of times a year resulting in serious degradation of surface water, adjacent wetland and mangrove ecosystem. Between 1986 and 1996, two and half million barrels of crude spilled or leaked into delicate riverine environment resulting in wholesale devastation of fish stocks that most villagers relied on.[33] Below is the statistics highlighting the magnitude of oil spillage experienced by the oil producing communities between 1976 and 1998.

Table 6.1: Oil Spill Data for Nigeria

S/NO	Year	Number of Spill Incidents	Quantity spilled (barrels)
1	1976	128	26,157.00
2	1977	104	32,879.25
3	1978	154	489,294.75
4	1979	157	694,117.13
5	1980	241	600,511.02
6	1981	238	42,722.50
7	1982	257	42,841.00
8	1983	173	48,351.30
9	1984	151	40,209.00
10	1985	187	11,876.60
11	1986	155	12,905.00
12	1987	129	31,866.00
13	1988	208	9,172.00
14	1989	195	7,628.161
15	1990	160	14,940.816
16	1991	201	106,827.98
17	1992	367	51,131.91
18	1993	428	9,752.22
19	1994	515	30,282.67
20	1995	417	63,677.17
21	1996	430	46,353.12
22	1997	339	59,272.30
23	1998	390	98345.00
Total		5724	2,571,113.90

Source: The Department of Petroleum Resources, Lagos, Nigeria. Also, see www.iosc.org/papers/IOSC 2005.

As depicted above, it is obvious that the incidence of oil spillage is enormous, and partly responsible for this is the flagrant violation of environmental laws by the operating oil companies as well as poor enforcement of the laws by the Federal Government of Nigeria. Bamidele Ajakaye argued that, 80 per cent of the oil spillage in Nigeria resulted from wilful vandalization and sabotage by the oil producing communities.[34] This assertion is corroborated with the survey carried out by Nigerian AGIP Oil Company (NAOC) on the causes of oil spills along the major pipelines between 1998 and November 2002. Out of the 46 cases recorded in Ogodo/Brass, sabotage was responsible for 45 while equipment failure was responsible for one. In Akiri/Ebocha, all the 25 cases resulted from sabotage, while in Tehidaba/Brass sabotage was responsible for 28 cases out of the 29 recorded.[35] It is important to note that sabotage, corrosion, equipment failure and operational error remains the yardstick used in assessing the causes of oil spillage, and as enumerated above, equipment failure accounted for minimal cases over substantial cases recorded by sabotage. However, to support the view that oil spillage and associated pollution is not monocausal (sabotage), insensitivity and carelessness on the part of the oil and gas companies have been identified as a major factor.

Gas flaring in the Niger Delta has a devastating effect on the flora and fauna of the environment thereby endangering the lives of the local people. Indeed, gas flaring is widely believed to have contributed more emissions of greenhouse gases than all other sources in sub-Saharan Africa combined.[36] Meanwhile, continued gas flaring causes acid rain which corrodes homes and other local structures, many of which have zinc-based roofing. In a bid to prevent the deterioration some people embraced the use of asbestos-based materials which is stronger in repelling acid rain deterioration. Unfortunately, this only contributes to their own declining health and the health of their environment, because exposure to asbestos increases the risk of lung cancer and mesothelioma.[37] Gas flaring is also a major contributor to air pollution and biodiversity depletion in the Niger Delta. Consequently, the natural environment became depleted; socio-economic activities paralyzed; and poverty, malnutrition and diseases become synonymous with an oil-rich region.

Be that as it may, the Niger Delta is enmeshed in an insurgency and conflicts of unprecedented proportions. The contradictions of wealth generation amidst poverty has generated anger, frustration and hostility to the state and the Multinational Oil Corporations (MNOC).[38] Consequently, violent protests, disruptions of oil production plants, seizures of oil platforms, installations and equipments, kidnapping of MNOC staff, confrontation with state security forces and militarization of the region become the order of the day. As it would be recalled, agitation took the form of writing petitions but this was generally ineffective but succeeded only in obtaining some social amenities from the

MNOC.[39] However, through the activities of Ken Saro Wiwa, the grievances of the oil producing areas were elevated to the national consciousness.[40] Through his Movement for the Survival of Ogoni Peoples (MOSOP), he championed the cause of his Niger Delta people who are said to live in one of the most densely populated and polluted parts of Africa·

Saro Wiwa demanded $10 billion in accumulated royalties and environmental damage compensation, and a greater say in future oil explorations. But Saro Wiwa was ignored by the federal government and the oil companies. He then organized a protest that managed to shut down virtually all oil production in Ogoniland.[41] This was a severe blow not only to the oil industry but also to the system of corruption and patronage it had spawned and the military reacted with predictable brutality. The struggle continued and by November 10, 1995, Saro Wiwa and eight other Ogoni activists were hanged by the Abacha government on trumped charges of incitement to murder.

The next major outbreak of violence occurred in 1998 when several Ijo groups tried to duplicate MOSOP's strategies by declaring Ijo territory off limits to the Nigerian military and demanding a stop to all oil extraction.[42] In response to these actions, the Nigerian military saturated the Delta and Bayelsa States with up to 15000 soldiers and commenced a series of attacks that resulted in death of hundreds of civilians. Ijo militants retaliated by shutting off oil well heads in their areas. Consequently, from the period up till now the region evolved armed militants that the government has not been able to contain. In fact, the emergence in 2004 of Asari Dokubo, an Ijo youth leader who retreated into the creeks to wage war against the government and the oil companies, increased oil prices in the international market. Asari Dokubo accused the government and its officials of bunkering oil; he agitated for control of the Niger Delta oil resources by the people of the region. His activities and campaign angered the Obasanjo-led administration such that by September 2005, he was arrested and detained by the Federal government.

In reaction to his arrest was the formation of Movement for the Emancipation of the Niger Delta (MEND). The first MEND attack was launched four months after its formation. They demanded for the release of Asari Dokubo and Diepreye Alamieyeseigha, the Bayelsa state governor charged with corruption. Their activities ranged from massive destruction of oil blocks to kidnapping of oil workers as well as killing of the security operatives of government.

Notes

1. S. Azaiki, *Oil, Gas and Life in Nigeria* (Ibadan: Y-Books Limited, 2007), 87.
2. Ibid.
3. Ibid., 88.

4. D. Omoweh, "Oil and Politics of Environmental Degradation in the Niger Delta of Nigeria: A Historical Perspective," in *Nigeria and The Wider World in the 20th Century*. Ed. Y. Akinyeye (Ibadan: Davidson Press, 2003), 182.
5. Ibid.
6. Ibid., 183.
7. Ibid., 184.
8. Azaiki, *Oil, Gas and Life in Nigeria*, 99.
9. Ibid.
10. Ibid., 100.
11. O. Ezewezili, Chairperson of Nigeria Extractive Industries Transparency Initiative and Former Minister of Education. Cited in J. Ikubaje, *Corruption and Anti-Corruption: Revenue Transparency in Nigeria Oil Sector* (Lagos: Joe Tolalu and Associates, 2006), xiii.
12. J. M. Mbaku, "Bureaucratic Corruption and the Crisis of Political Reform," in *Multiparty Democracy and Political change: Constraints to Democratization in Africa,* eds. John Makum and Julius Ihonvbere (Burlington, USA: Ashgate, 1998), 67-69.
13. H. Goran and M. Bratton, *Governance and Politics in Africa* (Boulder: Lynne Rienner, 1992), 178.
14. Nkrumah Kwame, *Handbook of Revolutionary Welfare* (London: Panaf Books, 1968), 65.
15. J. Ikubaje, *Corruption and Anti-Corruption: Revenue Transparency in Nigeria's Oil sector*, 16.
16. See, "An Introduction," in M. Robinson, ed., *Corruption and Development* (London: Frank Cass, 1998), 6.
17. J. Stiglitz, *Making Natural Resources into Blessing rather than a Curse in covering Oil: A Reporter's Guide to Energy and Development* (New York: Open Society Institute, 2005), 13-19.
18. K. Terry, *Understanding Resource Curse in Covering Oil: A Reporter's Guide to Energy and Development* (New York: Open Society Institute, 2005), 22.
19. Ezekwesili Obiageli, "Public Comments on Corruption in the Oil industry as the Chairperson of Nigeria Extractive Industries Transparency Initiative," in *Transparency Watch* of July 2006, Berlin.
20. Mallam Nuhu Ribadu (Former EFCC Boss), See for details, *Daily Independent*, 15, August 2006.
21. Ibid.
22. J. Ikubaje, *Corruption and Anti-Corruption*, 42.
23. P. Okojie and A. Momoh, "Corruption and the Crisis of Development in Nigeria," a paper presented at the conference on "Redesigning the state? Political Corruption in Development Policy and Practice," Manchester Metropolitan University, 2005.
24. Ibid.

25. Ayagi Ibrahim, *The Trapped Economy* (Ibadan: Heinemann Books, 1990), 73.
26. See *West Africa Reports*, 9 June, 1986. 14.
27. See The Punch newspaper (Lagos) 1 October, 2008, 5.
28. Azaiki, *Oil, Gas and Life in Nigeria*, 133.
29. Ikubaje, *Corruption and Anti-Corruption,* 90.
30. Tope Akintola, "Environmental Degradation in Nigeria's Niger Delta," Online publication in www.africanconservation.org.
31. Ibid.
32. See for details, 1995 World Bank Bulletin, cited in J. Oribabor, *Nigerian Oil Industry and the Impacts of Wastes in the Niger Delta* (Makurdi: Maxil Books Limited, 2003), 14.
33. *Daily Independent*, Section B2, 22 January, 2007.
34. Bamidele Ajakaye, Managing Director National Oil Spills Detection and Response Agency. Television interview, 18 August, 2009.
35. For details, see *Business Day Newspaper*, 16 December 2002, 42.
36. See for details, 2002 World Bank Bulletin, cited in J. Oribabor, *Nigerian Oil Industry and the Impacts of Wastes in the Niger Delta,* 18.
37. A. Ejim, "Niger Delta: Issues Arising," *The NEWS*, 01, June, vol. 32, 2009, 9.
38. A. Ikelegbe, "Civil Society, Oil and Conflict in the Niger Delta Region of Nigeria: Ramifications of Civil Society for a Regional Resource Struggle," *Journal of Modern African Studies* Vol.39, no.3 September (2001), 440.
39. Ibid.
40. Ibid.
41. See for details, *Daily Independent*, B2, 22 January 2007.
42. Ibid.

• Chapter 7 •

POTENTIAL OF MICROBIAL STRAINS FOR RECLAMATION AND CLEAN-UP OF POLLUTED NIGERIAN SOILS

Timothy O. Adejumo

Introduction

Fungi and bacteria are capable of breaking down hydrocarbons and organic contaminants in oil-polluted soils and are effective means of disposing hazardous wastes. This involves transformation of simple/complex chemical compounds into non-hazardous, water-soluble products. The lignin-degrading white-rot fungi (*Pleurotus tuber-regium, P. pulmonarius, Lentinus subnudus*), *Rhizopus nigricans* and *Pseudomonas aeruginosa* and *P. fluorescence* have been found to degrade diverse toxic, recalcitrant and persistent environmental pollutants in Nigeria. Studies show high potency of these microorganisms and the possibility of using them in bioremediation of petroleum refinery and petrochemical waste waters. The advantages of bioremediation include its scale ability, cost effectiveness and simplicity. This chapter discusses the mechanisms, impact and the extent of the use of microorganisms, especially fungi and bacteria in the breaking down of environmental pollutants in Southern Nigeria.

Crude oil contamination of soils in the Niger Delta, Nigeria has been variously reported to be a perennial problem. This is as a result of the development of petroleum industry into new frontiers. It is obvious that inevitable spillage do occur, especially during routine operations (extraction and processing), while there are also records of acute accidents during transportation.[1] The use and disposal of hazardous substances also result in contaminated lands which have been found to be a potential threat to human and animal health as well as environmental effects. These areas are potentially ridden with environmental problems, ranging from devastation of agricultural settlements, land and water (crippling income generation through farming and fishing), through death of thousands of domestic animals, disastrous fire out-breaks, and various illnesses to deaths.[2] The contamination changes both the physicochemical and biological properties of the soil, and also the air is constantly being polluted, the inhalation of which may lead to several chronic respiratory diseases and deaths of livestock, poultry birds and humans. Environmental crises may lead to political unrest, with the people of the area feeling exploited and calling for reparations. More studies into oil pollution problems, a regular occurrence, are necessary.

Efforts to reclaim or clean up polluted sites are based on the risk of adverse health, environmental effects caused by the contamination as well as enabling the sites to be redeveloped for use.[3] The traditional methods of remediation include digging up the contaminated soil and removing it to a landfill, or the cap and contain method, while other approach is the complete destruction of the pollutants or to transform them to harmless and less toxic substances by high-temperature incineration and chemical decomposition including base-catalyzed dechlorination and UV oxidation.[4] Although these methods can be very effective at reducing levels of a range of contaminants, but each of them has been found to have many negative aspects. They have been found to be technologically complex, costly for small-scale application, lack of public acceptance, site workers and nearby residents are prone to contamination by incineration method.[5] To remove the pollutants from the soil in a rapid, cost-effective, ecologically responsible and sustainable approach is a growing concern in environmental management, the use of microbial strains is one of the ways of achieving this and this is addressed in this chapter.

Bioremediation of Pollutants

Bioremediation is a process whereby environmental contaminants are degraded or transformed by living organisms under controlled conditions to harmless or less toxic forms, or to levels below concentration limits established by regulatory authorities.[6] The living organisms are mainly microorganisms: the naturally occurring fungi, bacteria or plants which detoxify substances that are dangerous to human health and/or the environment. The microorganisms involved may either be indigenous or imported to a contaminated site (bioaugmentation).

Many organopollutants released into the environment are persistent and are known to have toxic, carcinogenic and or mutanogenic effects.[7] They include the polycyclic aromatic hydrocarbons (PAH), polychlorinated biphenyls (PCBs), pentachlorophenols (PCP), trinitrotoluene (TNT), 1,1,1-trichloro-2,2-bis (4-chlorophenyl) ethane (DDT) and benzene, toluene, ethylbenzene, and xylene (BTEX). Table 7.1 shows the contaminants that are suitable for bioremediation with specific examples, mode of degradation and potential sources.

Ecological and physiological conditions of microbial strains

Various fungi and bacteria use some crude oil fractions as a sole carbon source and change them to non-toxic compounds such as carbondioxide.[8] Lignolytic fungi have been found to produce extracellular enzymes with very low substrate specificity and this makes them appropriate for breaking down a wide range of organopollutants (PAH, PCBs, TNT, DDT) that have similarities in structure to lignin.[9] Some aromatic molecules found in hydrocarbons (oil and petrochemicals) bear the same moieties as those found in lignin or humic materials (plant-based materials). These enzymes have been referred to as lignin-modifying enzymes (LMEs) which include lignin peroxidase, manganese dependent peroxidase, and laccase. Biodegradative abilities of white rot fungi (like *Phanerochaete chrysosporium, P. sordida, Trametes versicolor, T. hirsutus, Pleurotus ostreatus, Bjerkandera* sp.) are remarkable. This is because they contain all the three LMEs and are therefore able to breakdown and mineralize several environmental pollutants into nontoxic forms including those that are usually regarded as very recalcitrant to biodegradation. The fungi require carbon co-substrate such as potato pulp, wheat straw, peat, bark, wood chips,[10] sawdust, and straw.[11] Also, the branching, filamentous mode of fungal growth allows for more efficient colonization and exploration of contaminated soil.

Microbes will adapt and grow with an excess of oxygen, and in anaerobic conditions, with the presence of hazardous compounds or on any waste stream where the main requirements of energy source and a carbon source are present. This adaptability of microbes and other biological systems enable them to be used to degrade or remediate environmental hazards. Vidali[12] classified them into various groups: Aerobic bacteria like *Pseudomonas, Alcaligenes, Sphingomonas, Rhodococcus*, and *Mycobacterium* that are known for their degradative abilities on pesticides and hydrocarbons, both alkanes and polyaromatic hydrocarbon compounds are used as the sole source of carbon and energy.

Anaerobic bacteria: they are used for bioremediation of polychlorinated biphenyls (PCBs) in river sediments, dechlorination of the solvent trichloroethylene (TCE), and chloroform.

Methylotrophs: Aerobic bacteria that grow utilizing methane for carbon and energy. The initial enzyme in the pathway for aerobic degradation, methane monooxygenase, has a broad substrate range and is active against a wide range of compounds, including the chlorinated aliphatics trichloroethylene and 1,2-dichloroethane.

Table 7.1: Some contaminants potentially suitable for bioremediation

Class of contaminants	Specific examples	Aerobic	Anaerobic	More potential sources
Chlorinated solvents	Trichloroethylene Perchloroethylene		+	Drycleaners Chemical manufacture
Polychlorinated biphenyls (PCB)	4-Chlorobiphenyl 4,4-Dichlorobiphenyl		+	Electrical manufacturing Power station Railway yards
Chlorinated phenol	Pentachlorophenol (PCP)		+	Timber treatment (Wood preservatives) Landfills
BTEX	Benzene Toluene Ethylbenzene Xylene	+	+	Oil production and storage Gas work sites Airports (Aviation fuels) Paint manufacture Port facilities Railway yards Chemical manufacture
Polyaromatic hydrocarbons (PAHs)	Naphthalene Antracene Fluorene Pyrene Benzo(a)pyrene	+		Oil production and storage Gas work sites Coke plants Engine works Landfills Tar production and storage Boiler ash dump sites Power stations
Pesticides	Atrazine Carbaryl Carbofuran Coumphos Diazinon Glycophosphate Parathion Propham 2,4-D	+	+	Agriculture Timber treatment plants Pesticide manufacture Recreational areas Landfills

Source:[13]

For bioremediation to be successful, there have to be right microbes (bacteria or fungi, which have the physiological and metabolic capabilities to degrade the pollutants), biodegradable contaminant in the right place with the right environmental factors for degradation to occur, biodegradation occurring in the site naturally.[14] Factors such as temperature, moisture, oxygen accessibility and nutrients influence the rate of biodegradation of crude oil in soil. Many reports indicate the importance of nutrient addition to contaminated soil to enhance biodegradation.[15] Phosphates and nitrate salts are the most common nutrient additives.[16] Degradation occurs when the microbes and the contaminants are in contact. Mobile bacteria exhibit chemotactic response, sensing the contaminant and moving toward it, while others like fungi grow in filamentous form toward the contaminant. Mobilization of the contaminant can be enhanced by utilizing surfactants.

Decomposition mechanism of pollutants by microbial strains

The major metabolic pathways in the biodegradation of hydrocarbons by bacteria and fungi start with the initial step of oxidation of the substrate by oxygenases, for which molecular oxygen is required. Alkanes are subsequently converted to carboxylic acids that are further biodegraded via ß-oxidation (the central metabolic pathway for the utilization of fatty acids from lipids, which results in formation of acetate that enters the tricarboxylic acid cycle). Aromatic hydrocarbon rings generally are hydroxylated to form diols; the rings are then cleaved with the formation of catechols which are subsequently degraded to intermediates of the tricarboxylic acid cycle. Fungi and bacteria form intermediates with differing stereochemistries. Fungi, like mammalian enzyme systems, form *trans*-diols, whereas bacteria almost always form *cis*-diols (many *trans*-diols are potent carcinogens, whereas *cis*-diols are not biologically active). Since bacteria are the dominant hydrocarbon degraders in marine environment, the biodegradation of aromatic hydrocarbons results in detoxification and does not produce potential carcinogens. The complete biodegradation (mineralization) of hydrocarbons produces the non-toxic end products carbon dioxide and water, as well as cell biomass (largely protein) which can be safely assimilated into the food web.[17] The aliphatic and some aromatic fractions are the most biodegradable components and the resins and asphaltenes are believed to be resistant to biodegradation.[18] However, the bacterial degradation of PAHs with more than three rings, often referred to as high molecular weight PAHs, are often challenging to mineralize by microorganisms. This is related to their low solubility and high affinity to solid surfaces and organic liquids, which generally limits their bioavailability to microorganisms present in soils.[19]

In white-rot fungi, the ligninolytic enzymes catalyze the degradation of pollutants by using a non-specific free radical mechanism.[20] When an electron is added or removed from the ground state of a chemical it becomes highly reactive, allowing it to give or take electrons from other chemicals. This provides the basis for the non-specificity of the enzymes and the ability to degrade xenobiotics, chemicals that have never been encountered in nature.[21] The main reactions that are catalyzed by the ligninolytic enzymes include depolymerization, demethoxylation, decarboxylation, hydroxylation and aromatic ring opening. Many of these reactions result in oxygen activation, creating radicals that perpetuate oxidation of the organopollutants. Once the peroxidases have opened the aromatic ring structures by way of introducing oxygen, other more common species of fungi and bacteria can mineralize the products intracellularly into products such as CO_2 and other benign compounds.[22] The same mechanisms that give these fungi the ability to degrade lignin are also used to degrade a wide range of pollutants such as DDT, TNT, PCBs and PAHs.[23]

Some authors[24] have tested the biodegradation of two PAHs, nitrobenzene and anthracene, by a white rot basidiomycete, *Trametes trogii* and found that high activities of laccase and Mn-dependent peroxidase during both the trophophasic and idiophasic growth stages in the fungal cultures, suggesting that the enzymes are responsible for PAH degradation and that complete degradation of PAHs by most white rot fungi likely involve cytochrome P450 enzymes as well as ligninolytic enzymes.[25] The application of spent-mushroom compost (SMC) to contaminated soil utilizes the high levels of laccase and Mn-dependent peroxidase and high carbon contents to break down complex organopollutants[26] have been found to be a great recycling tool and potentially a sustainable approach to bioremediation.

Advantages of bioremediation

- Bioremediation is cheaper/cost effective than other clean-up methods;
- Biodegrades and detoxify hazardous contaminants while other techniques merely transfers pollutants from one medium to another;
- Generally a greater public acceptable waste treatment process for soil pollutants;
- Residues from the treatment are harmless, and include carbon dioxide, water, and cell biomass;
- Bioremediation can be coupled with other physical or chemical treatment methods;
- Useful for complete destruction of a wide variety of contaminants thereby get rid of relocation of contaminants from one medium to another;
- Eliminates long-term liability with regulatory encouragement; and
- Environmentally safe, remove threats to human health and the environment during transportation, since bioremediation can be carried out on site, and site disruption is minimal.

Disadvantages of bioremediation

- It is time-consuming; it often takes longer than other treatment options, such as excavation and removal of soil or incineration.

- Limited to those compounds that are biodegradable, not suitable for soils containing heavy metals, radionuclides, complex (high molecular weight) PAHs and chlorinated hydrocarbons.[27]
- Incomplete transformation.
- Biological processes are highly specific, meaning that one has to do treatability studies on a small scale before the actual clean up of the sites.
- It is difficult to extrapolate from bench and pilot-scale studies to full-scale field operations.
- Contaminants may be present as solids, liquids, and gases.
- Microbial metabolism of contaminants may produce toxic metabolites.

Environmental impacts of oil exploration

The environmental deterioration in the Niger delta has resulted in declining soil quality, soil erosion and depletion of vegetation cover in the Tropical Rain Forest belt of Southwestern Nigeria. The mineral exploration, particularly in the oil industry, has adversely affected land, water and air quality. The major atmospheric pollutants from the refineries are nitrogen oxides, carbon, sulphur and liquid effluents. Oil exploration in Ogoni land have recently gained international attention in recent years. There were 1,360 oil spill incidents on the mangrove swamps and offshore areas which constitute the most productive biological habitats of the Niger Delta between 1976 and 1983, while by 1994 the incidence of oil spills had exceeded 2,000.[28]

Crude oil exploration leads to landscape disturbances/ aesthetic deterioration of the landscape, land degradation and ecosystem destabilization: land surface devastation (including erosion), land subsidence, disruption of drainage systems, deforestation, excessive water draw down, and contamination of the water table. Processing and transportation leads to gas leakages, oil spills, noise and pollution of the air, soil and water: thermal loading of water, increased storage and consumption in CO_2 and CO, Ozone layer depletion, acidification of air, soil and water, weather modification, toxicity hazard to plants and animal, death of terrestrial and marine life, loss of crops and livestock, impairment of atmospheric visibility, vehicular accidents, damage to buildings and machinery, nervous disorders, respiratory diseases, cardiovascular illnesses, cancers and food poisoning.[29] However, increased industrial activities and socioeconomic development brings about more employment opportunities, commodities that used to be imported are now produced locally leading to increased material wealth. Alongside with the above social benefits, there is air pollution which is as a result of industrial effluents.

Efficiency of some microorganisms in the bioremediation of polluted soils

Several reports[30] claim that growth of microorganisms and reduction of total petroleum in crude oil contaminated soils indicated the high degradative potentials of the fungi and bacteria employed (See Table 7.2). White-rot fungus, *Pleurotus tuber-regium* on rice straw brought about increased organic matter content (5.19%), carbon (2.99%) and available potassium, (0.97 meq/100 g) of 1- 40% concentration of soil polluted with engine-oil after six months of incubation compared to 4.41%, 2.56% and 0.66 meq/100 g respectively in the control.[31] There was also bioaccumulation of zinc and nickel, lower nitrogen (0.16%), phosphorus (9.32 ppm) and pH (5.93) observed in incubated soil, while higher values of 0.32% nitrogen, 11.42 ppm phosphorus and pH 6.94 were obtained in the control. The authors[32] showed that *P. tuberregium* is able to decontaminate crude oil contaminated soils of Uvwiamgmughe in Delta State by reducing the various petroleum hydrocarbons to varying degrees. The contaminants instead of reducing the growth of the fungus improved their growth, while phyto assessment tests show that the addition of poultry litter, sawdust and banana leaf blades to fungal remediation restored the contaminated soil. Reports from authors[33] found that there was significant improvement on the growth of cowpea (*Vigna unguiculata* L. Walp) when polluted soil with crude oil, automotive gasoline oil and spent engine oil was amended with sawdust and waste cotton and bioaugmented with *P. pulmonarius* after one month of incubation as compared to planting on polluted soil with no amendments and bioaugmentation.

The ability of indigenous fungus, *Lentinus subnudus* to mineralize soil contaminated with 1-40% concentrations of crude oil was tested.[34] Organic matter and carbon were higher than the control at all concentrations of crude oil in inoculated soils for 3 months. Nutrient contents were generally higher after 6 months of incubations except potassium levels which were not significantly different from the control, while the Total Petroleum Hydrocarbon (TPH) in crude–oil contaminated soils; the highest rate of biodegradation was at 20% concentration after 3 months and 40% after 6 months of incubation after 90 days of incubation.[35] The authors also reported that concentrations of Fe, Cu, Zn and Ni recovered from straw/fungal biomass complex increased with the increase of engine-oil contamination and bioaccumulation. The author[36] reported that the 4 fungal isolates obtained from soils collected from Niger Delta region were assayed using four crude oil types [Bonny light, Dubr oil and Escravos light (Nigerian crude oils) and Arabian light (an Arabian crude oil)] showed potential for bioremediation: *Rhizopus nigricans* recorded the highest efficacy in crude oil degradation (18% weight loss in crude oil after 90 days of incubation) followed by *Aspergillus flavus* (14%) and *Penicillium notatum* (11%), while *A. niger* showed the least of 9%.

Table 7.2. The use of microorganisms for bioremediation of polluted soils in Nigeria

Microorganisms	Substrate	Nature of Pollution	Location	Source
		Fungi		
Pleurotus tuber-regium	Rice straw	Engine oil Polluted Soil	Ibadan, Nigeria	37
P. tuber-regium		Crude oil polluted soil		38
P. tuber-regium	Poultry litter+ Co-substrates	Crude oil Polluted Soil	Uvwiangmughe, Delta State	39
P. pulmonarius	Sawdust+ waste cotton	Crude oil, gasoline oil and spent engine oil		40
Lentinus subnudus,	Rice straw	Oil polluted	Ibadan, Nigeria	41
L. squarrosulus	Rice straw	Engine oil Polluted Soil	Ibadan, Nigeria	42
Rhizopus nigricans *Aspergillus flavus, A. niger* *Penicillium notatum*		Four crude oil types		43
		Bacteria		
Pseudomonas aeruginosa	Mineral medium	Bonny light crude oil	Irri, Niger Delta	44
P. aeruginosa *P. fluorescence*	Mineral salt	Refinery effluent- Phenol	Nigeria	45
P. putida	Inorganic nutrients	Crude oil polluted soil	Nigeria	46
P. fluorescence *Acinetobacteria iwofii,* *Bacillus subtilis,* *Arthrobacter globiformis*	Fertiliser	Crude oil polluted soil	Niger Delta	47

Four strains of *Pseudomonas aeruginosa* biodegraded Bonny light crude oil of Irri town in Niger Delta in 15 days (ranging from 4.9% to 29.6%), and that degradation rates and specific growth rates varied significantly between 0.049 and 0.351 day^{-1} and 0.017 and 0.028 $hour^{-1}$ respectively,[48] while major peak components of the oil were also reduced between 6.5% and 70.6%. *P. aeruginosa* and *P. fluorescence* were also tested for bioremediation potential on Warri Refinery and Petrochemical effluent with respect to phenol biodegradation in a batch reactor.[49] The authors reported that *P. aeruginosa* completely mineralized phenol at 60th hour of cultivation, while 75% (23 mg/l) of phenol was degraded by *P. fluorescence* and complete degradation at 84th hour of fermentation. The oil pollutant was completely biodegraded with the soil bacterium, *P. putida* at week 9 and the inorganic nutrient ions PO_4^{-3} and NO_3^{-1} were significantly utilized,[50] while there were no significant changes in the concentrations of oil and inorganic nutrients in the control. Also, the percentage germination and growth profiles of cress seeds (*Lepidium* sp.) planted as evidence of the recovery of the oil-impacted soils were poor in the control (27.5%) with pronounced abnormal morphology when compared with the results obtained for the sterile soils inoculated with the bacterium (98.8%). The authors[51] also reported that *P. fluorescence*, *Acinetobacteria iwofii*, *Bacillus subtilus* and *Arthrobacter globiformis* introduced into typical Niger Delta fertilized soil with 5% crude oil pollution level gave the highest number of leaves, % crop emergence, plant biomass, microbial population and degradation of petroleum hydrocarbon compared to treatments that had received crude oil without the microbes and fertilizer.

Landfarming and Phytoremediation methods

Remediation by Enhanced Natural Attenuation (RENA), a land farming treatment technology for intervention in petroleum hydrocarbon contaminated soils where contaminated soils, sediments and sludge's are periodically turned over or tilled into the soil to aerate the waste.[52] This process spanning 10 weeks was used for a farm settlement at Rumuekpe in Rivers State State in the Niger Delta Area that was contaminated by crude oil.[53] The authors reported that a TPH concentration of 1.1004×10^4mg/kg of the sandy soil was achieved after spiking and tilling and a reduction in the TPH level from 300 mg/kg after 8 weeks, to 282 mg/kg after 10 weeks. The nitrogen and phosphorus levels of the sandy soil were 24.6 and 22.8mg/kg respectively, suggesting that the nitrogen and phosphorus levels could no longer support biodegradation at the residual TPH levels of 282mg/kg and 22.8mg/kg after spiking and tilling respectively, which further reduced to 0.12mg/kg and 1.7mg/kg respectively after 10 weeks. The Total Hydrocarbon Utilizing Bacteria (THUB) increased from 3.0×10^4 cfu/g to 8.55×10^4 cfu/g and finally reduced to 5.38×10^4 cfu/g, while the Total Heterotrophic Bacteria (THB) reduced from 1.22×10^8 cfu/g to 5.98×10^5 cfu/g.

Two plant species are being used in Ogbogu, Niger Delta to clean up spills, improve the quality of drinking water, soil conditions and the health of their surrounding environment.[54] The first stage of the cleanup involves indigenous *Hibiscus cannabinus*, an annual herbaceous plant originally used for pulp production. This plant has high rates of absorbency and can be laid down on top of the water to absorb oil. The oil saturated plant material is then removed and sent to a safe location where the hydrocarbons can be broken down and detoxified by microorganisms. The second stage of the bioremediation involves *Vetiveria zizanioides*, a perennial grass species, having a deep fibrous root network that can both tolerate chemicals in the soil and can also detoxify soils.

Ways of adopting bioremediation for solving environmental problems

Bioremediation is a scientifically intensive procedure which must be tailored to the site-specifc conditions. The basic steps that may be necessary for any bioremediation project includes compliance analysis, site characterization, method selection (feasibility studies), remediation proper and end of project analysis.[55] Compliance analysis requires examination of the contaminated site in the light of government regulation and the required action plan. Here, the environmental protection agencies (EPAs) such as Federal Environmental Protection Agency (FEPA) and Department of Petroleum Resources (DPR) of NNPC oversee such regulations in Nigeria. The second stage: site characterization, is a challenging and difficult aspect of a bioremediation project; it involves analysis of the physical characteristics and nature of the soil. Parameters such as cation exchange capacity, inorganic matter content, acidity (pH), aeration status, hydraulic properties, geologic and hydrogeologic factors, meteorological and climatological data of the soil may be evaluated.[56] Feasibility study for possible bioremediation approach and bioassays for pollutant degradation are carried out under control laboratory conditions. The importance of bioassays is that they integrate the combined effects of the mixtures of all the chemical components of the crude oil and other chemicals present in the polluted soil, including their bioavailability.[57] This confirms complete or near complete removal of the contaminants. The use of this type of biological test method is recommended for the evaluation of the effectiveness of remediation process.

Conclusion

The knowledge on bioremediation capabilities of white-rot fungi is growing rapidly from laboratory studies to its utilization in the field. They provide an environmentally-friendly, economical approach that should be intensified and adopted. The introduction of indigenous mushroom species like *Pleurotus tuber-regium, Lentinus subnudus, L. squarrosulus* to polluted soils with engine-oil, automotive gasoline oil and spent engine oil amended with rice straw, poultry litter, sawdust and banana leaf blades after months of incubation, brought about a reduction of total petroleum hydrocarbons in crude oil contaminated soils, increased organic matter content as well as bioaccumulation of zinc and nickel, lower nitrogen, significant improvement on the growth of cowpea (*Vigna unguiculata* L. Walp). While non white-rot fungus, *Rhizopus nigricans* has also been shown to have high efficacy in crude oil degradation. Among the bacteria, strains of *Pseudomonas aeruginosa, P. fluorescence* and *P. putida Acinetobacteria iwofii, Bacillus subtilus* and *Arthrobacter globiformis* have showed good bioremediation potential, with high percentage germination and growth profiles of crops planted as evidence of recovery of the oil-impacted soils. Bioremediation may be a promising biotechnology for enhanced recovery and rapid rehabilitation of crude oil-polluted agricultural lands upon pollution. Hence, fungi and bacteria can be employed in decontaminating environment polluted with various types of oil. Considering a generally large subsurface microbiota there is considerable interest for the prospect of degrading hazardous contaminants *in situ* by stimulating selective microbial populations (biostimulation) or by the addition of organisms to contaminated sites (bioaugmentation). Stimulation of an indigenous population of white rot fungi and some bacteria is likely to enrich for species that are well adapted to their environment, whereas the deliberate addition of more microorganisms into such an environment may be compromised since the introduced organisms are not as likely to be able to compete. Evaluation, characterization, and utilization of microbial communities associated with *in situ* bioremediation of subsurface and groundwater contamination is a technological necessity for environmental restoration and assessment. Further research is therefore advocated to investigate the efficiency of these methods on crude oil contaminated soils, their toxicity on other indigenous microorganisms and synergy between the species. Satellite imagery combined with the use of Geographical Information Systems can be employed to quickly identify and track spilled oils. To hasten the cleanup of spills, regional cleanup sites along the problem areas could help contain spills more quickly. To make these tasks feasible, more funding must be provided by the stakeholders of the oil industry.

Notes

1. I. A. Okoh, S. Ajisebutu, G. O. Babalola, and M. R. Trejo-Hernandez, "A Study of the potentials of a *Burkholderia cepacia* strain (RQ1) in the biodegradation of heavy crude oil," (Maya) *Int. Microbiol.* 4, 2001: 83–87.
2. A. F. Gesinde, E. B. Agbo, M. O. Agho, and E. F. Dike, "Bioremediation of Some Nigerian and Arabian Crude Oils by Fungal Isolates," *Int. Jor. P. App. Scs.* 2(3) 2008: 37-44.
3. M. Vidali, "Bioremediation. An overview," *Pure Appl. Chem.* 73(7) 2001:1163–1172.
4. C. A. Reddy and Z. Mathew, "Bioremediation potential of white rot fungi," In: *Fungi in Bioremediation*. G. M. Gadd (Ed.) Cambridge, U.K.: Cambridge University Press. 2001, 52-78.
5. Vidali, "Bioremediation. An overview," 1163.
6. J. G. Mueller, C. E. Cerniglia, and P. H. Pritchard, "Bioremediation of Environments Contaminated by Polycyclic Aromatic Hydrocarbons," In *Bioremediation: Principles and Applications*, Cambridge: Cambridge University Press, 1996, 125–194,.

7. S. Hamman, "Bioremediation capabilities of white rot fungi," BI570 – Review Article Spring, 2004: 1- 12.
8. C. E. Cerniglia, "Biodegradation of polycyclic aromatic hydrocarbons," *Biodegradation* 3, 1992: 354-368.
9. D. P. Barr and S. D. Aust, "Mechanisms of white rot fungi use to degrade Pollutants," *Environ. Sci. Technol.* 28, 1994: 78A-87A; E. Lang, I. Eller, R. Kleeberg, R. Martens, and F. Zadrazil "Interaction of white-rot fungi and micro-organisms leading to biodegradation of soil pollutants," In: *Proceedings of the 5th International FZK/TNO Conference on contaminated soil* 30th Oct. – 5 Nov. 1995. Maastricht; and Hamman "Bioremediation capabilities of white rot fungi," 1.
10. A. Zeddel, A. Majcherzyk, A. Hutterman, "Degradation of polychlorinated biphenils by white rot fungi *Pleurotus ostreatus* and *Trametes versicolor* in a solid state system," *Toxicol. Environ. Chem.* 40, 1993: 255-266; R. T. Lamar and J. A. Glaser, "Field evaluation of the remediation of soils contaminated with wood-preserving chemicals using lignin-degrading fungi," In R. E. Hunchea, A. Leeson, L. Semprini, S. K. Ong (Eds.), *Bioremediation of chlorinated and polycyclic aromatic hydrocarbon compounds*. Florida: CRC Press, 1994: 239-247
11. M. Otte, Y. Commeau, R. Samson, and C. W. Greer, "Enhancement of pentachlorophenol biodegradation using organic and inorganic supports," CRC Press, 1999: 35-45.
12. Vidali, "Bioremediation: An overview," 1166.
13. Vidali "Bioremediation: An overview," 1165.
14. R. Boopathy, "Factors limiting bioremediation technologies," *Biores. Technol.* 74, 2000: 63-67.
15. J. Braddock, M. Ruth, P. Catterall, J. Walworth, K. McCarthy, "Enhancement and inhibition of microbial activity in hydrocarboncontaminated arctic soils; implications for nutrient-amended bioremediation," *Environ. Sci. Technol.* 31, 1997: 2078-2084; and J. E. Lindstrom, R. C. Prince, J. C. Clark, M. J. Grossmann, T. R. Yeager, J. F. Braddock, and E. J. Brown, "Microbial populations and hydrocarbon biodegradation potentials in fertilized shorelines sediments affected by the T/V Exxon Valdez oil spill," *Appl. Environ. Microbiol.* 57, 1991: 2514-2522.
16. B. A. Wrenn, J. R. Haines, A. D. Venosa, M. Kadhodayan, and M. T. Suidan, "Effect of nitrogen source on crude oil biodegradation," *J. Ind. Microbiol.* 13, 1994: 279-286.
17. R. M. Atlas, "Petroleum Biodegradation and Oil Spill Bioremediation," *Marine Pollut. Bull.* 31, 1995: 178-182.
18. R. M. Atlas, "Microbial degradation of petroleum hydrocarbons: An environmental perspective," *Microbiol. Rev.* 45, 1981:180-209 and J. Oudot, L. Dupont, S. Haloui, M. F. Roquebert "Biodegradation Potential of Hydrocarbon-assimilating Tropical Fungi," *Soil Biol. Biochem.* 25, 1993: 1167-1173.
19. J. J. Ortega-Calvo, "Microbial Physiology of PAH Biodegradation Under Bioavailability Restrictions," Ca. 2004.

20. S. B. Pointing, "Feasibility of bioremediation by white-rot fungi," *Appl. Microbiol. Biotechnol.* 57, 2001: 20-33; and. M. Law, W. N. Lau, K. L. Lo, L. M. Wai, S. W. Chiu "Removal of biocide pentachlorophenol in water system by the spent mushroom compost of *Pleurotus pulmonarius*,". *Chemosphere* 52, 2003: 1531-37.
21. Pointing, "Feasibility of bioremediation by white-rot fungi," 30.
22. Hamman, "Bioremediation capabilities of white rot fungi," 6.
23. E. M. Ogbo, J. A. Okhuoya, "Biodegradation of aliphatic, aromatic, resinic and asphaltic fractions of crude oil contaminated soils by *Pleurotus tuber-regium* Fr. Singer- a white rot fungus," *Afr. J. Biotechnol.* 7(23), 2008: 4291-4297.
24. L. A. Levin, A. Viale, A. Forchiassin, "Degradation of organic pollutants by the white rot basidiomycete *Trametes trogii*," *Int. Biodeterioration and Biodegradation* 52, 2003: 1-5.
25. K. L. Lau, Y. Y. Tsang, S. W. Chiu, "Use of spent mushroom compost to bioremediate PAH-contaminated samples," *Chemosphere* 52, 2005: 1539-46.
26. K. L. Lau, Y. Y. Tsang, S. W. Chiu, "Use of spent mushroom compost to bioremediate PAH-contaminated samples," *Chemosphere* 52, 2005: 1539-46; and H. P. Hestbjerg, A. Willumsen, M. Christensen, O. Andersen, C. S. Jacobsen, "Bioaugmentation of tar-contaminated soils under field conditions using *Pleurotus ostreatus* refuse from commercial mushroom production," *Environ. Toxicol. and Chem.* 22(4), 2003: 692-698.
27. Environmental Protection Agency, "EPA Guidelines (EPA 589/05) Soil Bioremediation," EPA, Adelaide, 2005: 1-9.
28. I. A. Ikhuorda and A. Avwunudiogba, "Environmental Impact of Oil Exploration on the Socio-economic Conditions of Nigerian Delta Communities: A Baseline Study of Nembe and Ekulama," A Paper Presented at a National Conference on Geography and Sustainable Development, University of Benin, Benin City. 1995.
29. I. O. Akinwumi, T. O. Oyebisi, and A. T. Salami, "Environmental Degradation In Nigeria: Implications And Policy Issues- A Viewpoint," *Intern. J. Environ. Studies* 58, 2001: 585-595.
30. O. M. Adedokun, A. E. Ataga, "Effects of amendments and bioaugmentation of soil polluted with crude oil, automotive gasoline oil, and spent engine oil on the growth of cowpea (*Vigna unguiculata* L. Walp)." *Sci. Res. Essay* 2(5), 2007: 147-149; C. O. Adenipekun, "Bioremediation of engine-oil polluted soil by *Pleurotus tuber-regium* Singer, a Nigerian white-rot fungus," *African J. Biotechnol.* 7(1), 2008: 055-058; C. O. Adenipekun, I. O. Fasidi, "Bioremediation of oil-polluted soil by *Lentinus subnudus*, a Nigerian white-rot fungus," *Afr. J. Biotechnol.* 4(8), 2005: 796-798; C. O. Adenipekun, O. S. Isikhuemhen, "Bioremediation of Engine Oil Polluted Soil by the Tropical White Rot Fungus, *Lentinus squarrosulus* Mont. (Singer)," *Pakistan J. Biol. Sc.* 11(12), 2008: 1634-1637; A. I. Okoh, "Biodegradation of Bonny light crude oil in soil microcosm by some bacterial strains isolated from crude oil flow stations saver pits in Nigeria," *Afr. J. Biotechnol.* 2(5), 2003: 104-108; S. U. Nwachukwu, "Bioremediation of Sterile Agricultural Soils Polluted with Crude Petroleum by Application of the Soil Bacterium, *Pseudomonas putida*, with Inorganic Nutrient Supplementations," *Curr. Microbiol.* 42(4), 2001:

231-236; T. V. Ojumu, O. O. Bello, J. A. Sonibare, and B. O. Solomon "Evaluation of microbial systems for bioremediation of petroleum refinery effluents in Nigeria," *Afr. J. Biotechnol.* 4(1), 2005: 31–35.

31. Adenipekun, "Bioremediation of engine-oil polluted soil by *Pleurotus tuber-regium* Singer, a Nigerian white-rot fungus," 056.
32. Ogbo and Okhuoya, "Biodegradation of aliphatic, aromatic, resinic and asphaltic fractions of crude oil contaminated soils by *Pleurotus tuber-regium* Fr. Singer- a white rot fungus," 4295.
33. Adedokun and Ataga, "Effects of amendments and bioaugmentation of soil polluted with crude oil, automotive gasoline oil, and spent engine oil on the growth of cowpea (*Vigna unguiculata* L. Walp)," 148.
34. Adenipekun and Fasidi, "Bioremediation of oil-polluted soil by *Lentinus subnudus*, a Nigerian white-rot fungus," 798.
35. Adenipekun, "Bioremediation of engine-oil polluted soil by *Pleurotus tuber-regium* Singer, a Nigerian white-rot fungus," 056.
36. Gesinde, Agho and Dike, "Bioremediation of Some Nigerian and Arabian Crude Oils by Fungal Isolates," 37.
37. Adenipekun, "Bioremediation of engine-oil polluted soil by *Pleurotus tuber-regium* Singer, a Nigerian white-rot fungus," 056.
38. O. Isikhuemhen, G. Anoliefo, O. Oghale, (2003). Bioremediation of crude oil polluted soil by the white rot fungus *Pleurotus tuber-regium* (Fr.), Sing. Environ. Sci. Pollut. Res. 10, 2003:108-112.
39. Ogbo and Okhuoya, "Biodegradation of aliphatic, aromatic, resinic and asphaltic fractions of crude oil contaminated soils by *Pleurotus tuber-regium* Fr. Singer- a white rot fungus," 4292.
40. Adedokun and Ataga, "Effects of amendments and bioaugmentation of soil polluted with crude oil, automotive gasoline oil, and spent engine oil on the growth of cowpea (*Vigna unguiculata* L. Walp)," 148.
41. Adenipekun and Fasidi, "Bioremediation of oil-polluted soil by *Lentinus subnudus*, a Nigerian white-rot fungus," 797.
42. Adenipekun and Isikhuemhen, "Bioremediation of Engine Oil Polluted Soil by the Tropical White Rot Fungus, *Lentinus squarrosulus* Mont. (Singer)," 1635.
43. Gesinde and Dike, "Bioremediation of Some Nigerian and Arabian Crude Oils by Fungal Isolates," 38.
44. Okoh, "Biodegradation of Bonny light crude oil in soil microcosm by some bacterial strains isolated from crude oil flow stations saver pits in Nigeria," 105.
45. Ojumu, Bello, Sonibare and Solomon, "Evaluation of microbial systems for bioremediation of petroleum refinery effluents in Nigeria," 32.
46. Nwachukwu, "Bioremediation of Sterile Agricultural Soils Polluted with Crude Petroleum by Application of the Soil Bacterium," *Pseudomonas putida*, with Inorganic Nutrient Supplementations," 232.
47. Ogbonna, Iwegbue, Sokari and Akoko, "Effect of bioremediation on the growth of Okro (*Abelmoshus esculetus*) in the Niger Delta soils," 304.

48. Okoh, "Biodegradation of Bonny light crude oil in soil microcosm by some bacterial strains isolated from crude oil flow stations saver pits in Nigeria," 107.
49. Ojumu, Bello, Sonibare and Solomon, "Evaluation of microbial systems for bioremediation of petroleum refinery effluents in Nigeria," 34.
50. Nwachukwu, "Bioremediation of Sterile Agricultural Soils Polluted with Crude Petroleum by Application of the Soil Bacterium, *Pseudomonas putida*, with Inorganic Nutrient Supplementations," 235.
51. Ogbonna, Iwegbue, Sokari and Akoko, "Effect of bioremediation on the growth of Okro (*Abelmoshus esculetus*) in the Niger Delta soils," 307.
52. T. I. Odu, "The effect of nutrient application and aeration on oil degradation in soil," *Environ Pollut*. 15, 1978: 235-239.
53. O. A. T. Ebuehi, I. B. Labibo, P. D. Shekwolo, K. I. Sigismund, A. Ladoki, I. C. Okoro, "Remediation of Crude Oil Contaminated Soil by Enhanced Natural Attenuation," *Technique. J. Appl. Sci. Environ. Mgt.* 9(1), 2005: 103-106.
54. J. Limson, "Indigenous Plants to the Rescue Environmental remediation in Nigerian oil regions," *Science in Africa*, 2002.
55. C. Bonaventura, J. Bonaventura, D. F. Bodishbaugh, "Environmental Bioremediation: Approaches and Processes," In *Ecotoxicity and Human Health* (Eds): F. J. De Serres, A. D. Bloom, Boca Raton, New York, London, Tokyo: CRS Lewis Publishers, 1995: 199-200.
56. J. L. Sims, R. C. Sims, J. E. Matthews, "Approach to bioremediation of contaminated soil," *Waste Hazard Matter*. 7, 1990: 1-17.
57. D. C. L. Wong, E. Y. Chai, K. K.Chiu, P. B. Dorn, "Prediction of ecotoxicity of hydrocarbon-contaminated soils, using physico-chemical parameters," *Environ. Toxoicol. Chem.* 18, 1999: 2611-2621.

• Chapter 8 •

THE ECONOMY OF ACCELERATED FOOD PRODUCTION, LAND CONFLICTS AND ENVIRONMENTAL DEGRADATION IN CENTRAL NIGERIA REGION

Mike O. Odey

Introduction

Accelerated food and agricultural production constitutes the main reason for the undue pressure on arable land in Sub-Saharan Africa and also represents one flash point of global environmental degradation which until lately has received very little scholarly attention. But elsewhere, such a scenario amounts to what is called "the tragedy of the commons." This chapter examines the levels and the nature of food production system in Central Nigeria that is popularly referred to as the "food basket of Nigeria." To be specific, it is a historical analysis of the implications of perverse agricultural production under traditional methods on the Benue environment over time. It further seeks to demonstrate how the environment in the Benue valley of Central Nigeria, especially farmlands in the savanna, can be used more efficiently at very minimal environmental costs to achieve food security without jeopardizing the environment into future risk. The most important question the chapter is answering is how to keep a fair balance between food production and the improvement of environmental resources in the region to meet demands of sustainable development. The conclusion of the chapter is that environmental insecurity in Central Nigeria cannot be isolated from similar instances in the country or be taken for granted but carefully interrogated to avoid bigger ecological disaster like what is currently going on in the Niger Delta. This is a deliberate diversion from the furor of the oil rich Niger-Delta in Nigeria which has received an undue attention from the international community at the expense of the potentially risky agricultural regions of Central Nigeria which we are also bound to make a case for.

In two related analyses, I have[1] argued that the impact of massive food production in central Nigeria leads to nowhere but creeping poverty, underdevelopment and land disputes. These are very common scenarios in most African societies. Here in the present analysis, the emphasis is placed on the economy of accelerated food production and the perversity of environmental degradation in Benue State The previous analysis cannot substitute the present. Rather, the first provides the aegis upon which the second argument now rests as provided here constituting another block in one complex edifice. For easy understanding of the issues being raised, a simplified version of the geographical background and ecosystem of the region is first all required.

In geo-political terms, Nigeria is divided into six zones with variegated ecosystems and environmental hazards, generally determined by human activities and engagement in available natural resources. Central Nigeria is an area of extensive and intensive agricultural production with several implications. According to Forde[2] and UNFAO,[3] central Nigeria is a frontier zone in the middle of Nigeria with a distinct economic base and is quite different from the forest and Sudan zone. The economy is mainly based on food crop production while export production is yet to be fully developed.[4]

The Benue area lies within the Southern Guinea Savanna and is characterized by coarse grasses and an avalanche of scattered trees. Much of the area is undergoing re-growth of vegetable due to extensive agricultural production. Thick vegetation is rare and dotted here and there as village forests, gallery forests and forest reserves. The soil conditions of the Benue region are more or less the same as the rest of the savanna tropical grass lands and the sub-tropical grass biomes-described by the early pedogenesists as "fairly homogenous, dominantly lateritic and very infertile, almost bordering on infertility .... Generally, the soils have low levels of organic matter ranging between 5 and 3%." [5]

With the abundance of kaolinite clay mineral, the soil is subjected to severe weathering condition and lack of stability in the surface soil due to hot and humid conditions throughout the year. The significance of this is the low inherent fertility of the soils when fertilizers are applied, and the soils do not have the capacity to retain nutrients for long before they filter out. The texture of the surface layers of the Benue soils vary greatly with the ferruginous tropical type dominating. The others are laterite in the basement complex, fadama soils on the alluvial plains and the lithosols on the hills and ridges at Anwase, Kyogen, Andibila hills as well as the slopes of Mkar, Haaga, Ngokur and Tsemker.

In central Nigeria, agricultural land is becoming increasingly inadequate to meet the demands of the burgeoning population. Land has become both a trigger and a dynamic factor in the consideration of violent conflict for two related reasons: agricultural production and depletion of life support system. This is because, much of the north central geo-political zone of Nigeria is regarded as the "food basket of the nation", and agriculture is the main occupation of the majority of the people but production strategies are very poor and are on a small scale, making agriculture largely unsustainable. Furthermore, food demands from all over the nation remain high especially under the prevailing circumstances of national food insecurity that has gripped the nation for nearly a year now. And unknown to most of the small-scale farmers is the unprecedented real and potential threats to life support systems under the widespread and accelerated agricultural production to meet such growing demands. The situation has assumed the elements of "the tragedy of global commons." For instance, right now, everyone is hell-bent in Central Nigeria on producing food for survival and for the local foodstuff trade in such a way that nobody would give that activity up for others or for posterity. Here, every individual and community prides themselves of their limitless access to the common heritage of land without realizing the limits of its carrying capacity. The trend is quite alarming and unless alternative ways and means of meeting the food needs of the people are devised, it would appear that the Malthusian specter is fast closing in on the area. Apart from poor agricultural practices, this is accentuated by massive migrations from other regions and within Central Nigeria where land related violence is quite common, especially during the post-colonial period.[6]

Conceptualizing the Land Question in Central Nigeria

Land is one of the most important gifts from Mother Nature, the most controversial issue in human conflict, and the most used and abused element by people. Next to human skill and labor, land is about the most important primary factor of production because even with availability of capital and whatever, development cannot take place in the thin air or space. Man's consequent relation to the rest of nature and the subsequent production of the means of subsistence is necessarily conditioned by his physical organization and the environment where he finds himself. Land here therefore refers to the totality of the environment or earth's surface, a country, state, or community with people. Geographically, it connotes soil, or ground where plants grow, through which river flows and the basis of many life support systems or human activities with wide implications. Land is a basis of life support systems (providing food, water, fodder, fiber and biotic materials), and a source of conflict and sink of greenhouse gases as well as a co-determinant of the global energy balances.[7] Furthermore, land is a storehouse of raw materials and mineral as well as the basis of terrestrial biodiversity, providing the biological habitats and gene reserves for plants, animals and micro-organisms.[8] The value of land depends on the socio-economic circumstances of its location; the extent of its use and laws surrounding its control. Thus, either it is useful or it constitutes an issue of conflict over which people lay claim and are ready to die for.

In the Benue valley, land tenure system is generally porous as elsewhere in Africa. It is both a community and individual property and regulated by simple rules. Every indigene of a community was entitled to a piece(s) of land with attached social responsibilities that cannot be violated for peaceful co-existence. In the Benue valley, land is considered to be divinely owned from which humans derive their source(s) of livelihood. Thus, even the poorest in society was entitled to portions of land for each crop. Usufruct rights were mostly left under the control of households and individuals because, as long as an individual cropping activities goes-on on any piece of land, no other individual could lay claim to it. However, as soon as he leaves it for any reason (fallow period or change to another plot), that individual ceases to be its *bona fide* owner. And with this simple flexibility and routine, land operations were maintained for centuries in the Benue Valley region.

However, after the advent of the British colonial rule in the Benue valley in the early 20th century, dramatic changes began to emerge over land holding in such a way that those productive activities and demographic dislocations continued to build-up together to accentuate the existing land disputes in the Benue valley. Furthermore, the concept and significance of land took a new dimension and implied that it was becoming scarce.[9]

The Capacity of Food & Agricultural Production in the Benue Valley/Central Nigeria

Before the creation of Benue State of Nigeria in 1976 and ever since then, agricultural activities constitute the most significant economic base of the people and it is on the increase. Over 80% of the population engage in food crop production as small-scale farmers on an avalanche of rich arable land spanning over 300 kilometers along the Benue river plains.[10] Well over 20 different crops are produced in the area for consumption and commercial purposes. These include yams, cassava, rice, soya beans, potatoes, beniseed, cocoyam, beans, millet, and sorghum. Others are groundnuts, maize, and tree crops such as oil palm, citrus, cashew, mangoes, cola-nuts, cotton and so forth as shown in Table 1 below. Most rural households in the region all engage in their production cycle and strategies to produce these crops simultaneously which is nearly more than have half the Nation's food needs. The implication of this is far reaching as clearly captured in the words of Nafis:

> Much of the environmental degradation witnessed today is due primarily to two groups of people-the top billion richest and the bottom billion poorest.[11]

The truth is that, if the rich nations of the Global North are guilty of atmospheric pollution and chemical hazards from their industrial smokestacks, carbon dioxide, sulphur dioxide and other types of contaminants that have become the bane of global warming, poor agrarian nations in the Global South are no less guilty of depleting the quality of their agrarian environment for the habitation of human life. But this appears not to be anybody's immediate concern in Central Nigeria.

The physical features, the location and the climatic conditions of Benue state in central Nigeria favor the production of this wide variety of crops. Thus 'the back bone of the state' economy is widespread agricultural production all over the state;[12] and this will likely continue to be the case for a long time to come as the state government has indicated in its current development policy of *our Benue our future* since 2008.But one significant issue here is, with what environmental consequences?

The Benue agricultural practice is intensive but predominantly subsistence under the control of rural peasant farmers using hoes and cutlasses. It is difficult to argue about the extent to which Benue agriculture is subsistence since much of the produce is not only consumed but sold in and out of the region. Apart from that, much of the Agricultural Extension services are master minded by the Ministry of Agriculture and Natural Resources, Benue Agricultural and Rural Development Authority (BNARDA) and their agencies which remove agricultural activities from the simple idea of "slash and burns." However, at the grassroots, each community and household is entitled to free land rights on which to cultivate their crops on rotational or shifting basis. Benue State spatial environment is well over 45,000 square kilometres out of which over 36% is under cultivation at the moment.[12] The heaviest areas under cultivation include; Vandekiya and konshisha (40.4%), Gboko/Buruku (38.7%), Ukum (28.2%), Oju (26.6%) and so forth. The best five cultivated crops are rice under 91,000ha, yams occupying over 32 000 ha, cassava occupies about 17.2 000ha and guinea corn using about 12 000 ha.Tables (8.1, 2, &3) below explains level of agricultural activities, the extent of land under cultivation and the areas mostly affected.

Table 8.1: Major Crops Produced in Benue State, Nigeria

S/No.	Crops	Local Government Areas of Production
1.	Cassava	All over the state with higher concentration in Okpokwu, Gboko, Vandekiya, Gwer, K/ ala and Otukpo.
2.	Yam	All over the state with higher concentration in K/ Ala, Otukpo, Gboko, Okpokwu and Vandekiya.
3.	Millet, Guinea Corn, Maize and Rice	All over the state, rice has higher concentration in Vandekiya, Gboko, Makurdi, Otukpo and Okpokwu.

S/No.	Crops	Local Government Areas of Production
4.	Soyabean	Gboko, Vandekiya,Katsina-ala, Makurdi,Gwer and Kwande.
5.	Groundnuts	All over the state
6.	Citrus	All over the state
7.	Beniseed	Gboko, Katsina-ala,Vandekiya,Makurdi,and Oju.
8.	Palm Produce	Okpokwu, Otukpo, and all over the state.
9.	Mango	All over the state
10.	Sugar Cane	All over the state
11.	Melon	All over the state with the highest concentration in the western zone.

Sources:

1. BNARDA (1986) *Crop Production Manual for Benue State.*
2. FOS, Lagos: *Annual Abstract of Statistics*, (1995).
3. Odey, Ph. D. thesis "A History of Food Crop Production ... Dialectics of Hunger and Rural Poverty," 2001, p. 285.

Table 8.2: Density of Land under Cultivation by LGA in Benue State in Central Nigeria

LGA	Total Area	Over 60%	35% - 59%	10% - 34%	Under 10 %	Reserved	Total	Area Under Cultiv..(Sq.Km)	% Area Under Cultiv.
Ado,Okpokwu & Ogbadibo	2357	8.0	19.6	40.1	29.8	2.5	100	563	23.9
Oju	1845	2.3	35.1	42.3	20.3	0.0	100	4.91	26.6
Otukpo/Apa	4092	2.9	18.1	48.3	28.0	2.7	100	866	21.2
Vandeikya/Konshisha	2652	29.0	30.8	31.2	8.0	1.0	100	1072	40.4
Makurdi/Guma	2185	2.5	9.2	46.9	41.1	0.3	100	536	24.5
Gboko/Buruku	3250	13.3	47.1	34.3	1.5	0.8	100	1257	38.7
Kwande/Ushongo	4051	5.5	22.5	35.1	36.4	0.5	100	893	22.0
Gwer/Gwer West	4017	0.2	10.9	51.2	36.6	1.1	100	699	17.4
Katsina–Ala/Ukum	5610	8.2	43.9	35.6	12.0	0.3	100	1837	28.2

Source: LRD Serial Photography and field surveys, 1968-74.

Table 8.3: Estimated Hectares of Major Crops by LGA ('000 ha.)

LGA	Corn	Yam	Cassava	Maize	Guinea Corn	Millet	Rice	Soya Beans	G/Nut	B/Seed	Cow Peas
Ado, Okpokwu & Ogbadibo	12.8	10.3	1.8	7.4	3.4	9.4	-	3.9	0.7	-	49.7
Oju	12.1	10.3	1.8	7.4	3.4	9.4	-	3.9	0.7	-	48.1
Otukpo/Apa	14.7	7.3	10.5	12.1	6.1	11.3	0.1	4.3	1.1	1.7	79.2
Vandeikya/ Konshisha	21.3	7.7	8.9	10.6	3.6	16.0	7.0	14.2	7.1	7.1	113.4
Makurdi/Guma	3.4	4.0	2.4	3.5	4.9	5.7	-	4.9	0.8	1.8	31.4
Gboko/Buruku	18.5	15.4	6.2	12.3	7.9	23.3	6.4	12.3	0.3	3.12	105.7
Kwande/ Ushongo	18.5	10.0	4.9	9.7	1.3	10.6	0.4	4.8	0.1	0.2	60.5
Gwer/Gwer West	5.3	7.8	8.4	7,7	2.4	6.8	0.1	2.0	0.2	0.5	41.2
Katsina-Ala/ Ukum	32.5	17.4	6.5	12.0	7.6	9.1	0.8	7.7	2.4	2.3	98.3

Source: MANR Estimates for 1978 and serial photography interpreted by LRD, 1968-74.

The Economy of Agricultural Production and Land Conflicts /Wars in Central Nigeria

There is no doubt that Central Nigeria has enough arable land for extensive food production for which the people in the region have engaged themselves over time, though not necessarily the food basket of the nation as often acclaimed. But against the backdrop of the potential for which the region is known, several emerging issues can be articulated. These include the scarcity of farmlands and conflicts over farmlands, intra-and inter-migration movements, environmental degradation due to excessive use of farm and forestlands and the unintended consequences arising from all these on both the long and short runs. Each of these will be interrogated and linked in the present section to see how they act and react upon each other to alter the environment over time. The first of these issues is wars/violent conflicts over land.

Most of the inter-group conflicts in central Nigeria are over land and shifting of boundaries between communities, chieftaincy problems, political matters and very little of religious conflicts. For the purpose of the present analysis, it is important to isolate land and boundary disputes as the most significant factor impinging on the environment in Central Nigeria. Conflicts over land in central Nigeria are wide- ranging and have made several appearances as inter-state, intra-state, inter and intra local governments areas inter community/village as well as inter- family disputes. There are too many instances of disputes and open conflicts over land and shifting of boundaries between and within groups in the Benue valley as part of the general phenomenon of "a veritable mix of bedeviling contradictions, not only on central Nigeria, but indeed the entire nation".[14]

One significant issue here is that land and boundary disputes in the Benue valley have a very long history and traceable to the pre-colonial period and is getting worse, involving constant massive migratory waves in search of new settlements. The long perspective has shown that no ethnic group was excluded from some form of migration and fighting other groups over land as demonstrated in a recent study in Central Nigeria.[15] A random selection of such cases is necessary here.

The Jukun Kingdom (Kwararafa confederacy) was perhaps the first group whose movement into the Benue valley in the 13th century was characterized by large-scale inter-group rivalries. When they arrived in the lower Benue valley, they engaged in large-scale territorial aggrandizement and displacement of people for over several centuries.[16] The emergence of the Jukun kingdom in the Benue valley has created a chain of reactions, accounting for series of migratory movements in the area and struggle over land since the 13th century onwards. The contemporary Jukun-Tiv conflicts are relics of the ancient wars.

The Tiv were perhaps the next to move in four successive generations from their cradle in Swem to the present location from the late 15th to the early 20th century. There after, the Idomas came in, although they argued they were the first to have settled in the Benue valley. The Etulo, the Igede, the Ebirra, the Afu, Akweya-Yachi, the Arago, the Yala and so forth came into the area at various times and settled in various locations, either displacing others or repossessing their own ancient land-marks.

Before the advent of British rule, conflicts over land and along ethnic/village boundaries were relatively few in central Nigeria. However, in 1910 (and later in 1916 Amendment Act) in the Land and Native Rights proclamation of Lugard, all lands right and control were confiscated from the indigenes and conferred on the British government. From this point, both autochthonous, unwritten land rights of the people and land policies of the former colonial power began and continued to co-exist without clear-cut reforms. Land assumed a new dimension of significance, officially and privately. Apart from this, fast population growth through migrations into central Nigeria due to new waves of political and religious crises in the North in the new democratic dispensations and intensification of agricultural production all combined to bring about increased utilization and pressure on land and to worsen its scarcity.

Like in other agrarian societies, land remains an invaluable asset for active agricultural production especially for high production of food in the region. But despite the abundance of arable land, population ratio to arable land is perpetually out of balance. Furthermore, long fallow periods are shortened and soil fertility is jeopardized. Indeed, no one can say that in central Nigeria, there are landless masses that were uprooted or alienated completely from their land and roaming on the streets of town or anyone dispossessed or forced out of his/her agricultural land as such.

In the Benue valley, land disputes are the most exasperating to the people and government, most violent and between very close relations with several losses of lives and property.[17] The best recorded boundary disputes were those of Ushongo-Konshisha, Gwer-Oju and Otukpo –Ohimini. One of the worst land disputes in recent time was that of Makurdi in 2004, the Benue state capital city and very recently in Katsina –Ala area. It happened against the background of indigene ship and settler in town and its environs with ferocious consequences that were almost unimaginable.[18]

Within the larger Central Nigeria, disputes over land have occurred and reoccurred between states: An instance of international boundary dispute was the one between Benue State and the Republic of Cameroon across Kwande local government area a few years ago. It was alleged that Swem shine and Hill the cradle of Tiv of Benue was taken over by Cameroon several years ago, which Nigeria and Cameroon have been trying to resolve. Furthermore, Benue State shares common boundaries with Enugu, Ebonyi, Cross River, Taraba, and Nasarawa and Kogi states within the north central geopolitical zone of Nigeria. Each of these states has had its rounds of conflicts with Benue State.[19] "Each of these conflicts has a long history, but there has been an escalation and expansion of conflicts in recent year."[20] Most of the problems leading to conflicts between states started as disputes over farmland or fishponds between neighboring villages that could not be resolved between them or their respective local governments.

From the web of our colonial history, the exploitation of the former territories for cheap raw materials and other natural resources in general and the Benue area in particular since early 20th century by Britain and now by the people themselves to eke out a living has undoubtedly resulted in environmental and biophysical damages. As bad as this is, even more serious complications have set in to compound the colonial and the post-colonial havoc. This has been reflected as an inseparable nexus between environmental crisis and the dialects of hunger and rural poverty.[21] From the colonial era, the Benue economy was subjected to the nature and the balance of ancient subsistence and the fragile market economy, especially the intensification of commodity production. But the forces of production were left untransformed.[22]

As clearly shown so far, land regimes in Africa are guided by unwritten customs and tradition and are specifically made to fit agrarian rhythms of life are susceptible to violent conflicts. For whatever reason, the use of land is unregulated and left at the whims of whoever was entitled to use it as he/she deems fit. Indeed, land is not scarce, but because it is very important for agricultural exploitation among the Central Nigerian population in the widest sense, its carrying capacity is likely to be thrown into jeopardy constantly. It is everybody's priority and simultaneously without practical rules of dealing with emerging problems such as ecology and biodiversity, infrastructure, urbanization or industrial development. It does not make much sense for peace building where land is both individual and group property on the one hand and the co-existence of native law/custom and the colonial brand of land laws under the colonial administration of Lord Lugard since 1902. Every family may be regarded as "farm unit" or "farm tenure" and geared towards subsistence survival and cash crop production at the same time to enhance their cash-nexus without the transformation of the basic production strategies.

The other factor is the decreasing land space and the rising population/influx of immigrants contending with the indigenous population for available arable land. This is linked to the phenomenon of "settler" verses "indigene" or the question of citizenship. Behind the issues of rejuvenation of ethnic consciousness lies the matter of rivalries for geographical locations between so-called "hosts" and "strangers" in the Benue valley and throughout Nigeria. The immediate consequences of land/boundary disputes are many and diverse. It led to loss of lives, houses, and property worth billions of naira. It has brought about social insecurity and the plight of refugees in their own country. Displacement of people has been widespread and for very long period of time. One significant effect of land/boundary conflicts in the lower Benue valley is that it brings up further argument whether the area is truly "the food basket of the nation." In other words, disputes over land have drastically reduced agricultural productivity, giving way to perennial famine, human deprivation[23] and ecological problems.[24] Next to loss of lives, the worst effect of land/boundary disputes in the area is the threat to food security (its availability, sufficiency and affordability for adequate diet for all concerned). This is because civil strife over land has brought down per-capita production to stagnation. On the long run, land scarcity/conflict adds to environmental stress and both of these on the existence food as well as environmental insecurity.

The Economy of Benue Agricultural Production and Environmental Insecurity in Central Nigeria

Several interlocking factors are responsible for the current environmental changes taking place in the Benue area. The most significant ones include continuous migrations from other parts of the country leading to rapid population growth and a wide variety of the increasing human activities especially agricultural production of the staples.

Historically, the Benue region is relatively peaceful and appears to be a haven for refugees from the troubled Northern parts of the country, fraught with religious crises. Furthermore, it has been clearly shown that agricultural lands in central Nigeria are vulnerable to all kinds of depletion such as erosion and reduction of fertility even before the bourgeoning population. For example, according to Briggs, Benue soil deterioration began to be noticed as far back as 1938 as a result of extensive agricultural cultivation.[25] This has been compounded by population pressure and overuse of same plots of land for long periods in the colonial period. Pressure on land and the destructive capacity of the people to vegetation supposedly for the economic development of the area are incompatible and unharmonised. This is because there has been no conscious effort to protect or replenish the depleted areas of the Benue soil or vegetation for purposes of sustainable agricultural production.

Thus, the significant issue directly related to the above is that of scarcity of land and worse still, its gradual loss of nutrient, wrong management of land resources because of traditional methods of farming, population movements as well as environmental and social insecurity in Central Nigeria area in a long perspective. According to one UNFAO publication[26] which is believed to be the most up to date and thorough satellite and aerial pictures involving 88 tropical countries, about one acre of land, estimated to be about the size of England and Wales, keeps disappearing per second per year, from the rich tropical regions of the world which were reputed to be housing the best of earth's heritage. Todaro has further observed that since the 1980s,most developing countries have continued to loose about 1.9 percent each of their arable lands annually with over 60 percent of the population of most countries in sub-Saharan Africa struggling to survive on agricultural soils that are fast loosing their nutrients. This includes Central Nigeria but with its perverse and crude agricultural production systems which put our rich ecosystems in danger of extinction. And the plausibility of this argument is quite diverse.

The common practice all year round in the Central Nigeria area is that arable lands and vegetations are cleared, burnt and cultivated for food crops. This is a routine that constantly subjects the area to the vulnerability of environmental crises which appears to be unknown to the poor rural farmers who ignorant about their plight. The traditional system of agricultural production has created tremendous environmental impact in several ways.

One of these is that increasing agricultural production on the same soils that are fast loosing their nutrients is creating chain responses on the environment. Apart from loss of soil fertility, it leads to loss of productivity. One other consequence of traditionally-driven agricultural practice is the continuous release of carbon dioxide and other greenhouse gases as a result of the regular *slash and burns* of vegetation and farmlands in the process of shifting cultivation as elsewhere in Africa. Undoubtedly too, what happens in the Benue agricultural plains is part of the phenomenon of the global warming. Until now, food production is through traditional methods as shown above and the havoc seems unending. For instance, it is practically impossible to persuade a Benue food producer not to cut down trees, which are his basic sources of energy or to stop him from bush burning which is his familiar way of agricultural practice unless these means/methods can be substituted. This is why government and people in the Benue region, as "the food basket of the nation", must begin to take initiative towards sustainable development, which is contingent upon sustainable environment.

Similarly, loss of land and nutrient of the existing patches of land as well as a great variety of the biodiversity of Central Nigeria area. What used to be regarded as natural vegetation and bushes that were untouched for years are no longer in existence and several plants and animals have quietly gone into extinction. Indeed, some of the species in the old forests in Benue region may never be known any more. The threat to the Central Nigerian environment as a result of agricultural activities is real and should be a matter of concern to all those who are familiar with the problem.

The changing face of the environment in Central Nigeria area as a result of the increasing agricultural production is linked to deteriorating levels of poverty because of the diminishing soil nutrients. It implies that the harder the poor peasant farmers struggle to produce food to survive, the farther they will drag the environment into depletion. Thus, in developing economies, the main environmental pressure is linked to poverty. With increasing imbalance between what people consume and what their habitat can provide like food resources, we have a good case in point.

> Poor people press on the land and forests, over-exploiting them to survive and undermining the resource base on which their well-being and survival depend.[27]

That is a very simple argument: for all the damage we inflict on Mother Nature in the area, we still remain poor, which makes no sense at all. So the doomsday looms quite large: who will tell the poor farmers that their common heritage cannot last forever?

The Benue river valley is currently washing off its alluvial plains more than ever before due to increasing activities that is not directly linked to food crop production alone. In the last ten to twenty years, there have been series of activities like burnt-brick (block) making which the vast majority of people prefer to use for building houses. This has increased soil erosion, instability of watershed and localized flooding of river bed and other nearby places.

Taming the Tragedy of the Commons for Sustainable Environment in Central Nigeria

The scenario of the global commons and the tragedy that usually follow it is a familiar terrain in environmental analysis and this has found its parallel among the agricultural populations in central Nigeria no doubt. The truth is demonstrated in the inseparable link between agricultural production and environmental damage as has been repeatedly underscored. Invariably, it points to a gradual damage of the human environment which must be repaired sooner or later. Perhaps, the best way to do so is all embedded in the idea of sustainable development. That is what the present section attempts to contribute to because, the preceding analysis on the agricultural economy and its impact on the environment in Central Nigeria is a good parallel of environmental tragedy in the offing which M. J. Watts captured succinctly:

> Let us not flatter ourselves over much on account of our human victories over nature. For each of such victory, it takes its revenge on us.[28]

Indeed, food crop production is uniquely significant in the overall national economic development, stability, security and source of power[29]. Food crop production is the backbone of the wealth of the nation. But that is just one side of the argument. Logically, if Benue region is regarded as "the food basket of the nation"–whether rightly or wrongly, then one can see the essence of endeavoring to work towards "sustainable development" as a way of dealing with the environmental insecurity in the area. This is line with the World Commission held in Brundtland on Environment and Development which conceptualized "sustainable development" as a framework of policy within which strong growth, necessary to overcome poverty can be achieved while adopting economic policies to take full account of environmental consideration.[30] It also means proper management of the global commons for development.[31] This must be addressed too in Central Nigeria to avert tragic implications of the reckless use of land, the people's common heritage.

The necessity of government intervention here is overriding: It is necessary to check environmental fluctuation, review the land tenure systems and improve the power sector by providing alternative sources of energy to avert land destruction and vegetation which are responsible for the reduction of ecological functions. There is also need to work out efficient alternatives to maximum capacity expansion for utilities like in Central Nigeria, so as to bring about progress without loss of the essence of the natural habitat through the process of manipulation and exploitation of ecosystems in accordance with the 1991 recommendations of the Stockholm initiative on Energy, Environment and sustainable development.[32] As far as this goes, there is also need for government to sensitize peasant farmers to be aware of the implications of their agricultural activities. There are several ways to do this:

> Through education about the ecological consequences of irresponsible actions and by bringing social pressures to Bear on members of the community who have not moderated their actions.[33]

Apart from this, use of farm and forestland and vegetation can also be regulated by way of placing penalties on offenders. In the Benue region, there is no such framework to cater for depletion of vegetation or land resources or bush burning - not even by the ministry of Tourism or ministry of Agriculture and Natural Resources. It has been suggested that agrarian reform division should be created in areas of heavy agricultural activities to coordinate agricultural activities, take inventory of land use and development, and also evaluate the implications of such activities on the environment. There is also need to keep in view a proper value of the environment and to put in focus the seriousness of the damage of natural resources.[34]

Similarly, to check land disputes and enhance environmental security in central Nigeria, there is need for uniformity in land law and ownership to be embarked by both the state and federal legislatures. This will bring a wide range of advantages: it will enhance the status of customary land tenure, improve investments on land as well as the improvement of the physical quality of land used for agricultural cultivation. Furthermore, it is long overdue but imperative to register all arable lands including virgin and free holding portions in the bush, family/community land in every part of the region as a way of overcoming conflicting claims over land.

In most of the Global South especially sub-Saharan Africa, one can only say that development is taking place if poverty, inequality, unemployment and other mysteries of human life are receding. But the problems linger: so, what moral rights have we to talk of sustainable development and sustainable environment when the depletion of the earth's resources and the degradation of its forest, soils, species, fisheries, water and atmosphere while peasant farmers remain in their abject poverty?[35]

One way to ensure sustainable development and sustainable environment is to strategically reduce the poverty of agricultural farmers to ease the pressure they exert on land. If this is successfully implemented, it will bring out another positive implication, namely improvement on the living standards of the majority of the people and enhanced consumption patterns. Although this may boomerang on the environment all over but that is on the long run unless people are linked to technological devices with limited equitable management of resources. But of course, that is still achievable.

Furthermore, there is need for voluntary restraint or willingness to curb the extent of damage inflicted on the environment by the rural farmers without slacking food production strategies. Here, what is required is technical and more specifically linked to poverty and underdevelopment but it can be done.

Most countries regard the conservation of most ecosystems on grounds of economic considerations to be the most compelling. Other arguments for this are ethical, aesthetic values and ecological issues.[36] In the Benue region, it is to be considered that the preservation of ecosystems is for a variety of economic benefits, such as use of biological species for diverse needs such as food, medicines, and industrial purposes as well as incorporation of genes from wild life stock to enhance performance of cultivation activities which is a popular technique in molecular biology.

In non-economic spheres, we shall also do well to preserve our biodiversity because of the multiple essential ecosystems services from plants and animals for maintenance of gaseous composition of the atmosphere, climatic buffering, maintenance of soil, nutrient transfer to wild and cultivated plants, waste disposal and nutrient recycling, pest control, pollination, provision of food for vegetarians and all that.

These services are so beneficial and intricate that they cannot be substituted. What our ecosystems can do without the preservation of our biodiversity is not precisely known yet. And one can even suspect that rise in death toll may be attributed to the depletion of our flora and fauna, directly and indirectly. This argument is a corollary of the ongoing one in which world attention has been on the potential climate changes as a result of global warming due to the release in quantity and varieties of green house gases in the atmosphere in recent time. It summons us in this part of the world too to provide a perspective on the effects of environmental fluctuations arising from agricultural activities in the area.

All these are central to the general idea of sustainable development which has been modified to suit local circumstances. Underlying the whole idea is the development that is capable of meeting the needs of the present generation without compromising the needs of future generations. It means development without destruction, and development with the connotation of economic development.

Conclusion

In this chapter, we consistently argued that the fertile plains and the environment in central Nigeria have been subjected to so much over-use and misuse throughout the known history of the region due to ancient and modern migrations, traditional methods of farming, conflicts and wars over land. In aggregate terms, the immediate result is that, the area is one of the hottest regions in the county today. But the long term results on the environment are even more complicated and should be a matter of concern to experts, stake-holders and the Benue state government now more than before.

We have demonstrated that as long as land continues to play a key role in life support systems of the people, it will remain almost a magic force and a major moderating factor in the Central Nigerian history and environment which is predominantly an agrarian economy. Land will not only continue to generate wars and conflicts but will also pose a threat to environmental balance and security because more than ever before, land is becoming increasingly recognized as an economic basis or source of life in Africa. Unlike the notion that land was "gift of nature or ancestors," it has become a productive factor without any form of regulation. That is the dilemma and the prospect.

Furthermore, because the river Benue is a major waterway running through the Benue region providing alluvial deposits, up and down stream conflicts over land cannot be abated. This is because land regulates the flow of surface water including the river Benue and its tributaries as well as storage of groundwater resources, which are in short supply. Thus, the control of the waterways has wide implications on the multidimensional function of land as well as environmental security of the area which must not be taken for granted. It is against this backdrop that the main debates of this analysis were situated to discuss the implications and to draw attention to the immediate and long term issues involved in a very modest way.

The implications of the continuous scarcity of land against the background of its numerous functions, its diminishing nutrients, the dwindling ratio between burgeoning agrarian users and its limitedness constitute some of the daily worries that the agrarian population and environmentalists in central Nigeria have to contend with and find answers. Whatever the implications and solutions, they must capture the needs of everyone involved and in a long perspective to ensure the sustainability of the environment for posterity. These include adequate productive assets, socio-cultural, economic and the health needs of the poor. According to NEPAD, this simply underscores the fact that "a healthy and a productive environment is a prerequisite to sustainable development".[37] This also means that mere replenishment of the central Nigerian environment may not even give us all the desired results unless this is linked to total environmental accounting which encapsulates human capital; in particular, knowledge, experience and skills that can keep the environment invariably sustainable. That is hard work and commitment which will not only give the people the food they need at the moment but also keep them alive in a healthy environment as well.

Notes

1. M. O. Odey, "A History of Food Crop Production in the Benue Area, 1920–1995: The Dialectics of Hunger and Rural Poverty," PhD Thesis, University of Jos, Jos, Nigeria, 2001. Also see my latest argument on this issue in M. O. Odey, "Implications of Food Production Activities on Benue Environment and Economic Development," *African Journal of Environmental Law and Development Studies*, (AJEDS), Vol. 1 No. 1, 2004.
2. Daryl R. Forde and R. Scott, *The Native Economies of Nigeria*, London, Faber, 1902, quoted in E. O. Ademiyi, "Land Tenure as a Socio-Cultural Factor in Rural Development in the Middle Belt of Nigeria" cited in Onigu Otite and C. Okali, eds. *Reading in Nigerian Rural Society and Rural Economy*, Ibadan: Heinemann, 1990, pp. 172-184.
3. E. O. Ademiyi, "Land Tenure as a Socio-Cultural Factor in Rural Development in the Middle Belt of Nigeria" in Otite and Okali eds., *Reading in Nigerian Rural Society and Rural Economy*, pp. 172-184.
4. Quoted in Onigu Otite, et al., eds., *Community Conflicts in Nigeria: Management Resolution and Transformation*, Ibadan: Spectrum, 2001.
5. The Benue region is about 300 kilometers from the Atlantic coast on the Eastern part of the Middle belt of Nigeria between latitudes 7 and 10 degrees east; and between longitudes 6 and 8 degrees north of the equator. The land mass is over 45,000 sq. kilometers. The climate is characterized by tropical wet and dry seasons annually. There are about 7 months of rainfall with totals ranging from 150 to 200mm for most of the years. Temperature is generally high, fluctuating within the range of 23^{0}C and 30^{0}C to a high range of 37^{0}C in March or April. The area has good drainage from the Rivers Benue, Katsina-Ala and their tributaries. The flood plains are good for fadama agriculture during the dry period of the year. Further insight on the region can be derived from J. L. Nyagba , "Soils and Agriculture in Benue State," in D. Denga, *Benue State: A Compendium,* Calabar: Rapid, 1995.
6. M. O. Odey, "Implications of Food Production Activities on Benue Environment and Economic Development," *African Journal of Environmental Law and Development Studies,* (AJEDS), Vol. 1 No.1, 2004. The reader may also look at two other related sources: A. L. Owen and T. K. Pickering, *An Introduction to Global Environmental Issues,* London, 1994; and C. W. Kegley and E. R. Wittkopff, *World Politics: Trend and Transformation.* Boston and New York: Macmillan, 1999, p. 329.

7. H. H. Munkner, "Land Rights in Africa: Collective Use Right or Private Property?" in A. Wilcke eds., *Agriculture and Rural Development*, Vol. 3 No. 2 Frankfurt, 1996, pp. 10-14.
8. W. M. Sombrock and H. Eger, "What do we understand by Land Use Planning? A state of the art report," *Agriculture and Rural Development*, Vol. 4 No. 2, 1996, p. 3.
9. Chom Bagu, "Conflict in the Middle Belt: Perspective, Reality and Window of Opportunity," mimeo, Centre for Middle Belt Studies, Abuja.
10. Odey, "Implications of Food Production Activities on Benue Environment and Economic Development."
11. Michael P. Todaro, *Economic Development*, 6th Edition, Longman, 1998.
12. *Ministry of Information Bulletin*, Makurdi, Benue State, Nigeria.
13. Odey, "Implications of Food Production Activities on Benue Environment and Economic Development," 99-113.
14. Gluckman (1957), "Custom and Conflict in Africa" quoted in M. Howard *Contemporary Cultural Anthropology,* British Colombia University Press, 1989.
15. Albert Anloho, "A Study of Land Shortage and Rural Poverty in Southern Tivland, 1930-2005," MA Dissertation, Department of History, BSU, Makurdi, 2008.
16. Erim O. Erim, "The Early Foundations of the Kwararafa Confederacy 1333-1902 AD," *The Nigeria Magazine* vol. 55 No. 3, July-September, 1987, pp. 35-43.
17. T. Gondo, "Land Disputes in Benue State and Government Efforts in Resolving them," Paper delivered by Director General, *Bureau of Internal Affairs and Special Service*, Makurdi at Federal and State Security Meeting, August 20, 1997.
18. There are three related arguments on this, in Ibrahim James et al, eds., *The settler phenomenon in the middle Belt and the problem of National Integration in Nigeria,* Jos, Midland Press, n.d.; Jibo Mvendega, *et al, Ethnic Groups and Conflicts in Nigeria*, Vol. 4, The North Central Zone of Nigeria, Ibadan, 2001; and Yoroms Joses Gani, "Colonial Legacy, Violence and Democracy: The Case of The Middle Belt States" cited in P. Achakpa, *Uneasy Peace: A Report on Violent Conflicts* in Catholic Ecclesiastical Province of Abuja, Peace Commission/Caritas, Abuja, 2004.
19. Achakpa & Aliyu, *Uneasy Peace: A Report on Violent Conflicts*; and Gondo, "Land Disputes in Benue State and Government Efforts in Resolving them".
20. Achakpa & Aliyu, *Uneasy Peace: A Report on Violent Conflicts.*
21. Odey, "A History of Food Crop Production in the Benue Area, 1920-1995: The Dialectics of Hunger and Rural Poverty".
22. M. O. Odey, "Alternative Approaches to Food Security Strategies in Nigeria in the new Millennium" in P. Ogiji, ed. *The Food Basket Myth*, Aboki Publishers, Makurdi, 2004.
23. Odey, "A History of Food Crop Production in the Benue Area, 1920-1995: The Dialectics of Hunger and Rural Poverty."

24. Odey, "Alternative Approaches to Food Security Strategies in Nigeria in the new Millennium."
25. One of the earliest colonial records on this is Briggs, SWS *NAKMAKPROF (Fed). 1935-1939, 4/1/1553.* "Southern Tiv Areas, Farming Conditions in Experimental Farms."
26. UNFAO Report, 1992 quoted in Michael P. Todaro, *Economic Development*, 6th Edition, Longman, 1998.
27. Commission on Global Governance (Report 1995), *Our Global Neighborhood,* Oxford University Press, Ibadan, p. 29.
28. M. J. Watts, *Silent Violence, Food Famine and Peasantry in Northern Nigeria,* Berkeley: University of California Press, 1983.
29. Denga, *Benue State: the Land of Great Potentials: a Compendium*, p. 147. Also see Odey, "Implications of Food Production Activities on Benue Environment and Economic Development."
30. Commission on Global Governance, *Our Global Neighborhood,* Ibadan: Oxford University Press, 1995.
31. Association of Commonwealth Universities, (ACE, 1994), "People and the Environment: Preserving the Balance," *Report of Proceedings of the 15th Congress.*
32. Owenand Pickering, *An Introduction to Global Environmental Issue.*
33. Sorros quoted in Kegley and Wittkopff, *World Politics; Trend and Transformation*, 329.
34. Segun Famoriyo, ed., *Land Tenure and Agricultural Development in Nigeria,* Ibadan: The Nigerian Institute of Social and Economic Research, 1979.
35. Owen and Pickering, *An Introduction to Global Environmental Issues.*
36. M. G. Murray, "Conservation of Tropical Rain Forests: Arguments, Beliefs and Convictions," *in Biological Conservation,* 52, 1990, 17-26.
37. *New Partnership for Africa's Development* (NEPAD) October 2001, Abuja.

• Chapter 9 •

GROWING SLUMS AND IMPROPER SANITATION HABITS AS IMPEDIMENT TO DEVELOPMENT IN NIGERIA

Hauwau Evelyn Yusuf & Ahmad Bawa Abdul-Qadir

Introduction

Development in African countries has always been assessed based on economic advancement and technological development, with little or no attention given to environmental impediments. There is need for refocusing on this aspect of the life of a nation to be able to assess properly need areas and proffer lasting solutions or enhance development in all spheres.. The principal objective of this research was to find out how the lack of portable water and poor sanitation impact on people's lives. As a result, questionnaires were administered in three local Government Areas of Kaduna State representing the three senatorial districts of the state. The study was conducted in Kaduna State located in the North Western region of Nigeria; the state is heterogeneous in nature consisting of various ethnic groups. Christianity and Islam are the dominant religions practiced by the majority of the people in the state; this is not to say that other forms of religions do not exist as you find traditional worshippers scattered across the state. There are at least 16 major ethnic groups in the state, namely, the Gbagyi, Gwarok, Juu, Akulu Kamanton, Adarah, Atyap, Hausa, Ham, Koro, Nkurme, Ninzom, Kamuku, Kagoma, Ashalio and Fulani. The state covers a land space of 45,567 square kilometers and has an estimated population of 6,066,562, according to the 2006 census. The state has 1,586 primary schools, 182 post-primary schools and 12 post secondary institutions. It has 23 local government councils, 3 emirates, 32 chiefdoms and 345 districts. The state is endowed with six precious minerals, which include molybdenite, kaolin, columbite, aquamarine, sapphire and gold. It grows several cash and food crops, and the state houses a number of industries some of which are moribund. One of the four major refineries in the country is located in the state.

The chapter is divided into four sections. After this introduction, section two discusses the relationship between environmental factors, poor sanitation habits and sustainable development. Section three discusses efforts made at improving water supply and sanitation. Section four, which is also the concluding one, presents the findings of the study and offers some suggestions.

Environmental and technological factors, poor sanitation and sustainable development

Development issues have generally been assessed based on economic advancement and technological development. Until recently little attention was given to environmental impediment to development. The level to which environmental factors affect development might be underestimated but at the same time enormous as we shall see later in this chapter.

Threat to the environment and environmental crisis is a global phenomenon. Threat to the environment comes from extraordinary expansion of industrial production and its accompanied technological intervention into nature has been more profound and extensive.[1] As industries expand, the environment becomes more threatened. In the western world and other advanced economics of the globe, toxin in the environment, particularly in the air, water and foodstuffs produced "systematic and often irreversible harm". Since the toxin is often invisible, people are not immediately aware of the risk they create.[2] It could be recalled that before modern industry the overall impact of humanity on nature was relatively slight. "Nature dominated human life far more than the other way round."[3] Today, the onslaught on the natural environment is so intense that there are few natural processes uninfluenced by human activities.

Nuclear Power

The place of nuclear technology and nuclear energy in modern economy cannot be underplayed. The advanced and industrial nations of the world are in a different and higher pedestal than other parts of the world. Atomic accidents associated with nuclear production, especially in nuclear power plants, could affect a large number of people. Radioactivity, for example, knows no geographical boundary and can spread across borders. In the same vein acid rain caused by industrial pollution in one country may affect another country.[4] In 1992 for example, there were 428 reactors in 31 nations. There had been four severe accidents: Three Mile Island in the United States in 1979, Chernobyl and Ukraine in 1986, and Japan in 2011.[5]

Creation of Waste

The creation of waste globally is assuming an alarming proportion, that is, waste which cannot be disposed of in short term or recycled. Material waste includes goods that are thrown away, gaseous waste pumped into the atmosphere, for example, carbon dioxide released into the atmosphere by the burning of fuels like oil and coal in cars, power stations and gas released into the air by the use of aerosol cans, materials for insulation and air conditioning units.[6]

Ozone Layer

The threat to the ozone layer is increasingly being appreciated by both political leaders and scientists. By 1975 the chlorine concentration was 1.4 parts per billion.[7] In the 1990s it was enough to open a hole in the ozone layer each Antarctic spring (that is to say that the chlorine level had alarmingly risen). Risks such as the green house effect and global warming and the depletion of the ozone layers are real and global in nature. As a matter of fact it is held responsible for the unprecedented climatic changes being experienced in the world in recent times.

Health & Global environment

There is no doubt that the health of the citizens of the world is connected with the environment around them. There is a connection between the global ecosystem and pattern of health and diseases as every human activity takes place within the confines of the social and physical environment. Some form of environmental degradation produce health hazards. Air pollution, for example, is said to lead to asthma, impaired lung function, lung cancer and other respiratory problems. In fact, air pollution affects agricultural production which could in turn lead to starvation and gradual extinction of the human race in such places if immediate measures are not taken.

Third World countries like Nigeria are not immune from global environmental related crisis, even though they may have certain peculiarities. In the major cities thieves may be killed and their bodies dumped on the road side, bush path or road intersection, where their decomposing bodies cause serious health hazard. In some cities in Africa during war situations, one can step over dead bodies.. Ironically, reporting such decomposing bodies could get one into trouble with the police, the Health department and the news media.[8] It does not really matter what side effects this situation poses for the people living around in that environment; even when the reports are made there is usually no immediate actions taken.

Migration from the rural areas into these cities has continued unabated and this has continued to put pressure on the grossly inadequate social amenities provided. People defecate anywhere anyhow and once there is flooding a regular occurrence, cholera is likely to spread since many people, especially in the rural areas, obtain their water supply from the ponds and some of the seasonal rivers. The flooding problem is compounded by lack of proper physical planning of the cities, blockage of the drainage system either by dumping refuse into the system or erecting of building on waterways. It is very important to put in place strict measures to ensure that constructions are done correctly following town planning procedures, putting in place disposal and sewage facilities for every building. This will go a long way in solving the problem of flooding and growing refuse pyramids that have become the regular sight in all major cities and villages in the country.

Furthermore, the nation is faced with serious desert encroachment, deforestation, and acute gully erosion in some parts of the country. This type of erosion has cut off some communities from major cities making it impossible for agricultural products to get to the markets. Desert encroachment comes with its attendant movement of soil on a relatively large scale and encroachment on farmlands posing difficulties for farmers. On the other hand, the oil producing areas, particularly the Niger-delta region, encounters oil spillage either by means of pipeline vandalization or break down or the system of the major oil producing companies. When this happens, the rivers are polluted and the species in them are killed making it impossible for the people who are mostly fishermen to earn a living. Meanwhile, the spillage makes it equally. impossible to cultivate crops since the land surface is usually covered with oil and crops are likely not going to germinate on such soil.

A number of studies have been conducted on environmental sanitation and sanitary habits of different groups but the impact of such habits on development have not been given the prominence it deserves.[9] This is in spite of the general situation of sanitation in the world and the importance of sanitation as "prerequisite for improvement in general health standards, productivity of the labour force and quality of life".[10] Increasing access to safe drinking water and improved sanitation are components of development and poverty reduction. The social and economic environment and health benefits associated with these are enormous. Hygiene and sanitation related diseases like diarrhea, bilharzia, and cholera are among the highest causes of ill health and death.[11] Current statistics indicate that about 1.1 billion people in the world do not have access to proper drinking water and 2.6 have no proper sanitation facilities like toilets. In Africa the sanitation situation is even worse. More than 65% of the population of the continent lack proper sanitation facilities and death as a result of water related diseases are high.[12]

Since the 1980s, efforts have been made to improve on the sanitary conditions of the world population. This started with the declaration of 1980-1989 as the International Decade of Water and Sanitation. Much has been achieved during the decade though some of the achievements have not been sustainable probably as a result of the strategy adopted or the lack of commitment on the part of its initiators. Political instability and economic crises experienced by many countries, especially in Africa, affected whatever achievement that might have been recorded because as soon as a new government takes over they start new projects abandoning existing ones regardless of its importance and relevance to sustainable development.

There was also the World Bank water and sanitation program. This program brought to the fore the challenge of providing proper drinking water and sanitation facilities. Investments in this sector bring substantial economic returns and reduces costs in other sectors thereby impacting positively on the economic development of any nation. Water supply and sanitation are also found to be very important tools in poverty alleviation. This is because resources used in treatment of disease caused by bad water and poor sanitary conditions would be invested on development projects even on personal basis.

In recognition of the above MDGs were adopted by a number of countries in 2000. With eight goals identified as needing attention by 2015. The seventh MDG goal is environmental sustainability. This goal has three sub-targets which are:

1. Ending exploitation of environmental resources
2. Access to safe drinking water and basic sanitation
3. Significant improvement in the lives of at least 100 million slum dwellers.

The supply of safe drinking water is considered critical to hygiene and sanitation and fundamental to achieving the MDGs and subsequent attainment of sustainable development in general.[13] A target of having the proportion of people with no access to safe drinking water and basic sanitation was set for 2015.[14] The period from 2005 to 2015 was declared by WHO as the decade of water; the aim was to eventually establish the framework of providing full access to water and sanitation for all. Other initiatives include the establishment of household water treatment and safe storage network made up of over 100 organizations working all over developing countries.

Following from experiences gained, 2008 was declared by the United Nations as the International Year of Sanitation. During the year efforts were made to meet the challenges of sanitation and water supply. These efforts include the creation of a political platform to gain political support for the efforts at providing sanitation to the vast population that need these facilities. One such effort was a conference held in February 2008 in South Africa which sought to bring together ministers in charge of water and sanitation in Africa. Commendable efforts were made at this conference. But the basic problem of safe water supply and sanitation still exists. In most poor countries access to safe water remains poor. Significant improvement is required if the UN MDG is to be met.

The importance of water and sanitation has been recognized by countries and investments in the sector are increasing. Efforts are made to achieve the MDGs. In India, the state of Madhya Pradesh works in co-operation with UN-Habitat in trying to achieve the MDG on water and sanitation.

Nigeria recognizes the pivotal role of water in national development. This was clearly stated in Nigeria's Poverty Reduction Strategy Paper called National Economic Empowerment and Development Strategy (NEEDS).[15] The documents see water playing a crucial role in the well-being of Nigerians and in poverty alleviation. An integrated approach that looks beyond water supply and sanitation and looks at bigger picture is adopted. The policy thrust of NEEDS is anchored on the National Water Supply and Sanitation Policy of 2000.

Some achievements have been recorded over the last decade or so. The proportion of people having access to improved water resources increased considerably though there are differences between access in urban and rural areas. Inadequate water supply in urban areas is attributed to poor urban planning and poor management, while in rural areas it is due to lack of commitment to rural development on the part of governments at various levels. In addition the rural population has limited knowledge about the implication of improved water supply on the health of the people.

A combination of ignorance, poverty and leadership has made this so. There is a high level of illiteracy in the country, and this is even more pronounced in the rural areas. This has made it difficult for the rural populace to appreciate the intricate link between water supply and health.

The rate of poverty has compounded the water problem of the rural areas as many households cannot afford to sink boreholes in their houses. Even the local types of wells which is the common means of getting supply of water apart from ponds, streams and rivers are increasingly out of their reach due to poverty.

Furthermore, poor communication has added its own gross dimension. The rural populace is not enlightened by the leaders about the place of quality water used by the household in the health of the people. The mass media, opinion leaders and traditional means of communication have not been exploited by the leaders in this regard.

A number of other policy guidelines and action plans have been developed to enhance the achievement of the MDG.[16] There have been increased sectoral interventions in sanitation through the implementation of the National Environmental Sanitation Policy (NESP) and National Environmental Sanitation Action Plan (NESAP). There is also an on-going water sector reform program which will ensure higher level of service delivery.

Investigating Sanitation in Kaduna State

This chapter takes a look at the availability of portable water and sanitation facilities as it affects development in Nigeria with particular reference to Kaduna State. To this end data collected was limited to three Local Government areas of the state with the three senatorial districts represented. Multi-stage random sampling was adopted in selecting the local governments and the respondents in each local government.

Questionnaire developed for the purpose was administered to respondents. Using the In-depth interview method, descriptive statistics which include frequency and percentages were used in the analysis of the data collected. This is in addition to inferential statistics in form of Analysis of Variance.

Empirical findings

Finding reveals that little or no attention is given to the provision of portable water and sewage facilities in Kaduna State and even where available, people portray high level of ignorance as to usage and maintenance.

Table 9.1: Showing respondents by Sex

	Frequency	Percentage
Male	84	63.6
Female	54	36.4
Total	138	100

Source: Field Data in Kaduna State Nigeria, 2008.

Table 9.1 shows that there were 138 respondents out of which 84 representing 60.9% were males and only 48 that is 34.8% were females, six of the respondents (4.3%) did not indicate their sex.

Table 9.2: Sources of water in study area

	Frequency	Percentage
Tap	64	47.1
Well	52	38.2
Borehole	22	14.7
Total	138	100

Source: Field Data in Kaduna State Nigeria 2008.

Table 9.2 shows the different sources of water in the study area, 47.1% have access to tap water, 38.2% have only well water and 14.7% have borehole water.

This goes to show that the majority of the respondents (47.1%) have portable water supply in their area. If the majority, that is 47.1%, only have access to tap water and up to 38.2% get their water from wells, the question here is, to what extent are the wells in good condition, how often are the wells treated to make it safe for human consumption? It is therefore important to give special attention to improving the sources of water supply in the study area. Making sure that all wells are treated regularly and that the people using the wells do their best to keep the surroundings clean and free from filth.

Table 9.3: Availability of refuse disposal facilities in respondents' house

	Frequency	Percentage
Yes	100	80.6
No	38	19.4
Total	138	100

Source: Field Data in Kaduna State Nigeria, 2008.

Table 9.3 displays that majority of respondents (80.6%) have refuse disposal facilities in their homes while 19.4% do not have at all. The implication here is that there are people in the study area who do not attach any importance to the provision of refuse disposal facility in their homes.

The availability of refuse disposal facilities in individual homes does not signify proper use of such facilities and may account for the poor health conditions of the people of the study area. It should be noted that it is possible to have facilities but yet not use them appropriately, in the sense that one may pile up refuse in the dustbin and leave indoors for more than 24 hours without disposing; some things may decompose and create odors that are hazardous to the health of occupants of the house.

Table 9.4: Availability of refuse disposal facilities in study area

	Frequency	Percentage
Yes	52	41.9
No	86	58.1
Total	138	100

Source: Field Data in Kaduna State Nigeria, 2008.

Table 9.4 indicates that some parts of the study area have central refuse disposal facilities while others do not, only 37.7% have central refuse disposal facilities while the majority 58.1% do not have. This goes to show that refuse disposal facilities in the study area are grossly inadequate; it means that even if they have facilities in their various homes they do not have a central area where they could transfer their refuse. It is important to make adequate provision for refuse disposal in the society and create awareness of the importance of such facilities to both health and development so as to encourage people to properly dispose refuse and live healthy and improved life.

The availability of a central refuse disposal facility does not readily mean that every member of the society uses it; even if they use it, to what extent is the facility properly managed? To what extent does these facilities not constitute a higher health hazard to the generality of the area as a result of their not been regularly cleared? The enlightenment and awareness discussed above must form one of the foundations that must be included in the provision process to forestall the chances of such facilities constituting higher health hazard.

There is need to put in place a monitoring mechanism that will ensure regular clearing of such facilities. Community members should be given regular talks on the importance of using the facility and its implication on health and development of the society at large, the participatory development strategy should be employed so that the end users see themselves as stakeholders and as such do their best to make sure it works, from experience when the community's consent is not sought in some development policies getting their maximum cooperation is usually difficult. It is important to note that this will go a long way in reducing the growth of slums thereby encouraging investors and in turn lead to development.

Table 9.5: Link between availability of refuse disposal facilities and development in study area

	Frequency	Percentage
Yes	94	77.0
No	44	23.0
Total	138	100

Source: Field Data in Kaduna State Nigeria, 2008.

To some respondents there is a link between refuse disposal and development, while to some there is no link at all; however it is the majority that sees a link between development and proper refuse disposal. Their reasons for this are shown in Table 9.6: they believe that a clean environment enhances good health which in turn provides a healthy working force needed for optimum productivity.

Table 9.6: Link between availability of refuse disposal facilities and development in study area

Impact on development	Frequency	Percentage
It helps to make the Environment clean	34	25.0
Absence of refuse disposal causes illness	24	17.0
Absence of refuse disposal facilities makes society dirty and untidy	22	16.0
Diseases are spread easily in dirty environments	32	23.0
Causes under-development	26	19.0
TOTAL	138	100

Source: Field Data in Kaduna State Nigeria, 2008.

For further examination of the dependence of development on proper sanitary habits and clean environment, availability of portable water and development, cross tabulations of some dependant and independent variables were carried out; the results are as displayed.

From the findings, it becomes clear that there is a strong link between clean environment and development. Clean environment begets good health and good health creates room for a healthy workforce, a good basis for investment and in turn better room for development. For every prospective investor the assurance of a healthy workforce is a basic ingredient for successful production, it also encourages expansion enabling growth for the economy.

Table 9.7: Link between availability of portable water and development in study area

		Absence of Portable Water Can Affect Development in the Community?		
		Strongly Agreed	Agreed	Total
Sex Of Respondents	Male	64	20	84
	Female	44	4	48
Total		108	24	132

Source: Field Data in Kaduna State Nigeria, 2008.

108 representing 81.8% of respondents (male 64 and female 44) strongly agree that there is a link between availability of portable water and development, while 24 (18.2%) agree but not strongly, what is notable here is that all respondents see a link between development and portable water.

Portable water supply, just like clean environment, contributes to development as bad water leads to bad health conditions which affects the totality of the society while portable water enhances good health and development. Majority of the respondents see a link between none availability of refuse disposal facilities and development. Only 24 (20.7) do not see any link between them.

$H_{0:}$ Absence of portable water can affect the health of the people is not significantly different from the absence of portable water affects development.

H_1: Absence of portable water can affect the health of the people is significantly different from the absence of portable water affects development.

Table 9.8: Showing the test of hypothesis

ANOVA					
ABSENCE OF PORTABLE WATER CAN AFFECT THE HEATH OF THE PEOPLE?					
	SUM OF SQUARES	DF	MEAN SQUARE	F	SIG.
Between Groups	.101	1	.101	1.819	.180
Within Groups	7.429	134	.055		
Total	7.529	135			

Source: Field data in Kaduna State Nigeria 2008 (processed through the Statistical Package for Social Sciences).

The hypothesis was tested using one-way ANOVA at 1% level of significance and the H_0 was not rejected, that the absence of portable water can affect the health of the people as well as development. Hence there is no significant difference between the two. This implies that availability of water is essential to development as ill health leads to shortage of manpower which is a necessary factor in development.

Observations

- None availability of portable water is an impediment to development;
- None availability of proper refuse disposal facilities is also an impediment to development;
- Level of education of people does not necessarily determine sanitation habits as people in the rural areas at times have better sanitation habits than people in urban centers;
- Poor sanitary habits are impediments to development as the outbreak of diseases resulting from such habits leads in most cases to lose of manpower which in turn affects productivity; and
- There is a relationship between availability of portable water and health as well as development.

Taking a walk round Kaduna State, it is observed that polythene and paper bags are littered all over the place. It could be that the people are not aware of the hazard as well as the possibility of turning such waste or garbage to abundant wealth through the simple process of recycling that could change lives and make the society neater. Clearing up the mess will also provide jobs for many people.

Conclusion

The following recommendations are presented for possible implementation:

> Massive investment by government in the development of facilities for proper refuse disposal, either in form of pits or refuse vans located in strategic locations, which are to be taken away at specified intervals preferably on weekly basis is important.

Provision of portable drinking water: this could be in form of pipe borne water, boreholes and deep wells. Previous governments had done a lot in providing portable water for the people but there is still a lot to be done as large population of the state and the country at large have no access to portable drinking water.

> Public enlightenment on health and sanitation issues: the government may consider the reintroduction of public health inspectors, whose duties will include enlightenment on both health and sanitary issues

> There should also be proper town planning schedules in all parts of the nation rural or urban making sure for proper layouts with good drainage systems and refuse disposal areas, as against the indiscriminate building of residential houses on land areas marked as roads in layout areas.

In conclusion, there is a clear indication that more attention should be accorded the issue of proper sanitation habits and portable water in all societies so as to enhance development. There is need to give special attention to the effects of poor environmental conditions, bad sanitation habits and lack of portable water on the general development of the individual and the society at large. Poor sanitary habits and bad water affect the health conditions of the individual and in turn the society at large. There is urgent need for the growing industries the world over to pay attention to the final destination of industrial waste making sure that they do not constitute health hazard to the larger society. The possibility of recycling should be explored as this will probably take care of some of the refuse thrown all over. It is also hoped that further research would be conducted. This time on a wider scale (as Kaduna State is not the only state experiencing poor sanitary conditions and lack of portable water in

Nigeria) so as to get a clearer picture of the effects of poor sanitation habits and development. Every individual in the society needs to know that the state of their environment determines their development rate; for instance an environment that is filthy does not attract investors most especially international investors. Enlightenment workshops, if organized in urban and rural areas, will go a long way in improving the general sanitation habits of the people. As discussed earlier, the bottom up strategy should be used to enhance success. Meanwhile, the government, community leaders and all stakeholders must be committed to improving the living conditions of the larger community, ensuring that the provisions necessary for enhancing clean and healthy living is adequately provided.

Kaduna State has a large number of industries that can flourish more if it has a healthy environment and workforce. The surrounding environment of some of these industries are slums that can benefit from these industries if only these industries see themselves as stakeholders, Places like Kakuri where industries like Peugeot, Arewa Textile, Coca-cola and others are located should be given a face lift. The impact of these industries should be felt by the residents of the area.

Notes

1. Anthony Giddens, *Sociology* (Cambridge: Polity Press, 1993), 551.
2. M. Haralambos and M. Holborn, *Sociology Themes and Perspectives* (London: Collins International Textbooks, 2004), 87.
3. Haralambos and Holborn, *Sociology,* 87.
4. Ibid., 87.
5. Giddens, *Sociology*, 552.
6. Ibid., 553.
7. Ibid., 553.
8. P. F. Wilmot, *Sociology: A New Introduction* (London: Collins International Textbooks, 1985).
9. Pankaj Dwivedi and A. N. Sharma, "A Study on Environmental Sanitation, Sanitary Habits and Personal Hygiene among the Balgas of samnapur Block of Dindori Distict, Madhya Pradesh," *J. Hum Ecol*, 22, 1, 2007: 7-10 .
10. Ibid.
11. WHO/UNICEF, *Global Water Supply and Sanitation Assessment* (Geneva: WHO, 2000).
12. World Bank Project 2004, *Water Supply and Sanitation in Ethiopia.*
13. United Nations, Human Development Report. (New York: UNDP, 2004).
14. J. D. Sachs, *Investing in Development: A Practical Plan to Achieve the MDGS.* (New York: UNDP, 2005).
15. National Planning Commission, (MDG Report, NPC Abuja, 2004).
16. National Planning Commission, (MDG Report, NPC Abuja, 2005).

• Chapter 10 •

POST-COLONIAL ENVIRONMENTAL ISSUES IN AFRICA: GLOBALIZATION, SCIENCE, TECHNOLOGY AND THE LANGUAGE QUESTION

Ann Albuyeh

The wailing is for the Great River
Her pot-bellied watchers

Despoil Her . . .

– Christopher Okigbo, "Lament of the Drums"[1]

Introduction

African authors have used Europhone languages, especially English and French, to both celebrate nature and voice their particular environmental concerns.[2] Notwithstanding more recent controversy surrounding language choice in the context of African literature, epitomized by the stance of Kenyan writer Ngugi wa Thiongo[3] the fact that Christopher Okigbo, an Igbo who was killed in 1967 during the Nigerian War, was unapologetic about writing in the ex-colonial language, was not noteworthy during the transition period just before and after independence. Stating in an interview: "I think that I've been influenced by various literatures and cultures,"[4] Okigbo, who studied Classics at University College, Ibadan, explains the international references he incorporated into "Lament for the Drums." The poem's five sections integrate aspects of African oral tradition, Christian tradition, Roman mythology and Babylonian/Assyrian mythology.[5]

That a spiritual link to the natural world is an important part of traditional Africa's relationship to its environment is illustrated by fellow-Nigerian Ken Saro-Wiwa's description of the beliefs of the people of his Ogoni homeland in the Niger Delta. "The land is a god and is worshipped as such."[6] Saro-Wiwa was an environmental activist who gained worldwide fame and support for his Ogoni people during his fight against the Royal Dutch Shell Oil Company[7] and a complicit Nigerian government.[8] That such protests regarding the pollution in the Niger Delta have been internationalized through the use of the English language appears to have gone unremarked. Perhaps this is because there has been a long-standing tradition of protest in Europhone languages in Africa, beginning in the colonial period.[9]

Yet, however relevant language choice is to environmental concerns in the context of either artistic expression or protest, central to real environmental progress in Africa are questions regarding language and the debate over whose science and technology will have what role in Africa's future development. In particular, the value of traditional vs. western science and technology has been linked to the languages they are communicated in, and this debate is, thus, comingled with issues regarding the value of African vernaculars vs. Europhone lingua francas, often in the context of higher education. That the goal of developing African environmental resources from North to South, East to West, will ultimately require the participation of all Africans highlights the fact that the language choices made will constitute a complex and critical component in the direction of Africa's environmental future.

Early Lingua Francas Transmit Scientific and Technological Knowledge

Linguists Cécile Vigouroux and Salikoko Mufwene point out "one cannot make sense of *globalization* without connecting it to *colonization* .... [and] both colonization and globalization are as old as the dispersal of Homo sapiens out of East Africa forty to sixty thousand years ago,"[10] and caught up in the history of colonization and globalization is the spread of international lingua francas. Ashurbanipal, the last great monarch of the Assyrian Empire,[11] assembled a library of cuneiform tablets in the ninth-century BCE, of which 30,000 survive, including descriptions of contemporary scientific knowledge and technology.[12] In *Empires of the Word,* Nicholas Ostler states that the multilingual Assyrian Empire was, moreover, innovative not only for the first libraries, but also for the first lingua franca, the "first generalized use of a totally foreign language for convenience in communication."[13]

The Assyrian lingua franca, Akkadian, is a close relative of a more familiar international language of science and technology, Arabic,[14] which was carried to Africa centuries ago.[15] Introducing the concept of "Arabic Science," Egyptian scholar Abdelhamid Sabra[16] explains:

> This immense and long-lasting enterprise has been called 'Arabic,' firstly, because it owed its inception to Arab initiative and Arab

patronage and, secondly and more importantly, because the Arabic language was the medium in which it developed.

From its beginnings in the mid-eighth century, Arabic science involved scholars from many ethnolinguistic backgrounds. For the educated elite, as Sabra states "the Arabic language rapidly became an international language of science in a stronger sense than had been true of any other language."[17]

However, there can be a downside to the spread of scientific knowledge through a foreign lingua franca. Writing in *Science Magazine* in 2004, the geologist Scott Montgomery stated: "Languages of intellectual power---such as Greek in the Hellenistic period, Arabic in the 8th to 12th centuries, and Latin in medieval Europe—can divide the world into 'haves' and 'have nots.' This means the possibility for severe intellectual isolation."[18]

Discussing Arabic science, Dennis Overbye called communication and financial support "two of the main pillars of science."[19] Although a critical discussion of global finance is beyond the scope of this chapter, in the modern world also, the dominance of means of communication and the financing of science and technology are linked in ways which emphasize the "have" "have not" status of participants. In the Niger Delta, the English language and millions of dollars worth of oil drilling technology dominated the Ogoni and other farmers and their agricultural interests to their detriment. Other examples abound, from the policies of the World Bank to initiatives promoted by Microsoft. The globalization of communication and financial support in the twenty-first century has arguably greater repercussions for the continent of Africa and its environmental concerns than anywhere else, and the linguistic dominance of English is implicated in both.

English, the First Modern Global Language of Science and Technology[20]

Although there have been predictions that in the future, the international dominance of English could face challenges from e.g. Chinese,[21] it appears that for the present, the use of English as an international lingua franca will continue to grow, a fact most apparent among the world's scientific elite. For example, in his *The Beginner's Guide to Winning the Nobel Prize* (2005), Nobel Laureate Peter Doherty illustrates this point by stating: "It isn't necessary to be a Shakespeare . . . but anyone who wants to be recognized as a top scientist must be able to write clear, concise English."[22]

In fact, the percent of non-native speakers of English is increasing worldwide, and the international value of Global English can be predicted to increase too among the scientific and technological elite of all countries. The British linguist David Crystal reports that while approximately 400 million people speak English as a mother tongue, almost three times as many people speak it as a second or foreign language, i.e. over a billion speakers. Montgomery cites Graddol's 1999 statistic that "Overall the proportion of native speakers [of English] is decreasing worldwide, from 8% in 1950 to [about] 6.5% today, projected to 5% or less by [the year] 2050."[23] This reality has moved some to sidestep the language debate and focus instead on practical issues such as the acceptance of nativized [24] varieties of English in scholarly publications. One response to the perceived gate keeping of mother-tongue English speakers on e.g. editorial boards of scientific journals and technology websites is to call for an acceptance of international variety in vocabulary, syntax, etc. in the English of non-native speakers.[25] If on the elite end, the use of English is a necessary evil, surely the accommodation of variety across the global speech continuum is not too much to ask.

However, practical considerations aside, the spread of this global lingua franca and its dominant role in science and technology is being actively rejected by many, to a perhaps unprecedented degree. The cuneiform tablets do not tell us how science and technology in a foreign language were received by non-Akkadian speakers, nor do we know whether non-Arabic speakers felt left out of the scientific progress of their age, but certainly the "Globalization of scientific English,"[26] in the 21st century has become an international topic of concern, and in some quarters, even outrage.

In fact, the loudest protests do not emanate from scientists such as Montgomery, quoted above, but from educators, political leaders, and, it seems, especially, European linguists. Recently, whole volumes of essays devoted to the controversy have emerged with titles such as *The Dominance of English as a Language of Science*,[27] and *Linguistic Inequality in Scientific Communication Today: What Can Future Applied Linguistics Do to Mitigate Disadvantages for Non-Anglophones?*[28]

That both books are published in English is only one of the ironies inherent in this controversy. A second is that European criticism of linguistic imperialism seems to have only arisen, as Paulin G. Djité, a noted sociolinguist from Côte D'Ivoire points out, "when the English language started overtaking and overwhelming [other European's] own linguistic space and expression."[29] Although the plight of non-Anglophone Europeans should not be discounted, the question of the dominance of English and linguistic inequality in science and technology is worth examining within an African context because it is becoming clear that this debate could prove more important to Africans than to anyone, especially as they relate to the environment and development.

Referring to technology transfer and development, Djité states: ""Language is generally not perceived as playing a significant role in the causes of underdevelopment in Africa, and therefore not thought of or mentioned in trying to work out solutions to this situation. The absence of linguistic input in development planning in Africa is one of the key reasons why the majority of Africans are left 'on the edge of the road.'"[30]

Biodiversity and Indigenous Knowledge

In the comprehensive report on the environment in Africa published by the United Nations Environment Program in 2006, *Africa Environment Outlook 2: Our Environment, Our Wealth*,[31] the authors emphasize the past, present, and future role of Africa's environmental resources in the region's development. Of all the environmental areas examined, many of which also impact on biodiversity, such as freshwater, coastal and marine environments, forests and woodlands, biodiversity arguably has the greatest significance at regional and global levels in equal measure.

Coastal fishing village, Ghana
Source: Ann Albuyeh

Throughout the continent "African biodiversity is closely linked to nutrition and achieving food security."[32] For example, at least 669 species of fish are found in the Congo River, the second largest river system on Earth, many of which provide significant sources of protein for local populations living on its banks.[33] Moreover, in rural areas, a large variety of plant foods provide essential micro-nutrients and are the source for almost three-quarters of recorded protein consumption.[34] Although species of millet and sorghum, originating in Africa, are, unsurprisingly, important food crops, forests and woodlands also contribute food sources such as, for example, the leaves and seeds of the baobab[35] tree.[36]

Baobab forest, Senegal
Source: Ann Albuyeh

The baobab tree is also harvested for medicinal uses,[37] and, although non-profit groups such as The International Centre for Underutilized Crops often promote indigenous plants as food sources, the potential for African biodiversity to open up new medicinal applications is where traditional knowledge and the global science and technology nexus most famously intersect. A case in point occurred when a Southern African spiny succulent called hoodia,[38] which has reputedly been used for thousands of years by the San people of the Kalahari to stave off hunger during long hunting trips, attracted the attention of "Big Pharma" in the West.

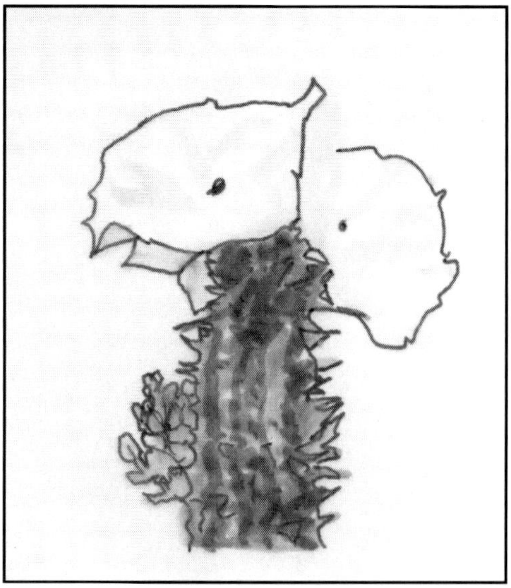

Hoodia gordonii
Source: Ann Albuyeh

Researchers were hoping that the appetite suppressant properties of a group of steroidal glycosides, "P57," isolated in the hoodia plant could provide an important ingredient for weight loss drugs, an especially lucrative area of research in the wealthiest countries.[39] A European court granted the patent for P57 to the South African Council on Scientific and Industrial Research (CSIR) which isolated and identified P57. CSIR licensed a UK-based company called Phytopharm, which in turn licensed pharmaceutical giant Pfizer to handle development and global marketing, followed by the global consumer goods company Unilever.[40]

The rest of this story would be "business as usual," especially big international business as usual, except that the San people demanded recognition of their contribution to global science and technology and inclusion in the accruing economic benefits. The CSIR and the South African San Council, set up in late 2001 to represent the Khomani, Xun and Khwe people, spent three years negotiating over these two issues. Having won a 1999 court case which transferred 100,000 acres of white-owned farmland on the edge of the Kalahari, the San people hired the same lawyer to successfully argue their landmark case regarding what ActionAid characterized as "a major case of bio-piracy."[41]

Minority and Endangered Languages

The case of the San people highlights the linguistic difficulties faced by speakers of minority languages in standing up to national and international bodies. The San people speak varieties of "click" languages much beloved by linguists the world over, but which have most often been treated with incomprehension and even contempt by speakers of neighboring languages, both colonial and indigenous. African click languages belong to a linguistic family known as Khoisan (Khoi-San) or Khoesan, of which the San languages are considered a subset;[42] the term Khoisan was adopted by linguists of South Africa's University of Witwatersrand to restore dignity to the languages which the common labels "Bushman" and "Hottentot" did not. Unfortunately, prejudice towards the click languages and their speakers continues.[43]

The fact that the clicks, consonants formed by clicking with the tongue placed in various positions, have numerous variants which can involve more common processes like aspiration,[44] provides these languages with the possibility of a very large, and to the rest of us exotic, sound inventory.[45] While no language is impossible to learn, even if adult learners experience greater problems, a language's difficulty is measured in direct relationship to its "genetic" distance, in terms of phonology, syntactic structure, etc., from the language(s) a person happens to already know. Although neighboring speakers of Bantu languages, including the Nguni language, Zulu, have borrowed a limited range of clicks,[46] the speakers of Khoisan languages have most often been the ones forced to adapt linguistically by learning the languages of their neighbors, and not the other way around.[47]

A significant problem for the San people and other linguistic minorities in Africa, as elsewhere, is the attrition rate in terms of number of speakers. While it is believed that San languages had been spoken in Southern Africa for some 8000 years, modern survivors are few and spoken by increasingly smaller populations.[48] San language expert, Anthony Traill, dramatically opens his chapter on Khoesan languages, thus, "The sociolinguistic story of the South African Khoesan . . . finds its place in the discussion of language death in Africa."[49] Languages may become extinct due to the death of their speakers or due to the shift of their speakers to other languages;[50] both have happened during the history of the San people. During the colonial era, there is ample evidence of the former.[51] In addition, throughout their history speakers of Khoisan languages have shifted to speaking the languages of those they have come into contact with, especially Bantu languages. Salikoko Mufwene likens the spread of Bantu languages into previously Khoisan-speaking areas in southern Africa to the spread of Arabic throughout the North.[52]

Where Biodiversity and Linguistic Diversity Overlap

The study of biology and the study of language have sought theoretical and methodological parallels since the nineteenth century when analogies between languages and species were first proposed and later rejected.[53] However, particularly in the last couple of decades, potential parallels have been revisited which focus on diversity. In her 2005 book, *Linguistic, Cultural and Biological Diversity,* Luisa Maffi points out that if the 1980's might be characterized as the decade of biodiversity, the 1990's were the decade which produced the concept "biocultural" diversity.[54] Sometimes called "biolinguistic" diversity, the emerging field has continued to grow, investigating:

> the parallels and correlations between biodiversity and linguistic diversity, the overlaps in the global distribution of languages and biodiversity, and the relationships between language, traditional knowledge, and the environment.[55]

Maffi credits recent interest in the potential link between language and the environment in part to the work of "ethnobiologists and ethnoecologists studying indigenous knowledge and use of local flora, fauna, and ecosystems."[56] In linguistics, various notions of "linguistic ecologies" have arisen, ranging from that espoused in Peter Mülhäusler's 1995 work, *The Interdependence of Lin-*

guistic and Biological Diversity to the version of "genetic linguistics" found in Salikoko Mufwene's 2001 work, *The Ecology of Language Evolution.*[57]

Although attempts to causally link biodiversity and linguistic diversity fail in the face of the ways in which language and species are not comparable on a number of levels,[58] it is probably more than a coincidence that many areas of prolific biodiversity in Africa also support a high incidence of linguistic diversity. Joslin Moore and fellow scientists quantitatively tested the hypothesis that cultural and biological diversity have similar geographical distributions in Africa.[59] Their study compares the distribution data of 3882 species from four vertebrate groups: mammals, birds, snakes, and amphibians and the contemporary distribution of 1686 languages in Sub-Saharan Africa.[60] Moore *et al* found a correlation between richness of languages and vertebrates strongest within tropical forests, consistent with Daniel Netttle's hypothesis that language diversity is aided where transportation and travel are difficult. Moreover, their results show that:

> Both species and language richness were most strongly correlated with the climate 1 variable, which correlates with wetness and observed growth. This fits with the idea that species richness is related to the total amount of resources available and the ideas of Nettle (1996) that more languages will develop in areas of low ecological risk.[61]

The findings of Moore *et al* support other international studies on biodiversity and linguistic diversity which highlight especially West Africa as an area with a very high correlation between the two, such as Steppe et al 2004 which focuses on plant life.[62] Africa is a continent where some 2000 languages co-exist, and the numbers are particularly high in the sub-Saharan region.[63]

"Ecolinguistics" and Indigenous Knowledge

Researchers in the field known as "Ecolinguistics"[64] often emphasize the fact that many of the world's most endangered languages are vast repositories of traditional knowledge. Such ecolinguists, therefore, promote the importance of endangered languages to world culture and world knowledge of the environment. In linguist Jon Orman's words: "According to the ecolinguistic standpoint the loss of these languages would represent an irretrievable loss of knowledge that would weaken the adaptational strength of our species." [65]

However, as seen in the case of the San people, the preservation of the traditional knowledge of people speaking endangered languages does not necessarily benefit them as much as Western scientists ready to exploit it. As Orman's discussion points out the ecolinguistic argument above can be seen as highly

Western-centric, although that is not necessarily its intention, to the extent that "Knowledge from indigenous languages has been imported chiefly to serve the needs of Western societies in the form of pharmaceuticals, etc."[66]

More troubling, if that were possible, is the extent to which ecolinguistic views have followed a strong Whorfian belief regarding the relationship between language and worldview. Benjamin Lee Whorf was an anthropologist studying Native American languages in the United States in the early twentieth century who popularized the concepts "linguistic relativity" and "linguistic determinism" which hypothesize that languages vary substantially and unpredictably and that the structural organization, vocabulary range, etc. of one's language determines one's understanding of reality.[67] Ecolinguists such as Peter Mühlhäusler have not escaped criticism for such a position. For example, Sue Wright criticizes "a central claim" of the ecolinguistics school which argues that:

> [C]ertain languages are perfectly adapted to communicating certain values and attitudes. For example, the languages of pre-industrialized societies are held to promote respect for the natural world. . . . Mühlhäusler [2001] argues that the languages of the big polluting states allow them to systematically underestimate and play down the consequences of dangerous practices.[68]

In a strong Whorfian view, translation is ultimately impossible, and, therefore, the worldviews of speakers of differing languages cannot be communicated. For example, Tove Skutnabb-Kangas appears to claim that the concept/language link is so strong that a population that shifts to another language will inevitably lose the richness of its own traditional knowledge:

> Traditional knowledge may indeed linger even after a native language is lost, but the richness and diversity of that knowledge cannot survive even one generation of language loss.[69]

Skutnabb-Kangas quotes a statement by Mühlhäusler that "It is language that allows people to become efficient users of the environment."[70] In fact, this deterministic view of language predates Whorf, and has been used against indigenous languages just as often as in their support.

Can African Languages Bridge the Science, Technology Divide?

If modern ecolinguists question whether speakers of "Standard Average European" languages, as Whorf referred to them, can comprehend the traditional knowledge and understanding of the environment preserved in indigenous languages, they are not alone in questioning whether some languages are incapable of expressing valuable concepts. Although many such linguists' focus is on the validation of indigenous languages (and pidgins and creoles), their arguments are, unfortunately, reminiscent of the linguistic prejudice which favored European languages, a view which they oppose.

No better example of this prejudice in the West is the infamous 1835 "Minute" of the British administrator and historian Thomas Macaulay, which addressed the question of the adoption of English instead of Sanskrit or Arabic in Indian education. Macaulay argued whether:

> when it is in our power to teach [English], we shall teach languages in which, by universal confession, there are no books on any subject which deserve to be compared to our own . . . whether . . . we shall countenance . . . medical doctrines which would disgrace an English farrier, --Astronomy, which would move laughter in girls at an English boarding school.[71]

Macaulay attacked Indian vernaculars, questioning, like the ecolinguists discussed above, the possibility of disseminating knowledge through translation:

> All parties seem to be agreed on one point, that the dialects commonly spoken among the natives of this part of India, contain neither literary or scientific information, and are, moreover, so poor and rude that, until they are enriched from some other quarter, it will not be easy to translate any valuable work into them.[72]

That it is now Macauley's Minute which moves us to laughter when we first encounter it masks the pernicious way in which his doubts about the capacity of Indian vernaculars, made over 150 years ago, are reflected in current statements about African vernaculars. Speaking as a linguist, a disturbing aspect of this research has been to find doubts raised, even by Africans, regarding the capability of especially minority African languages, of which the Khoisan languages are an example, to discuss and "do" science and adapt to twenty-first century technology. Particularly egregious are suggestions that any of the languages of Africa are too lexically or structurally deficient for this purpose. For example, J. A. I. Bewaji, writing in the *Journal of African Philosophy* in 2002, criticizes fellow Nigerian, Moses Akin Makinde and his book, *African Philosophy, Culture and Traditional Medicine* because "[Makinde] is of the opinion that logical grammar is underdeveloped in the indigenous languages of Africa."[73] Makinde, a respected philosopher of science, is certainly wrong about language, as are others who have made this suggestion. Makinde, whose goal is to support African philosophy, culture and traditional medicine, nevertheless states:

> Science has nonetheless advanced through well developed languages. Hence, the poverty of African languages has led to the poverty of scientific ideas and meaningful contributions to the development of philosophy, science, and technology.[74]

From Mandarin Chinese to Inuit Eskimo, which are extreme opposites on a continuum of phonological, grammatical and lexical types, and moreover "big" vs. "small" languages with regard to number of speakers, there is no such thing as a language which is "undeveloped" with regard to the logic of its structure. The structure of human languages is dictated by the physical structure and functioning of the human brain, which learns, processes and produces these languages. Despite the attraction this idea apparently has for especially speakers of majority languages, there is not one shred of evidence that there exist human languages which do not fully exploit the capabilities of the brain or are in any way less capable of complex thought.[75]

Even worse than doubts regarding their suitability for science, is the suggestion that some languages are not even useful for conveying knowledge. This is especially true in the context of African higher education where many of the subjects taught have historically followed western traditions, exemplified by the education of Christopher Okigbo, discussed earlier. Furthermore, as will be examined below, in such institutions the issue of what constitutes knowledge has most often been decided in favor of "modern," i.e. western, science, technology, etc., taught in the European lingua francas established in educational systems across Africa.

In the introductory chapter of *African Higher Education: An International Reference Book,* published in 2003, the Ethiopian scholar Damtew Teferra and his co-editor Philip G. Altbach list among other issues confronting "the development of vernacular languages into an instructional medium of higher education" the purported "poor vocabularies and grammatical conventions of indigenous languages that make it difficult to convey ideas and concepts."[76] Even discussions of language choice at all levels of education seem to raise doubts regarding the suitability of vernaculars. Questioning the issue of inter-translatability, Mohammed Abdulaziz asserts: "But it is unlikely that there is an African language that is as developed, say, as German or English, especially in modern areas of technology," although he states that Amharic, Somali and Swahili "are being developed very quickly."[77] Well-meaning but patronizing phrases such as the "Intellectualisation of African Languages" called for by Neville Alexander in 2007[78] or even "Modernizing the Languages of Africa" called for by Paulin Djité in 2008[79] compound this problem.

It is true that vast differences exist in vocabulary across languages, especially with regard to domains such as those of science and technology. Here Makinde, for example, is on solid ground when he states:

> One of the greatest problems of African thinkers today is to find the words in their respective languages or dialects that would catch precisely the meaning and reference of foreign words such as the scientific terminologies: atoms, electrons, molecules . . ." [80]

Historical linguistics teaches us that all languages throughout time have had to add to and adjust existing vocabulary to reflect changes such as technological breakthroughs. It is completely legitimate to explore vocabulary expansion and translation without being guilty of underestimating either non-western traditional scientific achievements and technological breakthroughs or the intellectual capabilities of those who use indigenous languages to pass on traditional knowledge. In fact, the African Academy of Languages (ACLAN), which introduced the "Year of African Languages" from 2006-2007, is sponsoring various lexicography projects whose goal is to "develop terminologies [in African languages] especially in science and technology."[81]

However, I have an additional suggestion, which I haven't seen anywhere, which is why not also do what English has mostly done, and just borrow the words which express desirable foreign concepts? English has borrowed literally tens of thousands of words from other languages, across the globe, and if we were able to definitively test this, I am convinced it would turn out to have the largest borrowed vocabulary of any language in the world, thanks especially to the British Empire.

And notably, among the domains in English which have most benefited from borrowing are science and technology. In fact, the words *science* and *environment* are not native to English but come from French, while the word *technology* comes from Greek, and all the words in Makinde's list of English scientific terms have been borrowed from foreign languages.[82] While especially European language purists decry the introduction of "Anglicisms," into their languages, the history of the English language itself teaches the usefulness of adopting ready-made words from somebody else. Foreign words like those mentioned above have been successfully "nativized," i.e. adapted to the pronunciation patterns and grammatical forms of the English language. In fact, the usefulness of the vast range of foreign words found in English is such that no one could imagine the language without them.

Languages of Instruction in Higher Education

As Djité's observation quoted above suggests, language inequality is a major stumbling block for achieving African development goals. One of the most important sources of linguistic inequality throughout Africa is formal education where European ex-colonial languages, African majority languages and languages privileged under colonial regimes dominate. As alluded to above, this is especially true in colleges and universities, whose choices regarding language of instruction, once established by colonial policy, have continued to favor European languages.

Even in countries with a population substantially in support of the use of vernaculars in education, practical impediments to language policy can thwart the establishment of African languages at the highest levels. Many Tanzanians are reported to have become disillusioned when Julius Nyerere's "Education for Self Reliance" policy did not lead to complete "Swahilization" of Tanzanian education, but left English in place at the tertiary level. Nyerere, who was a strong advocate of Kiswhahili, came up against harsh economic realities, which caused him to turn to the West for help and decreased the likelihood that Kiswhaili would be established as the medium of education at the college and university level in favor of the international language, English. Samuel Gyasi Obeng states: "Nyere had obviously misled the people by giving them a false hope about Kiswahili" whether this was what he intended.[83]

Despite such early efforts to replace Eurpean languages, Damtew Teferra, asserts that "Unfortunately, the impact of the prominence of Western languages such as French and English, has received little attention in the research on improving the quality of African higher education."[84] According to Teferra, other languages most used in higher education in Africa include Afrikaans, Arabic, Italian, Portuguese and Spanish, of which Arabic and, arguably, Afrikaans are the only African languages in this list.[85]

University of Ghana-Legon
Source: Ann Albuyeh

However, English has been making inroads into the domains of all of these languages, even Arabic. Ten years ago the Cairo weekly Al Ahram was reporting a perceived decline in students' proficiency in Arabic in the face of an "ever-increasing demand for English, ...the only way to get ahead in prestigious fields like medicine."[86] Nonetheless, African criticism of this state of affairs is on the rise. In fact, in 2006, the British linguist David Graddol asserted that "It is in Africa that the most heated debates about the place of English-medium education are now arising."[87]

Kenyan scholars, Ali A. and Alamin M. Mazrui in their 1998 book *The Power of Babel: Language and Governance in the African Experience,* called to "global Africans" to "continue with the struggle to 'centre' their own indigenous tongues in their lives." Mazrui and Mazrui warned:

> [A]n important source of intellectual dependence in Africa is the language in which African graduates and scholars are taught.... [T] oday, in non-Arabic speaking Africa, a modern surgeon who does not speak a European language is virtually a sociolinguistic impossibility. So is a modern physicist, zoologist . . . and so forth. . . . .

The call for the use of indigenous languages, in particular for the teaching of subjects involving science and technology, has dramatically increased since Julius Nyerere first raised the hopes of Tanzanians, post-independence. Such proposals now come from a wide range of sources, for example, from Nigerian traditional leaders, such as Oba Okunade Sijuade, the Ooni of Ife, southwest Nigeria in 2007, to the editors of The Herald newspaper in Harare, Zimbabwe, in 2008.

Linguists claim Nigeria alone has approximately 500 languages. In light of the challenges which such a large number of national and regional languages pose, currently only "big" languages, in terms of numbers of speakers, are making inroads into the dominance of ex-colonial languages. Not coincidentally languages such as Yoruba or Hausa in the West and Kiswahili in the East also function as a lingua franca their region.

It is these major languages which are able both to move into a potentially significant role in higher education and attract initiatives to incorporate their use in fields of science and technology. For example, Túndé Adégbölá, a professor at the University of Ibadan states that standard Yoruba "is used as a language of instruction in higher education and doctoral theses have been written in it." In addition, Adégbölá, who is also the executive director of the African Languages Technology Initiative, reported in 2007 on the development of a computer keyboard layout for Yoruba.

Languages in Cyberspace

In fact, technology and the Internet may be the back door through which indigenous languages traditionally left "on the edge of the road" can take advantage of the "modern" scientific advances and technological breakthroughs. While English has been touted as the dominant language in cyber technology, and everything down to the design of computer keyboards tends to reinforce this, the development of computer and Internet capability for indigenous languages running the gamut from varieties of Chinese to Eskimo is causing rapid changes in technology use and access.

The statistics compiled by Internet World Stats as of December 2008 not only show that Mandarin Chinese is catching up to English, but also indicate significant growth in Internet use on the African continent, as well. The African language which makes it into the top ten is Arabic.

Table 10.1: Top Ten Languages on the Internet

Language	Millions of users
English	464
Chinese	321
Spanish	131
Japanese	94
French	74
Portuguese	73
German	65
Arabic	41
Russian	38
Korean	37
All the rest	258

Source: Internet World Stats: www.internetworldstats.com

Nine African countries boasted Internet users in the millions at the end of 2008, with Egypt and Nigeria at the head of the list. It is not a linguistic coincidence that all the top-ten Internet-using African countries either have a significant Arabic-speaking population or, thanks to the British Empire, an English-speaking elite, or both. In addition, Internet users living in countries showing a relatively high "penetration" level, i.e. a high percent of users per population, such as Tunisia, Morocco, Egypt and Algeria, may be accessing the Internet in French.

Table 10.2. Africa's Top 10 Internet Countries

Country	Millions of Users	Penetration (% Population)
Egypt	10.5	12.9%
Nigeria	10.0	6.8%
Morocco	6.6	19.2%
South Africa	4.6	9.4%
Algeria	3.5	10.4%
Sudan	3.5	8.7%
Kenya	3.0	7.9%
Tunisia	2.8	27.0%
Zimbabwe	1.4	11.9%
Ghana	0.9	3.8%

Source: Internet World Stats Dec. 2008. www.internetworldstats.com

Given continuing infrastructure challenges as basic as reliable electricity, not to mention literacy issues among various populations, discussed below, the growth in Internet use over the past few years is remarkable. As Table 3 illustrates, both Egypt and Nigeria doubled Internet use within only two years.

Table 10.3. Egypt and Nigeria: Internet Statistics

Country	Year	Internet Users
Egypt	2006	5,100,000
	2008	10,532,400
Nigeria	2006	5,000,000
	2008	10,000,000

Source: Internet World Stats Dec. 2008. www.internetworldstats.com

Moreover, a number of initiatives designed to bring speakers of other languages into twenty-first-century communications hope to see these numbers swell even further.

Table 10.4. Sample of African Language – Technology Initiatives

Name	Description	Website
AfLaT	African Language Technology	http://www.aflat.org/
Alt-i	African Languages Technology Initiative	http://www.alt-i.org
ANLoc	The African network for Localization	http://www.africanlocalisation.net/

Bisharat	A language, technology, and development initiative	http://www.bisharat.net/introen.htm
PanAfriL10n	PanAfrican Localization Wiki	http://www.panafril10n.org/

Knowledge Production in Africa

As the adaptation of the expertise of the San people by research groups and pharmaceutical companies, discussed above, amply illustrates, even so-called modern scientists in practice appreciate what has usually been termed "traditional" or "indigenous" knowledge. Notwithstanding their undoubted superiority complex, for the scientists involved, profit would seem to have most strongly motivated their denial of the recognition of the importance of the knowledge of the San community. In other words, modern scientists, especially when it comes to the environment and biodiversity, have shown themselves only too happy to learn the knowledge of indigenous communities, as long as they do not really have to admit these people as equals. Thus, whereas Thomas Macauley surveying nineteenth-century India really thought that the people's traditional knowledge was useless, in the twenty-first century, the "scientific West" appears to want to cynically appropriate useful knowledge, while retaining "deniability." Happily, from the successful San court case conducted in post-apartheid South Africa to recognition of traditional knowledge in the United Nations sponsored *Africa Environmental Outlook 2* report, discussed above, there is hope for a change in this state of affairs. Moreover, the literally hundreds of websites in English such as those which currently tout the benefits of hoodia attest to the willingness of the world's technologically-savvy elite to wish to learn from the knowledge produced by communities like the San people.

A 2009 special volume of *Africa: The Journal of the International Institute*, called *Knowledge in Practice: Expertise and the Transmission of Knowledge* offers an important update on the evolving state and status of traditional knowledge on the continent. The editors, anthropologists, Kai Kresse and Trevor Marchand, adopt the term "endogenous knowledge," first coined by P. J. Houndtondjii in 1997 to refer to knowledge which is "an internal product drawn from a given cultural background." However, instead of conceiving of endogenous knowledge in opposition to e.g. western science, the recent scholarship explores "the relations between Africa's longstanding traditions of science and literacy with its ever-present traditions of orality and myth."

In multilingual Africa where a language may be spoken by millions or as few as a thousand, literacy is especially an issue because languages with few speakers are unlikely to invent or adopt a writing system. Speakers of these languages who are literate typically have learned to read in a "bigger" language. Thus, there are serious practical obstacles to literacy which are the inevitable result of the continued use of, especially the "smaller," minority languages. It seems safe to predict that the complementary roles of orality and literacy will be as important in Africa's future knowledge production and dissemination as they are today.

Furthermore, Kresse and Marchand are right in supporting the renaming of "traditional knowledge" as tradition may be conceived of as static and conserving of old, outmoded thinking. The West has exhibited a felicitous amnesia regarding scientific and technological innovations from Assyrians to Arabs and a predisposition to not give credit where credit is due. That this appropriation has caused global harm and imposed a false limitation on human development is especially obvious with regard to African biodiversity and environmental issues, where indigenous expertise is most conspicuous and minority languages are most prevalent.

Conclusion

The language issues which have preoccupied African authors such as Ngugi wa Thiongo over the last few decades are now coming into focus where considerations of language intersect with science, technology, and most recently, the environment. While cultural considerations like language choice with respect to African fiction should not receive less attention than they do, it is not an exaggeration to view decisions regarding language use in the context of science, technology and the environment as capable of nothing less than impacting the very future of the planet and its inhabitants.

The continent of Africa itself is both the best illustration of this fact and the place where the linguistic compromises required appear most likely to be successfully addressed. As the overlap between biodiversity and linguistic diversity on the continent show, accessing and developing knowledge "stored" in minority and even endangered languages is an important challenge which, as in the case of the San people, Africans are in a unique position to meet. Moreover, on a continent whose colonial legacy has been the dominance of foreign languages in areas of higher education and academic research, the linguistic means to access global knowledge on the part of African elites already exists. In addition, the power of the major indigenous lingua francas, such as Kiswahili and Yoruba, support their increasing role alongside international lingua francas from Arabic to English. Furthermore, present and future technological advances have the potential to "leapfrog" the sharing of information in unforeseen ways, as the increase in Internet use, and in an increasing number of languages, illustrates. Knowledge once stored on clay tablets by kings has evolved into a mind-boggling, unlimited cyberspace whose potential for bringing people "on the edge of the road" onto the so-called "information highway" we are only now beginning to appreciate.

Despite the dim view of the possibility of translating knowledge from one language to a totally different other, illustrated from Macauley to Mühlhäusler, the future of knowledge transmission worldwide rests on the successful accomplishment of precisely this. If the Nigerian poet Christopher Okigbo could call on oral and written traditions, Eastern and Western beliefs, ancient and modern convictions, and the wisdom handed down by his own ancestors to construct his art, neither science nor technology nor our understanding of our environment need be limited by the accident of one's mother tongue.

At the heart of the social dimensions of knowledge lies language. The "linguistic cooperation" of translation from one language to another, not to mention the sharing of useful terms, as in the suggestion above with regard to borrowing scientific vocabulary, already is a normal and accepted fact of life throughout multilingual Africa, as any stroll through a crossroads marketplace makes manifest. However, when the Europeans appropriated both *knowledge* and *progress* as their own during the modern period, they also, as illustrated in Macaulay's "Minute," promoted their languages as privileged purveyors of both. It is time that models of successful linguistic cooperation, that also include languages such as English, support the integration of scientific, technological and environmental development in ways that will benefit both the new "global" world we live in and local populations from North to South, East to West on the African Continent. The contributions of great and small are critical. To quote a Hausa proverb:

dà yayyafī kṑgī yakàn cika. With (small) showers a river is filled.

Notes

1. Christopher Okigbo, *Collected Poems* (London: Heinemann, 1986), 66.
2. Nigeria alone boasts a wide range of Anglophone literature expressing social, political, and environmental concerns, encompassing not only the poetry of Christopher Okigbo, quoted above, but the literary creations of internationally acclaimed authors such as Chinua Achebe, Ben Okri, and Wole Soyinka.
3. Ngugi wa Thiong'o famously bade "farewell to English" in 1986. Ngugi wa Thiong'o, *Decolonizing the Mind: The Politics of Language in African Literature* (London: James Curry, 1986).
4. Robert Serumago, "Interview with Christopher Okigbo" (July, 1965) in Donatus Ibe Nwoga, ed., *Critical Perspectives on Christopher Okigbo* (Washington, DC: Three Continents Press, 1984), 249.
5. Christopher Okigbo, *Collected Poems*, xxv—xxvi. Okigbo, raised a Catholic, professed to being able to integrate the beliefs of his ancestors with his imported faith in Collected Poems, he states:: "I do not feel that as a Christian I have ever been uprooted from my own village gods. We have a goddess and god in our family, our ancestral gods." Christopher Okigbo, *Collected Poems*, xx.
6. Ken Saro-Wiwa, *Genocide in Nigeria: The Ogoni Tragedy* (Port Harcourt, Nigeria: Saros International Publishers, 1992, resissue 2005), 12.
7. Royal Dutch Shell is part of the global Shell Oil Company, with over 100,000 employees working in more than 100 countries. The corporation was listed as number one , i.e. the world's largest, by the Fortune 500 in 2009. Its revenues of $458bn rival that of the nation of Saudi Arabia. http://www.shell.com/home/content/aboutshell/at_a_glance/. BBC News, "Shell Beats Wal-Mart as Top Firm," July 9, 2009. http://news.bbc.co.uk/go/pr/fr/-/2/hi/business/8142515.stm.
8. The government of Sani Abacha sentenced him to death by hanging with other Ogoni activists in 1995. Craig W. McLuckie and Aubrey McPhail, eds., *Ken Saro-Wiwa: Writer and Political Activist* (Boulder, CO: Lynne Rienner Publishers, 2000).
9. Ann Albuyeh, "Wars of Words: Enlisting Colonial Languages in the Fight for Independence in Africa," in *The Struggle for Representation: Narrating Wars and Peace in Africa,* ed. Toyin Falola and Hetty ter Haar (Rochester, NY: University of Rochester Press, forthcoming).
10. Cécille B. Vigouroux and Salikoko S. Mufwene, eds., *Globalization and Language Vitality: Perspectives from Africa* (London: Continuum Press, 2008, 1.
11. The Assyrian Empire came into existence some three thousand years ago and at itspeak included the peoples of modern-day Iraq, Syria, Palestine, Israel, and Armenia, Ibid.; Hooker, Richard. *World Civilizations.http*://wsu.edu/~dee/MESO/ASSYRIA.HTM.

    1996 (updated 1999); Elizabeth Wyse, et al., eds., *Past Worlds: The Times Atlas ofArcheology* (Maplewood, NJ: Hammond, 1988).

12. Nicholas Ostler, *Empires of the Word: A Language History of the World.* (New York: Harper Collins Publishers, 2005).
13. Ostler, *Empires of the Word*, 34.
14. Ibid, 35.
15. Modern Semitic and Afroasiatic linguistic descendents include languages from Morocco to Nigeria to Somalia The branches of this wide-ranging family include; Berber (e.g. Tuareg), Chadic (e.g. Hausa), Egyptian (surviving in the liturgical language of the Coptic church), Semitic (e.g. Egyptian Arabic). Cushitic (e.g., Somali), and Omotic (e.g., Wolaytta, spoken in Ethiopia). Ibid, 35—7; Bernard Comrie, ed., *The Word's Major Languages* (Oxford, Oxford University Press, 1987); Raymond G Gordon, Jr.., ed. *Ethnologue Languages of the World, Fifteenth Edition* (Dallas: SIL International, 2005). http://www.ethnologue.com.
16. Abdelhamid Sabra, "The Exact Sciences," in *The Genius of Arab Civilization: Source of Renaissance,* ed. John R. Hayes (New York: New York University Press, 1975), 121.
17. Ibid.
18. Scott Montgomery, "Of Towers, Walls, and Fields: Perspectives on Language in Science," *Science* Vol. 303, no. 5662 (February 2004): 1333.
19. Dennis Overbye, "How Islam, Won, and Lost, the Lead in Science," *New York Times,* October 30, 2001.
20. As is clear from the discussion above, Arabic, and even Akkadian, can be considered earlier global languages of science and technology. However, in the modern era, English is the first such international lingua franca, and, moreover, is more truly "global" in the sense of being used by scientists and others in more countries, farther from its core area, exemplified by the use of English in Japan.
21. Ulrich Ammon, ed., *The Dominance of English as a Language of Science: Effects on Other Languages and Language Communities* (Berlin: Mouton de Gruyter, 2001).
22. Peter Doherty, *The Beginner's Guide to Winning a Nobel Prize* (New York: Columbia University Press, 2006): 240—241.
23. David Graddol and Ulrike H. Meinof, eds., *English in a Changing World AILA Review 13,* 1999) 57-68, cited in Montgomery "Of Towers, Walls, and Fields: Perspectives on Language in Science," 1334.
24. "Nativized" here refers to the linguistic term which commonly describes a variety of an international language which encompasses phonological, syntactic, morphological, lexical, and sometimes pragmatic, adjustments which have developed in communities of speakers who have other mother tongues. For example, nativized varieties of English have typically resulted from the language contact situations arising because of the spread of the British Empire. These varieties include socially well-respected local standards, such as West African Standard English.
25. Ammon, *The Dominance of English as a Language of Science.*
26. Montgomery, "Of Towers, Walls, and Fields: Perspectives on Language in Science," 1333.

27. Ulrich Ammon, ed., *The Dominance of English as a Language of Science: Effects on Other Languages and Language Communities* (Berlin: Mouton de Gruyter, 2001).
28. Augusto Carli and Ulrich Ammon, eds., *Linguistic Inequality in Scientific Communication Today: What Can Future Applied Linguistics Do to Mitigate Disadvantages for Non-Anglophones?, AILA Review,* Vol. 20 (Amsterdam: John Benjamins, 2007).
29. Paulin G. Djité, "From Liturgy to Technology: Modernizing the Languages of Africa," *Language Problems and Language Planning* 32:2 (2008): 141.
30. Paulin G. Djité, "The Nexus between Education, Learning, and Language" (paper presented at the UNESCO/UNU 2008 Conference on "Globalization and Languages: Building on our Rich Heritage" Tokyo, Japan, August 27-28, 2008, 1. www.unu.edu/globalization/2008/files/UNU_UNESCO-Djite.pdf.
31. UNEP, *African Environment Outlook 2* (Nairobi, Kenya: United Nations Environment Programme, 2006).
32. Ibid, 232.
33. L. J. Champan, "Fishes of African Rainforests: Diverse Adaptations to Environmental Challenges," in *African Rainforest Ecology and Conservation*, eds. H. S. Moreland, et al (New Haven, CN: Yale University Press, 2001), 263—90, reported in UNEP, *African Environment Outlook 2*, 240.
34. FAOSTAT, FAOSTAT—*FAO Statistical Data Bases,* Food and Agricultural Organization of the United Nations, 2005, http:faostat.fao.org, reported in UNEP, *African Environment Outlook 2,* 232.
35. The baobab tree *Adansonia digitata* is native to Africa. (A related species, also called "baobab" or "bottle tree," *A. gregrii,* is found in Australia (*Encyclopedia Britannica Online: Academic Edition).*
36. G. Kranjac-Berisavljevic, et al "Securing Food in the Hungry Season: The Role of the Baobab Tree," in *International Symposium on Underutilized Plants for Food Security, Nutrition, Income and Sustainable Development, ISHS Acta Horticulturae 806,* eds., H. Jaenicke, et al., Arusha, Tanzania: ISHS, 2009. http://www.actahort.org/books/806/; *PhytoTrade Africa.* http: //www.phytotradeafrica.com/products/baobab.htm.
37. *PhytoTrade Africa.* http: //www.phytotradeafrica.com/products/baobab.htm.
38. *Hoodia gordonii*, *African Environment Outlook 2,* 40.
39. Ibid., 40.
40. Ibid., 40. However, due to difficulties adapting the hoodia extract to the liquid form that Unilever's weight loss product "Slimfast" is usually produced in, the company dropped this line of research. Ben Deighton, "UPDATE 2: Unilever Abandons Phytopharm Weight-loss Product," Reuters (Nov. 14, 2008). http://reuters.com/article/rbssPharmaceuticals%20-%20Generic%20&%20Specialty/. Despite (or perhaps because of) controversy over whether hoodia products can help people lose weight, a huge number of websites, such as http://www.hoodia-weightloss-products.com can be found on the Internet. A recent Google search for "hoodia weightloss products" yielded 640,000 hits.

41. Anthony Barnett, "In Africa the Hoodia Cactus Keeps Men Alive. Now Its Secret is 'Stolen' to Make Us Thin," *The Observer (London)*, June 17, 2001.
42. The membership in the Khoisan family is problematic, something recognized for decades. Khoisan has been used as a cover term "to refer to those African languages which use clicks as regular speech sounds and are not obviously members of one of the other families." Bonny Sands, "Comparison and Classification of Khoisan Languages," in *Language History and Linguistic Description in Africa,* eds. Ian Maddieson and Thomas J. Hinnebusch (Trenton, NJ: Africa World Press, 1998), 75. Although many of the click languages in Southern Africa appear to be indeed related, there are languages usually attached to this family which are far off in Tanzania which are more doubtful. G. Tucker Childs, *An Introduction to African Languages* (Amsterdam: John Benjamins, 2003); Bernd Heine and Derek Nurse, eds. *A Linguistic Geography of Africa* (Cambridge, UK: Cambridge University Press, 2008).
43. A young South African working on a cruise ship out of San Juan told me she and the other white South Africans onboard regularly clicked at each other, ridiculing these languages, after I had naively said something positive about them. (The fact that the Parisian spa attendant dismissed Canadian French as "sounding horrible" during the same cruise shows that language prejudice is unfortunately alive and well everywhere.)
44. Pronounced with a puff of air, as in English where the /t/ at the beginning of the word *ton* has a puff of air, while the /t/ in *stun* does not.
45. The linguist G. Tucker Childs reports that the click language !Xũ has "the world record" for the largest sound inventory with 141 segments, compared to Zulu and English which have approximately 40, an inventory characterized as "large, but not unusual." An impressive total of 48 clicks and 47 "regular" consonants are used to produce words in !Xũ.
46. Ibid., 60.
47. Obviously other differences such as grammatical structure and vocabulary, referred to below, are important as to whether a language is considered difficult, but this brief discussion of clicks (probably the most salient aspect of these languages to others) suffices to make the point regarding problems of inter-ethnic communication.
48. A. Traill, "The Khoesan Languages," in *Languages in South Africa,* ed. Rajend Mesthrie (Cambridge UK: Cambridge University Press, 2002), 36.
49. A. Traill, "The Khoesan Languages."
50. Languages like Latin are also called "dead," but they may be characterized (at least in one or more varieties) as having evolved into offspring languages.
51. The /Xam language died out about 170 years after the first conflicts between the /Xam people and the frontier farmers of the Cape Colony in the mid-eighteenth century. N. Penn, is quoted by Traill as reporting that "their societies [were] shattered by warfare, starvation and disease; the women and children enslaved; the men all but exterminated by the genocidal hatred of their enemies." N. Penn "the /Xam and the Colony," (paper presented to the Bleek and Lloyd 1870—1991

Conference, University of Cape Town, 1991, as reported in A. Traill, "The Khoesan Languages," 37.

52. Salikoko S. Mufwene, *The Ecology of Language Evolution* (Cambridge: Cambridge University Press, 2001), 186.

53. Charles Darwin appears to have been the first, in his *On the Origin of the Species: By Means of Natural Selection:* "It may be worthwhile to illustrate this view of classification by taking the case of languages" appealing to paradigms found in contemporary comparative Indo-European philology. Charles Darwin, *On the Origin of Species,* ed. Gillian Beer (Oxford: Oxford University Press, [1859] 1996), 342. However, then as now, "The new intellectual alliance between biology and linguistics was, understandably, welcomed more warmly by linguists than by biologists." Roy Harris and Talbot J. Taylor, *Landmarks in Linguistic Thought I: The Western Tradition from Socrates to Saussure,"* 2nd ed. (London: Routledge, 1997), 188. Linguistic theories which e.g. looked at language change as a kind of Natural Selection, such as those of F. Max Müller and August Schleicher, were discredited.

54. Luisa Maffi *Linguistic, Cultural and Biological Diversity, Annual Review of Anthropology 29,* (2005): 600.

55. Ibid., 600.

56. Ibid., 601.

57. Peter Mühlhäusler, "The Interdependence of Linguistic and Biological Diversity," in *The Politics of Multiculturalism in the Asia/Pacific,* ed. David Myers (Darwin, Australia: Northrern Territory University Press, 1995) 154—61; Salikoko Muflwene, *The Ecology of Language Evolution.*

58. Some of the linguists discussed above would not seem to support this statement.

59. Joslin Moore et al., "The Distribution of Cultural and Biological Diversity in Africa," *Proceedings of the Royal Society of London 269*(2002), 1645.

60. Ibid., 1647. The species were those compiled by the Zoological Museum of Copenhagen and the languages from Ethnologue 1996, www.ethnologue.com. Ex-colonial languages such as English, French and Portuguese and widespread lingua franca such as Arabic, Hausa, and Swahili were excluded.

61. Ibid., 1651. Daniel Nettle, "Language Diversity in West Africa: An Ecological Approach," *Journal of Anthropological Archeology* 15, (1996): 403—438.

62. Other areas were: MesoAmerica, the Andes, the Himalyas, and South Asia/Pacific. Luisa Maffi *Linguistic, Cultural and Biological Diversity,* 609.

63. See, for example, Raymond G Gordon, Jr.., ed. *Ethnologue Languages of the World, Fifteenth Edition* (Dallas: SIL International, 2005). http://www.ethnologue.com. Linguists generally use mutual-intelligibility as the rule of thumb for determining whether speech systems are separate languages or only dialects of the same language. However language exists in a continuum. For example, between Paris and Madrid, there is such variety among spoken dialects that the political border between France and Spain can seem like an arbitrary determiner between what is called "French" and what is called "Spanish" in a way that is not at all apparent when one looks at the two standard languages. Obviously, in Africa, the same

practical problem exists so linguists understand the counting of number of languages to be an inexact endeavor.

64. The term *Ecolinguistics* was introduced in 1972 by Einar Haugen in *The Ecology of Language* (Stanford, CA: Stanford Univeristy Press) to refer to the arrangements societies would need to make if languages are to be preserved.
65. Jon Orman, *Language Policy and Nation Building in Post-Apartheid South Africa* (New York: Springer, 2008), 64.
66. Ibid., 64.
67. Benjamin Lee Whorf, *Language Thought and Reality: Selected Writings of Benjamin Lee Whorf,* ed. J. B. Carroll (Cambridge, MA: MIT Press, 1956) (Original works written 1927-1941). See also the earlier work of Franz Boas and Edward Sapir on their views as to the relationship between language and culture. The "Sapir-Whorf" hypothesis is a common name for early versions of the theory.
68. Sue Wright, *Language Policy and Language Planning,* (Basingstoke, UK: Palgrave Macmillan, 2004) 220.
69. Tove Skutnabb-Kangas, *Linguistic Genocide in Education—Or World Diversity and Human Rights?* (Mahwah, NJ: Erlbaum, 2000), 259.
70. Ibid., 259.
71. Barbara Harlow and Mia Carter, eds., *Archives of Empire, Volume I. From the East India Company to the Suez Canal* (Durham, NC: Duke University Press, 2004), 231.
72. Ibid., 230.
73. J. A. I. Bewaji, "Language, Culture, Science, Technology and Philosophy *Journal on African Philosophy* 1.1 (2002): 8.
74. M. Akin Makinde, *African Philosophy, Culture and Traditional Medicine* (Athens, Ohio: Ohio Univeristy Center for International Studies, 1988): 19.
75. See, e.g., Ewa Dąbrowska *Language, Mind and Brain: Some Psychological and Neurological Constraints on Theories of Grammar* (Washington, DC: Georgetown University Press, 2004) for a discussion of how the brain processes and orders sounds and words into meaning.
76. Damtew Teferra and Philip G. Altbach, *African Higher Education: An International Reference Book* (Bloomington: Indiana University Press): 13.
77. Mohammed H. Abdulaziz, "The History of Language Policy in Africa with Reference to Language Choice in Education," in *Towards a Multilingual Culture of Education,* ed. Adama Ouane (Hamburg, Germany: UNESCO Institute for Education, 2003), 181-91.
78. Neville Alexander, "The Role of African Universities in the Intellectualisation of African Languages," *JHEA/RESA* 5, no. 1 (2007): 29—74.
79. Paulin G. Djité, "From Liturgy to Technology: Modernizing the Languages of Africa."
80. Makinde, *African Philosophy, Culture and Traditional Medicine,* 19.
81. African Academy of Languages, ACALAN. http://www.acalan.org.
82. *Oxford English Dictionary* (Oxford: Oxford University Press, 2009).

83. Samuel Gyasi Obeng, "'For the Most Part, They Paid No Attention to Our Native Languages:' The Politics about Languages in Sub-Saharan Africa," in *Political Independence with Linguistic Servitude: The Politics about Languages in the Developing World,* eds., Samuel Gyasi and Beverly Hartford, (Hauppauge, NY: Nova Science Publishers, 2002), 85.
84. Teferra and Altbach, eds., *African Higher Education: An International Reference Book,* 111.
85. Damtew Teferra and Philip G. Altbach, "African Higher Education: Challenges for the 21st Century" *Higher Education* 47 (January 2004): 30.
86. Amira Howeidy, "From Right to Left," *Al-Ahram Weekly On-line* 448 (23-29, September 1999) http://weekly.ahram.org.eg/1999/448/feature.htm.
87. David Graddol, *English Next: Why Global English May Mean the End of English as a Foreign Language* (London: British Council, 2006): 83. http://www.britishcouncil.org/files/documents/learning_research_english_next.pdf.
88. Ali A. Mazrui and Alamin M. Mazrui, *The Power of Babel: Language & Governance in the African Experience* (Oxford: James Curry, 1998): 64.
89. Ibid., 64-65.
90. Abiose Adelaja, "Nigeria's Native Languages to Promote Science Application," *My Africa* (May 5, 2007): 2. http://myafrica.wordpress.com/2007/05/05/605/.
91. Sifelani Tsiko, "Africa: Use Indigenous Languages in Science" *The Herald* (Harare) (January 7, 2008). http://allafrica.com/stories/200801070713.html.
92. Ibid.
93. Túndé Adébölá. "Hitting the Right Tone." *ICT Update: A Current Awareness Bulletin for ACP Agriculture: Issue 40 Language Technology* (December 2007) http://ictupdate.cta.int/en/layout/set/print/Feature-Articles/Hitting-the-right-tone.
94. Of course, as I myself am aware, having spent thousands of dollars on computers, software, etc. since I bought my first Apple in 1983, the question of the spread of the technology needed to access "cyberspace" involves a huge profit for firms like Apple and Microsoft and their famous billionaire CEO's and "dot.com" executives. Personally, I do not see any way around this because, e.g. currently cash-strapped governments across the globe are not going to be able to take over all aspects of Internet communication.
95. Denise E. Murray, "The Language of Cyberspace," in *Language in the USA: Themes for the Twenty-first Century,* eds. Edward Finegan and John R. Rickford (Cambridge: Cambridge University Press, 2004): 475; BBC News, "Inuit Language Finds Home on Net" (2004/11/03) http://news.bbc.co.uk/go/pr/fr/-/2/t/hi/technology/3975645.stm.
96. Internet World Stats figure as of 12/2008.
97. Paulin Houndtondji, ed., *Endogenous Knowledge: Research Trails,* (Dakar: Council for the Development of Research and the Social Sciences in Africa (CODESRIA) 1997), 17.

98. Kai Kresse and Trevor H. J. Marchand, eds., *Special Issue: Knowledge in Practice: Expertise and the Transmission of Knowledge, Africa: The Journal of the International African Institute, 79, no.1* (2009), 2.
99. Gordon, *Ethnologue Languages of the World,* http://www.ethnologue.com.
100. The fact of having a writing system, no doubt, reinforces the popular perception, discussed above, that majority languages are superior to minority ones.
101. The possibility of language shift by speakers of small languages is always a factor. However, I do not necessarily agree with Bhola's 2002 assertion that "for languages to survive they have to be written." H. S. Bhola, "Reclaiming Old Heritage for Proclaiming Future History: The Knowledge-for Development Debate in African Contexts," *Africa Today* 49, no. 3 (Fall, 2002), 17.
102. Kai Kresse, "Knowledge and Intellectual Practice in a Swahili Context: 'Wisdom' and the Social Dimensions of Knowledge," in Kai Kresse and Trevor H. J. Marchand, eds., *Knowledge in Practice: Expertise and the Transmission of Knowledge,* 148—67.
103. Charles H. Kraft and Marguerite G. Kraft, *Introductory Hausa* (Berkeley, CA: University of California Press, 1973), 264.

• Chapter 11 •

AFRICAN WRITERS AND THE METAPHOR OF ENVIRONMENTAL DEGRADATION: AN ECOCRITICAL PERSPECTIVE ON NIYI OSUNDARE AND TANURE OJAIDE'S SELECTED POEMS

Ademola O. Dasylva

Introduction

Two ideas informed the basis for this chapter: (i) conflict arising from economic/industrial activities, the subsequent environmental degradation, general industrial pollution and the struggle for survival by the victims of environmental degradation; (ii) conflict arising from a near-cultural miscegenation and psychosocial disequilibrium, including issue of identity, etc., foregrounded by historical events that pre-date Africa's colonial history to the present. The second category of conflict as identified in this chapter is both ideological and psychological, as well as rooted in the subconscious. Therefore, it is a conflict that requires a conscious psychological and ideological application for its resolution. In the case of Nigeria, like many African countries, the experience, so far, has become an extremely repeated expensive mistake.

Using a combination of Anthony Vital's suggested "African ecocriticism" and the interpretive approach, this chapter sets out to examine, among other things, how African writers respond to the serious issue of environmental degradation and its implications on the people's psyche. Let me quickly add that this chapter was originally conceived differently, before a colleague of mine called my attention to some recent publications on the possibility of an African ecocriticism. However, the one I found quite fascinating and which inspired the new direction that this chapter has taken was Anthony Vital's essay.[1] African ecocriticism, explained the critic, entails taking into consideration, "the ways modernity in African contexts transforms human relations with nature and, as a result, the impact of societies on natural environments," (87) the purpose of which is to complement the efforts that enable, "social worlds to find more equitable, sustainable, and healthy ways of inhabiting their place- as well as strengthen historical self-understanding" (87) Besides, according to the critic, "it will need, ... to work from an understanding of the complexity of African pasts, taking into account the variety in African responses to currents of modernity that reached Africa from Europe... , but now influence Africa from multiple centers, ... in the present form of globalizing economy." (88)

Vital's case for an African ecocriticism is in view of observable critical gaps not only in ecocritcism which neither includes nor engages in anything that has to do with the consequences of European colonialism of Africa but also postcolonial criticism which is not consistent with ecocritical preoccupations and strategies. He suspects that the two critical schools rather than emphasize areas of convergence and tap into them, or explore how best they could be exploited to human advantage, have continued to widen the delineation. African ecocriticism, therefore, as a variant of ecocriticism acknowledges both the fundamental peculiarities of Africa's original pastoral (green) setting and how it had impacted the psyche of the people, on the one hand, and on the other hand, how Africa's colonial history, and the so-called "modernity" which has its origin in Europe and North America, and later in Asia had adversely affected the ecological as well as the psycho-social components of the African society.

Earlier in 1994, Julia Martin[2] who unarguably wielded a significant influence on Anthony Vital had earlier proposed "reading for nature with an awareness of colonial history."(1) Similarly, like many African critics have argued in favor of African feminism, Vital, it seems, finds it convenient to locate and situate African ecocriticism at the point where ecocriticsm as understood by many today, and postcolonial critical canons do converge. In other words, the brand of African ecocriticism that is being proposed takes into account an exploration of the possibility of ecocriticism that fits into postcolonial discourses. In which case the thrust of his argument can be summarized thus; for ecocriticism to be truly global and relevant to the Africa-project, it must take into cognizance the totality of Africa's psychosocial and ecological constituents. Vital's African ecocriticism finds expression in J.M. Coetzee's *Life & Times of Michael K* in his article under reference. It is in light of the above that, in adopting the African ecocritical approach, the present study interrogates selected African writers' subordination of their postcolonial discourse to a literal and non-literal celebration of nature. In addition, it further proposes an understanding of the textuality of the different forms of violence foregrounded by ecological issues of degradation and death in the selected works.

In order to achieve the objectives of this study, the selected works of African writers I like to examine include, but not limited to, Niyi Osundare and Tanure Ojaide. Besides, whenever it is found expedient, I shall be making a passing reference to such other writers as Ayi Kwei Armah and Ken Saro Wiwa. This is with a view to analyzing the extent to which the writers' understanding, and subsequent interpretation, of the complexity and contradictions which characterize Africa's pasts, including its colonial history. I also intend to include the variety in African writers' responses, including protestation, and subversive dropping[3] which seem to strengthen Africans' historical self-understanding for sustainable, and healthy ways of inhabiting their space, to currents of modernity which have impacted negatively natural environments and the African psyche.

Environmental Degradation and the Survival Instinct: A Conflict of Interests

Perhaps I need to clarify one or two issues on the idea of "survival" as intended in this chapter. Agents of environmental degradation and geophysical denudation of the magnitude that is being described here are largely big-time businessmen and women with interest or huge investment in industries. The sole aim of any capitalist-propelled industry is to make profit. Osundare describes this as "industries for profit." The industrial toxic wastes are dangerous to both ecological equilibrium and human survival. The scenario in the Niger Delta region of Nigeria, for example, has to do with the activities of oil companies, and the periodic environmental degradation through oil spillage that is now a common occurrence, and atmospheric pollution occasioned by a continuous gas flaring. The people of the Niger Delta of Nigeria are protesting against these activities which render their land barren and their homes desolate. A collective protestation of affected communities soon culminates in armed struggles, which is a natural/instinctual agitation for survival. I call it, redemptive violence. At the same time, the multinational companies, owners of the plants for crude oil exploitation or prospecting are also busy doing every thing possible to remain in production, and as such protect their investment. This is also another dimension to survival conflict. Susie O'Brien refers to this phenomenon which is an aspect of resilience theory as "a seemingly paradoxical correspondence" (4).

Since African ecocriticism, according to Vital, is largely concerned with the impact of societies on natural environment, and especially, "the ways modernity in African contexts transforms human relations with nature and, as a result, the impact of societies on natural environments,"(87) my immediate preoccupation is situating "modernity" not in a historic holistic sense of Africa's contact with the Arab world, and later the trans-Atlantic slavers, but located within a manageable scope of western capitalism and its attendant problems. This is, in my opinion, a logical thing to do rather than focus on "colonialism" which to a large extent is a by-product of capitalism. Western capitalism promotes individualism; it is sustained on the maximization of profit. On the other hand African communalism, had thrived on individuality, and was, largely, nature-friendly. This was largely so until the colonial period. Capitalism, like a predator, had brought among other things, industries, heavy-duty equipment, for purposes of exploitation of natural resources, the consequences of which include, but not limited to, human and ecological death, albeit, by installments, through toxic wastes, and lethal gas emission.

Therefore, Osundare published his collection of poems, *The Eye of the Earth*,[4] to call attention to nature's beauty and abundant resources which were pure, full, and satisfying. The poet argues further that nature, originally, had determined the socioeconomic activities, promoted sound health, and guaranteed longevity of life. According to the poet, at first, a symbiotic relationship between humankind and nature had sustained existence, such that humanity, animals and plants' lives, hardly experienced infection or diseases arising from environmental degradation and pollution. The introductory lines read:

> A green desire, perfumed memories,
> a leafy longing lure my wanderer feet
> to this forest of a thousand wonders
> ... my nose one charmed universe
> of budding herbs and ripening roots ...
>
> ... Behold, cornfields flourish around your foot
> elephant grass fallows the land
> for unborn harvests.
> Swell the grain
> With living water from your rocky arteries
> Fatten the tuber,
> So the hoe does not scoop a sterile clod ...
>
> ... And thus Oke Eniju
> where coy cobs rocked lustily
> In the loin of swaying stalks ...[5]

The poet apparently reminiscences his childhood days when the Ikere-Ekiti community, in the south western Nigeria, where he was born, had lived solely on subsistent farming and depended on what nature had in abundance to offer. Through a poetic (imaginary) journey into his childhood past, he could perceive the natural sweet smelling perfume the rich green of the forest vegetation offered. He remembered how the "cornfields" had flourished, and the fallow system was put in place to check the abuse of the tilled lands. During the fallowing period, the recuperating soil was characterized by the presence of tall, "elephant grass" while the soil regained its fertility for a future harvest- "unborn harvests". It was an indigenous method by which the community used to prevent soil aridity- "sterile clod." Besides, the tuber (presumably, yam, coco-yam and cassava) farms were watered by the natural waters from the hanging rocks. The concern of the poet here is that gone were the days the days when the society had structures that guaranteed environmental protection and, the society had, in turn, benefited tremendously from nature. The poetic trip that was undertaken by Osundare was consequent on his desire to escape the apparent caustic environmental degradation which is quite alarming and nobody seems to care anymore about ecological or environmental stability, hence his longing for his childhood days when the humankind and Nature had symbiotic relationship and had benefitted mutually.

In other words, in a vein similar to the one described above, at the wake of the industrial revolution in England, John Keats[6] had reacted to the decay that characterized the England of his time for which he desired to escape "on poesy wings" into the greenland (world) of the singing "nightingale." His confession of "envy" for the joy the "winged dryad" exudes, is also similar to Osundare's who expresses his desire for the abundant resources of nature and chooses to wander or escape, as it were, from the life-threatening environmental decay that daily confronts him. As already pointed out above, he does this by means of metaphoric journey (flight) into memory and imaginativeness of his childhood days, when cornfields had flourished everywhere and the green field and forest have not been displaced by paved concrete and roaring industrial machines; when the village rivers that supplied fresh, living water have not been polluted, and there was abundance of subsistence farm produce to meet the need of every home.

Informed by a similar pathos, Tanure Ojaide's "When green was the lingua franca,"[7] expresses no less a dazzling dream-like experience as the persona reminiscences on what the rustic simplicity of a typical African community was like in his childhood days, and the response of the society to nature's beckoning:

> My childhood stretched
> one unbroken park,
> teeming with life.
> in the forest green was
> the lingua franca
> with many dialects. (p. 15)

The poet refers to "one unbroken park" suggesting an undefiled ecology, natural vegetation that was not devastated, where there was a symbiotic relationship between the people and nature;

> Everybody's favourite,
> water sparkled...

I remember *eruvwudjaorho*
such a glamorous fish
But denied growing big. (p.15)

Like Osundare's "living water," Ojaide's community water had sparkled. The "Earthworm" referred to in the poem is an appellation given to women of easy virtue, who naturally depend on their beauty and power of seduction to *eke* a living. They have a penchant for easy-does-it life; they shun hard work and would not have anything to do with any struggling man, but are ready to go with any man who is considered to have "arrived." The persona's society acknowledges and encourages hard work, and does not approve of life lacking in integrity. Earthworm is an apparently fragile, quite delicate, "spineless" creature that could break into pieces at the least pressure, or simply melt at the slightest contact with any saline substance. Some earthworms are of a medium-size snake, and at first could be mistaken for snakes because they have a semblance of snakes, however, they are harmless and do neither have the strength nor the temerity of a snake:

Earthworm, communal name
for the kind of women seeking
to flourish in soft spots.
Uwara, beauty that defies
tyranny of *Akpobrisi,*
forest manic and recluse-
what flesh or fibre fails
to capitulate before charm?

Snails and *koto* lured me
to tear through tangles
that seasoned my soles
to defy every distance.
Urhurhu grapes coloured
my tongue scarlet,
the*owe* apple fell to me
as cherries and breadfruit
on wind-blessed days.
The cotton tree made me
fly for tossed-out fluffs;
the gum tree took fingerprints
before invisible policemen.
*Ikere*froglets fell from skies
that covered the land
with tropical sheets; .(p.12)

The emphasis here is the fact that in the society of the poet's childhood everybody had equal opportunity to life and living.

> the skipper-fish overflew
> culverts into fisher's ambush.[8]

But even at that, unsuspecting skipper-fish's freedom is encroached upon by human predator's snare, "fisher's ambush," which traps the exuberant fish which is merely exercising its agility.

In the persona's "paradise" of his childhood, the strong had co-existed with the weak, the rich with the poor, while justice (equity) was both visible and accessible to everyone, no matter the status in the society.

> Undergrowth kept as much
> alive as overgrowth,the delta
> alliance of big and small,
> market of needs, arena
> of compensation for all... (p.13)

I must quickly add, however, that the fact that English pastoral poets, like Christopher Marlowe, celebrated similar beauty and abundance of nature's endowed gifts that were offered freely to humankind, as in, "The passionate shepherd to his love;" or powerful lyrics of the English romantics as in John Keats', "Ode to Autumn," is indicative of the fact that Osundare's or Ojaide's celebrated experience was not peculiar to Africa alone, it was universal.

Notwithstanding the celebration of beauty and the life-sustaining endowments with which "nature" is blessed, and regardless of the celebrative mood that Osundare and Ojaide's poetry conjure up, the former is quick to identify inherent contradictions that nature is associated with. Natural disasters do occur that humans are neither directly responsible for nor are they capable of providing answers for: tornadoes, Hurricane Andrew, Katrina and Tsunami, among others. Although scientific claims have traced some of these occurrences remotely to human activities, for example, the depletion of the ozone layers and its implication on glacier-reaction and flooding, minning activities and disasters associated with them, etc., the persona in Osundare's "Waiting laughters," believes that there are disasters occasioned by natural phenomena. They include, a dormant volcano suddenly erupting, lethal gas oozing out from the earth crust, land tremor, earthquake, tornadoes, incessant downpour leading to flooding, or a prolonged dry season leading to scourge and famine, as in Ethiopia, Eritrea, and Darfur, etc. Nature's phenomena do also contribute to general decay and ruin of the Earth. In the poem below, Osundare calls attention to the significance of the rain as a key life-sustaining gift of nature, but that it has its ugly side to life and living too. He traces the nature's rain-making process to climatic changes that occur, beginning with the hot weather, the rising vapor, the high humidity in the atmosphere, the moisture-laden tropical cumulus cloud- "seminal smoke" and the subsequent heavy torrential downpour often characterized by thunder and lightning, giving way to more rains, overflooding, "unbridled deluge", and unquantifiable disaster to life and property:

> The rain. The rain
> Truth of the rain's seven ingots
> In the womb of the forge
> And seminal smoke which leaps
> Above the roof, plodding skylanes
> Before taunting thunder's raw temper
> To a wild, unbridled deluge.[9]

Talking of the unquantifiable disaster the rains could cause, the poet reminds his audience, "But sometimes when it Rains," the persona observes, and highlights the extent of damage to lives and property, as rain storms blow off roof tops, fell trees, flood the farmstead, and destroyed growing farm produce:

> But sometimes when it rains
> And an angry thunder raps earth's ears
> Sometimes when it rains
> And a heartless storm beheads
> The poor man's house
> Like some long-convicted felon,
> Sometimes when it rains
> You wonder who sent the skies weeping
> Sometimes when it rains
> And an impregnable mahogany falls
> Across your farmward path
> Sometimes when it rains
> and a streamlet swollen with watery pride
> drowns your fields and tender tubers ...[10]

The inherent contradictions in nature's phenomena is more telling in Osundare's "Moon" in *Moonsongs,* which could be a blessing, when it is the Queen of the "fruitful" and "ripening" autumnal seasons with the night's multi-fragrant boughs:

> Moonfire too is deciduous:
> Its conifers weep their leaves
> Like yellow tears

And in the dappled darings
Of autumn
When yearning winds press lyrical lips
On the apple's demurring bosoms
Skyfields are sweaty tracks
For the garnerer's unwavering feet

There is gold chatter
In the bower of busy barns
The air leans low with the fragrance
of active ripening
(*Moonsongs*, 21)

The context here is the seeming conspiracy between the full moon and the weather in dry season resulting in a lot of human activities. The dry season is associated with harmattan. It begins with the panting north-easterly dry cold wind which moves towards the south-west, laps up every visible moist and wetness on its path. Then, the forest foliages turn brown with dryness that compels the fruits to full ripeness, while farmers are busy harvesting and gathering farm produce into barns. During the dry season, there is apparent low humidity in the atmosphere while in the air hangs the fragrance of ripening fruit and late flowers.

But the poet quickly warns that the same benevolent moon can also be an inclement agent of disaster. In the quoted verse below, the poet describes the moon as a despot, apparently because the poet knows that changes in the ocean tides are caused by specific Luna movement during the year- "moon's diurnal fingers", which in turn, bring about changes in weather, causing rains, thunderstorms, and flooding and more disasters upon natural vegetation, animals and other creatures, etc.:

... Oh the poetry it all!
... raging fire quickens the river of the moon
October-red
(*Moonsongs*, 65)

A despot devastating life;
The storm has seized the sky
seized the sky like a long-awaited despot
henpecked several seasons
by the moon's diurnal fingers

The hurricane harvests a shadow
of sullen sugar

rattled roots and oceans
of capsizing branches
a hawk telescopes the range
... divining orphan chicks
andoffals of assassinated henhood ...[11]

Again, the humankind is not spared by the entangling despotic rule of the moon, as some human maladies, including "lunacy," are associated with the calendaric movement of the moon. The poet, while using a metaphoric projection of the despotic moon, wonders if the moon reigns in the head of some African rulers who, for selfish reasons, compromise their office, mortgage the life-sustaining resources, "rocks" and the future, "the stars" of the people. Forest trees were felled and carted away, lakes are polluted with toxic wastes, or simply chartered out to the multinationals in the following lines:

And Man;
A madding moon
Has sold the stars
Sold the rocks
There is a bickering banter

In the budget of the sky
The moon plundered the gold
Drained the diamonds
And battered its silvery ore

That has felled the forests
Laundered the lakes
Harassed the hills
A yellowing chill stalks

Now the moon has crowned our silence
Gripped our songs
Laid a frenzied ambush[12]

In effect, Osundare's *The eye of the earth* and *Moonsongs* and Ojaide's *Delta blues and home songs,* among others, express the African poet's concern on the consequences of environmental degradation and general pollution. But beyond this, as I have pointed above, they represent the poets' social vision which transcends mere nostalgic feelings of the glorious agrarian past of the indigenous African society.

Conflict informed by ideology or lack of it

The gravity of the damage to Africa's eco-system through exploitation and plundering of her natural resources, the assault to Africa's psyche, the mutilation of Africa's personality and identity, which had gone on for over seven hundred years of trans-Atlantic slave trade and, later colonialism, are unquantifiable, and might take much longer to rectify. This is in light of the introduction and dominance of capitalist economy which thrives on individualism (as opposed to individuality and communalistic subsistence economy which had existed in Africa's cultural subsoil), and maximization of profit (a subtle term for exploitation), cash crops and mining activities, as well as planting of industries, etc. The consequences which include, complete disoriented and dehumanized society through capitalist economic activities- environmental degradation, atmospheric pollution, diseases and accelerated death. They foreground the so-called "modernity" and "civilization". Niyi Osundare describes the scenario thus:

> I have heard
> the thud of sleepy boots plodding
> toilwards in dreary dawns
> I have seen
> busy hands rouse slumberous yard
> into a hive of humming demons. (p.37)

The "slumberous yard" that is suddenly roused into a hive of "humming demons" describes the activities of heavy industrial equipment like bulldozers used in clearing the natural vegetation, felling of trees, digging, cutting, embankment, and excavation, as in mining activities or mineral prospecting, or planting of industries. The poet's diction, "humming demons", predictably, registers his protestation and displeasure at the devastation of an otherwise peaceful, serene and harmless (slumberous) natural vegetation by heavy-duty industrial machines, etc.

The capitalist economy is also characterized by a "classed" society and apparent inequality, in which case, a gulf exists between the very rich and the very poor. The capitalist system deliberately creates the poor in order to guarantee factory hands, etc., that must service the system. Each time the poor seem to level up, for example, in the United States, the system recruits more cheap labors by means of visa lottery. It is the reason why "penuried" that is, pauperized lives, "spent in ghetto dungeons." The word "penuried" which is a deliberate coinage, assumes a passive voice. This is significant in the sense that it connotes a class of people who are made poor by the capitalist system. Similarly, the word "spent" has dual grammatical functions in the rank-shifted sentence structure. "Spent" the past tense of the verb "spend," meaning, to spend one's life (or live) in a ghetto dungeon. And "Spent" a qualifier, as in, "a spent cartridge," meaning a used cartridge; "a spent life," meaning an exhausted life, or an empty life that does not add value or meaning to existence. Both possible meanings bear relevance to the topicality of the poem.

I have seen
penuried lives, spent, in ghetto dungeons
I have seen
native executhieves hold fort for alien wolves

...Here
there. (*The eye of the earth*, 45)

Similarly, reference to "native executhieves," a coinage by the poet, is a contraction of two words, "executive" and "thieves". It refers to home-grown tyrants with comprador mentality; these are African rulers, not leaders, who "hold fort ..." to facilitate the exploitation of the nation's natural resources by the multinationals; who give undue state protection to the "alien wolves" who degrade the environment and pollute the waters with toxic wastes from the factories, and annihilate aquatic lives by their mining activities and oil spillages. These way they make life unbearable for the communities. These "rulers" lend support to the "wolves" because of what they, as individuals, stand to gain, at the expense of the people they govern:

Everywhere

a lake is killed by the arsenic urine
from the bladder of profit factories
a poisoned stream staggers down the hills
coughing chaos in the sickly sea
the wailing whale, belly up like frying fish,
crests the chi-ling swansong of parting waters[13](emphasis mine)

The aftermath of the large scale industrialization, mining, oil prospecting, etc., is death by installments of the community as symbolized in the death of the "lake" from which the people *eke* their livelihood. The lake "is killed", due to the huge volume of lethal wastes (arsenic urine) emptied into it from the factories. The cycle seems to complete as the poisoned stream also empties its acidic content into the sea, leading to a total annihilation of existing aquatic lives, the "sickly sea" is no longer safe for creatures like the whales regardless of their massive size, "the wailing whale"which also died like a common fish as a result of the pollution.

Talking of atmospheric pollution, which heavy vehicular traffic cause on daily bases in Nigeria, Okinba Launko in *Ire! ... and other poems for performance* (1998:8) appeases Esu Laaroye, the trickster god and patron of all confusions, "Dweller on the fatal cross-roads/ Confuser at Oke-Ado market/ Ruler of Bere roundabout/King of the cross-roads/ Whose favourite fun/ Is a traffic jam/ Let my road be smiling today."

In the introductory chapter of Ayi KweiArmah's *Fragments*, the narrator describes the station as ill-kept, with hanging cobwebs on the ceilings, the walls defaced and murky, the electric bulbs dusty and, as such, poorly lit. He expresses dismay at the destruction of friendly green vegetation and the authorities' preference for the construction of commercial banks, as well as the opening up of every available space to commercial activities, all of which speak volume about the psychological state of the new Ghanaian society, the destruction of the natural vegetation, and the subsequent environmental degradation.

Osundare consciously pursues the same theme of exploitation and marginalization of the poor by the rich in the wake of introduction of capitalist economy, in "They too are the earth":

> They too are the earth
> The swansongs of beggars sprawled out
> In brimming gutters
> Under snakeskin shoes and Mercedes tyres
> They too are the earth
> The sweat and grime of
> Millions hewing wood and hurling water[14]

Similarly, in Ayi Kwei Armah's *Fragments,* the narrator captures very vividly the tragedy of the emergent independent African state by showing that the failure of Baako, the protagonist is measured by his deliberate choice not to compromise with a passive, decadent society that is governed by a corrupt regime. The author also identifies the elite ruling class, which Osundare refers to as 'native "executhieves,"' as well as the inherent contradictions in his postcolonial Ghanaian society in the wake of the emergent "cargo-culture."

Ocran, an old teacher of Baako, sees beyond environmental degradation, he sees cultural pollution and a warped national psyche that characterize the emergent society, as he warns Baako:

> Nothing works in this country. What can you expect? The place is run by this so-called elite of pompous asses trained to do nothing. ... If you come back thinking you can make things work in any smooth, efficient way, you'll just get a complete waste of your time. It's not worth

bothering about.[15]

The same issue of environmental degradation of a once beautiful natural habitat engages Ojaide when he refers to his childhood communal world as paradisal, "This share of paradise, the delta of my birth," and bemoans the complete disappearance of the beauty, the comfort, the freshness, the fruitfulness, the security, the abundance that nature had offered humanity of his childhood days. However, he is dismayed by the extent of the incurable wound, "Reels from an immeasurable wound," his land of birth has sustained through crude oil exploitation and accidental oil spillages, "Barrels of alchemical draughts flow/ From this hurt to the unquestioning world," a familiar occurrence occasioned by the activities of "a cabal of brokers," the menace of which no one seems to be interested in checking, and which leads to a general decay and death. The scenario that characterizes the reality of the poet's community today.

Like John Pepper Clark's fish that is forced out of its natural habitat in "Night Rain", the persona in "Delta blues" finds his hitherto natural habitat completely disabled; rather than support life and living, gradually, it gravitates to a phenomenon that suffocates and kills him; "now crushes my body and soul.... there's no refuge east or west, north or south of this paradise." His despair at this juncture becomes discomforting and more pungent as he laments the irretrievable loss of that "paradise" of his childhood years;

> My nativity gives immortal pain
> masked in barrels of oil-
> I stew in the womb of fortune.
> I live in the deathbed
> prepared by a cabal of brokers
> breaking the peace of centuries
> & tainting not only a thousand rivers[16]

Osundare's persona, rather than give up on the seemingly irreparable loss and foolish killing of the Earth (or nature), through economic activities, he chooses to dissuade the society from habitual abuse of nature, and cautions on plundering the Earth in "Ours to plough not to plunder":

> The earth is ours to plough and plant ...
> Let wheatfields raise their breadsome hands
> to the ripening sun
> let legumes clothe the naked bosom
> of shivering mounds
> let the pawpaw swell and swing
> (*The eye of the earth*, 48)

Resolving the Contradictions

In *Pedagogy of the oppressed* (1980:32), Paulo Freire defines oppression as an act which prevents a person or people from being more fully human:

> Behavior which prevents the restoration of the oppressive regime cannot be compared with acts which create and maintain it. One cannot compare it with acts by which few men deny the majority their right to be human.(32)

It is in light of this that one may begin to appreciate Freire's recommendation to resolve the apparent contradictions, by way of organized "redemptive violence". According to Susie O'Brien, "The imperative to survive impels oppressed individuals to processes," and quoting Elizabeth Crosz, 'not of remediation but of self-transformation (28)'" O'Brien states further that this entails "the capacity to mobilize elements already latent in the present. ... Survival in these terms isn't predicated on dominance, but on resilience."[17] It earned for South Africa, the majority black rule in 1994. Freire's remark:

> It is only the oppressed who, by freeing themselves, can free their oppressors. The latter, as an oppressive class, can free neither others nor themselves. It is therefore essential that the oppressed wage the struggle to resolve the contradiction in which they are caught ...(32-33)

Therefore, for a positive attitudinal change for the purpose of putting a check on further degradation, and encourage a reclamation and restoration of the devastated environment in Africa, it calls for a gradual process of psychic retrieval and an ideological repositioning through self-knowledge and the approximate consciousness. This is necessary to bring about a collective effort at combating the agents of degradation and denudation, especially, the multinationals and their home-grown despotic collaborators. This position has become imperative in view of the fact that situation such as this compels political struggles for the purpose of overcoming oppression. Oppression here is in the context of what Elizabeth Grosz describes as, "system of harm and injustice that privilege the bodies and activities of some at the expense of others" (28). The problem of self identity and self knowledge although often taken for granted, is a crucial one that must be resolved as a first step to self freedom which must be given a collective expression.

For example, it is evident that Armah's Asante-Smith in *Fragments*, like Ralph Ellison's invisible protagonist in *Invisible Man*, has lost his identity and focus hence the apparent contradictions. The invisible man confesses with shame, "I realized that I no longer knew my own name ... I was in the grip of some alien personality lodged deep within me". (195, 203). It would be recalled that the search for a lost identity has engaged Sophocles's king Oedipus, and Ola Rotimi's king Odewale, respectively, until they each get this resolved at the end of the two plays. Similarly, the quest for self identity, self knowledge has kept the invisible protagonist running for over 400 pages of the narrative. Finally, he declares, "If I couldn't help them to see the reality of our lives I would help them to ignore it until it explodes in their faces."[18]

Ojaide's persona in "For the sake of freedom" must have had similar idea in mind in his description of the activities of the predators and the scavengers cohabiting together- Hyena is a predator, the vulture is a scavenger. Again, in his description of the National Assembly (House that oil windfalls built), he intones that there are those that represent predators and those who are mere scavengers;

> hyena and vultures faithfully cohabit
> in the House that oil windfalls built. (Delta blues, 62)

Their inaction is described as sitting for days debating in flowery speeches without action;

> they sit for days defraud with speeches
> and leave for orgies the humanimals. (Delta blues, 62)

The national assembly predators and scavengers become more visible when it is time to loot the nation's treasury and sharing among themselves what should have been spent on general welfare, job provision, improved healthcare delivery, etc., so the result is hundreds of jobless youth roam the streets:

> those sacks they carry out of carpeted rooms
> are pumpkins meant for the commonwealth
>
> they starve youths jobless in the streets
> the occupation forced on the able-bodied
>
> pilfering – Abuja is full of them
> Lagos is full of them Area Boys[19]...

Self realization requires a level of consciousness and sound ideological inclination. It is this awareness that informs Tanure Ojaide's resolve after the nine Ogoni leaders, including Ken Saro Wiwa, were "murdered" on the order of the former Nigerian military Head of State, late General Sani Abacha. The poem is dedicated to the nine heroes. It reads in part:

> Nigeria sleeps in a makeshift grave.
> ... the next world will be brighter for me and my compatriots.
> A gunful of children broke the tetrarch's legs
> & the elephant that once pulled the forest along for a path has
> Fallen
> ... If Nigeria wakes from the grave after the murders,
> let the people cast general, staff, and cap
> into a marble grave in their born-again memory,
> there is no other way to live free here than kill
> or be killed. You can tell from our stone country.
> ("Sleeping in a makeshift grave," 24)

In the preface to *Invoking the Warrior Spirit* (1998: ix), Ojaide writes:

> I am interested in the current generation of Africans gaining inspiration from traditional African institutions, practices, and philosophies that can help overcome contemporary adversaries... I got to know of Sankofa, the Akan mythical bird, which symbolizes going back in time and bringing to the present, past values needed now.

In his poem titled, "Mbira o mbira o mbira o", Ojaide reiterates his faith in the effect of art for the purpose of creating general awareness, mass mobilization and forging a positive revolution for a desired change:

> If the song doesn't cause insomnia in thrones,
> Why are we, for opening our mouths, thrown into jail
> To relax the tortoise heads of Ken Moi and Ban Dawi?
> One single shot may not knock the leopard down,
> But forays of barbed words will mortally wound the demon.[20]

In "Ivwri: Invoking the warrior spirit" the poet is emphatic; "Some cries pierce the soul, others stifled to shame the torturer..." (5)

Conclusion

From the foregoing, it is pertinent to note that African writers have been part of the on-going campaign against abuse of nature. They have vehemently been opposed to environmental degradation. More than often, African writers have drawn attention of their target audience to such topicalities as nationalism, and the challenge of modernity, which includes environmental degradation. However, until very recently, critics, it seems, have hardly considered the issue of environment as a major thematic preoccupation that equally engages African literature.

This study through the deployment of African ecocriticism addresses how different African writers have responded to ethical issues that relate to environment and humanity; and how through such writers there has been a growing awareness about destroyed habitats, polluted rivers, careless mining activities: oil prospecting and exploiting, etc., all of which destroy the ecology. It would be recalled that it was for this cause that Ken Saro Wiwa, a writer of note, an environmentalist, was hanged along with eight other Ogoni leaders. In view of the effect of literary works on society, more African writers should engage in this struggle, because if humankinds fail to put an end to the menace of environmental degradation and pollution, pollution and degraded environment will certainly put an end to humankinds.

Notes

1. Anthony Vital, "Toward an African Ecocriticism: Postcolonialism, Ecology and Life and Times of Michael K," *Research in African Literatures*, Vol. 39, No.1 (Winter 2008). 87-106.
2. Julia Martin, "New, with Added Ecology? Hippos, Forests and Environmental Literacy," *Interdisciplinary Studies in Literature and Environment:ISLE* 2.1 (1994): 1-11
3. Odia Ofeimun, a Nigerian poet, refers to this form of writing/poetry as "subversive litanies" in his poem titled, "How can I sing?"
4. Niyi Osundare, *The Eye of the Earth*, Ibadan: Heinemann, 1986.
5. Ibid., 3, 14 & 15.
6. John Keats, "Ode to a Nightingale."
7. Tanure Ojaide, *Delta Blues & Home Songs*, Ibadan: Kraftgriots, 1998.
8. Ibid., 12-13.
9. Niyi Osundare *Waiting laughters*, 3.
10. Osundare *The eye of the earth*, 36.
11. Osundare, *Moonsongs*, 66.
12. Ibid., 37.
13. Osundare, 46, 50.
14. Ibid., 45.
15. Ayi Kwei Armah (1977) *Fragments*, London: Heinemann Educational Book, pp.116,118.
16. Ojaide, "Delta blues," 21.
17. Susie O'Brien, "Superfluous Words: Ecological and Cultural Resilience in *Things Fall Apart* Postcolonial," *Text*, Vol 5, No 1 (2009), pp. 3-4.
18. Ralph Ellison, *Invisible Man,* (1969) London: Penguin Books, p. 411.
19. Ojaide, *Delta Blues*, 62.
20. Ibid., 1.

Part III

ECOLOGY, SCIENCE AND TECHNOLOGY AND CONFLICTS IN THE POSTCOLONIAL ERA

• Chapter 12 •

ECOLOGICAL ROOTS OF SOCIAL CONFLICT IN KENYA: PASTORALISM, LAND, AND THE DEVELOPMENT PARADOX

Shadrack Wanjala Nasong'o

Introduction

Globally, at the beginning of the decade of the 2000s, five of the ten countries generating the most refugees were in the Eastern African region alone. Estimates at the close of the 1990s indicated that 29 percent of Rwanda's population lived as refugees in neighboring countries, and 12 percent of Eritrea's. The refugee crisis was compounded by the civil war in the Sudan, with over one million displaced persons, state collapse in Somalia, and a variety of internally displaced groups on account of social conflicts in Uganda and Kenya. Accordingly, three countries in the region, including Kenya, Ethiopia, and the Democratic Republic of the Congo ranked among the top-ten refugee hosting countries worldwide.[1] Emphasis on intervention in these conflicts in the region, and in Africa generally, has historically centered on negotiating competing political viewpoints. This chapter argues, however, that there is an environmental dimension to conflict in Kenya in particular and in Africa generally that has either not been understood or effectively incorporated into the development of conflict prevention and management strategies.

The chapter contends that the causal variables that help explain the occurrence of social conflict, defined as antagonism and tension that escalate into open hostilities, fighting, and deadly clashes between social groups be they class, ethnic, racial, or religious, can be divided into three: underlying causes, intermediate causes, and precipitating causes. Whereas the precipitating and intermediate causes of social conflict in Africa and elsewhere have been adequately covered in available literatures, the underlying causes have not. This chapter posits, in regard to Kenya and the Eastern Africa region in general, that most social conflicts have the ecological dynamic as their underlying causal variable. The chapter delineates and elaborates the ecological roots of social conflict in Kenya focusing on conflict over the "White Highlands", the Shifta war, the ethnic clashes of the 1990s, and the Kikuyu-Maasai conflict in Narok. It concludes that unless this ecological dimension is taken into account in designing tools and techniques for conflict prevention, the management of social conflict will remain overly problematic.

Mapping out the Conceptual Boundaries

Harold and Margaret Sprout[2] were among the first post-war social scientists to emphasize the importance of geography in explaining political behavior. They contended that most, if not all, human activity is affected by the uneven distribution of human and nonhuman resources. The Sprouts rejected unidimensional geopolitical theories in favor of an ecological perspective because it provides a holistic view that takes account of the physical and nonphysical features of the environment. The holistic nature of the ecological perspective is captured by the definition of the concept of ecology by Webster's Third New International Dictionary according to which, ecology is the totality of the pattern of relations between organisms and their environment. It is a branch of science concerned with the interrelationship of organisms and their environment, especially as manifested by natural cycles and rhythms, community development and structure, interaction between different kinds of organisms, geographic distributions, and population alterations. The environment, as Anyumba[3] posits, consists of all the surroundings of an individual organism or community of organisms, reaching up to the entire biosphere, the zone of the earth that is able to sustain life.[4] Surroundings on the other hand, are the totality of living and non-living materials that play any role in an organism's existence, what the organism feeds on, and the organisms that can feed on it. According to Anyumba,[5] cultural factors, what the Sprouts call the social milieu, may be considered as part of the environment since culture interacts with, determines, and impacts on the environment, or physical milieu, in a great number of ways.

The interface between culture and environment is multifaceted and, as Anyumba rightly notes, generates environmental conflict in the form of a conscious antagonistic state of relationships between individuals, members of society or governments in connection or contestation over one or several aspects of the environment, at the local, regional, or national territory for a given span of time. However, the environment-conflict linkages cannot be captured in a simplistic "stimulus-response" model, but in one that allows for "culturally mediated behavior."[6] Accordingly, Dabelko and Dabelko[7] argue that antecedent and intervening social and political conflict variables characterize would-be environmental conflict. Attempts to isolate environmental causes of conflict, therefore, can be misleading in conceptualizing the complex causal relationships behind conflict. The two scholars argue that there is evidence of environmental scarcity serving as an underlying yet strong cause of intrastate conflict in the form of ethnic clashes due to environmentally induced population movements, and civil strife stemming from environmental scarcity that affects economic productivity and therefore livelihoods, elite interests, and state capacity to confront these challenges.[8]

The total effect of human activity on the environment in a particular ecological region, according to Homer-Dixon,[9] is mainly a function of two variables. First, it is the product of total population in the region, and physical activity per capita. The second variable is the vulnerability of the ecosystem in that region to those particular activities. Activity per capita in turn, is a function of available physical resources—non-renewable resources like oil and minerals, and renewable ones like agricultural land, forests and water—and ideational factors including institutions, social relations, preferences, and cultural values among others. In particular, environmental effects may cause social consequences that could lead to conflict. For instance, agricultural land degradation could lead to large-scale migration, which in turn could result in ethnic conflicts between migratory groups and indigenous populations, with important feedback loops to the ideational factors and back to the activity per capita and population.

The Arab-Israeli conflict in the Middle East, for instance, is not simply a fight enflamed by nationalist sentiment; it is, to a great extent, also a contest over natural resources, especially control over water, given its limited availability in the region. As Cooley[10] rightly observes, water is not necessary for life but rather life itself, "... long after oil runs out, water is likely to cause wars, cement peace, and make and break empires and alliances... as it has done for thousands of years."[11] Indeed, the constant struggle for the waters of the Jordan, Litani, Orontes, and Yarmuk rivers among others was a principal cause of the 1967 Arab-Israeli war. It is noteworthy that Israel consumes roughly five times as much water per capita as each of its less industrious and less intensively farmed neighbors. Accordingly,

> In 1967 Israel went to war against Syria and Syria's ally, Egypt, partly because the Arabs had unsuccessfully tried to divert into Arab rivers Jordan River headwaters that feed Israel. During that war, Israel captured Syria's Bariyas River, the last of the important Jordan headwaters not under Israel's control.[12]

Similarly, the sensitive nature of exploitation of the River Nile waters brought the countries of Burundi, Egypt, Kenya, Sudan, Tanzania, and Uganda to the negotiating table to re-negotiate an agreement over the use of the Nile waters signed between the British colonialists and the Egyptians in 1929. Some Kenyan and Tanzanian leaders had vowed to proceed to utilize the waters of Lake Victoria, the source of the Nile, without regard to the 1929 Agreement and irrespective of the implications of their activities on Egypt's water needs. The Egyptians declared the move tantamount to a declaration of war, hence the effort at peacefully re-negotiating a new framework between the Nile Basin countries.[13]

Three other historical examples are in order here. First, Honduran efforts to reverse the flow of Salvadorans crossing the common border in search of arable land helped precipitate the "soccer war" in 1969. Second, Bangladeshis fleeing overcrowding and famines moved into the less populous Indian states of Assam and Tripura where they encountered both organized and spontaneous violence. The migrations and resulting violence were the source of great tensions between the two countries. Third, severe soil erosion and near complete deforestation in Haiti have deprived much of the Haitian population a means of survival. This catastrophic environmental degradation contributed to the economic and political hardships that led thousands to attempt a perilous boat ride to the U.S. This migration helped create tensions between the two states and among domestic constituencies in both countries that culminated in the September 1994 U.S. military intervention in Haiti.

Overall, there are two basic schools of thought in regard to the environmental or ecological roots of social behavior. The first school is one of environmental optimists, commonly referred to as Cornucopians. The second is one of environmental pessimists commonly known as neo-Malthusians. As their name suggests, Cornucopians are environmental optimists who do not worry very much about protecting the stock of any single resource, because of their faith that market-driven human ingenuity can always be tapped to allow the substitution of more abundant resources to produce same end-use service. As Julian Simon, the ultimate Cornucopian notes, "[t]here is no physical or economic reason why human resourcefulness and enterprise cannot forever continue to respond to impending shortages and existing problems with new expedients that, after an adjustment period, leave us better off than before the problem arose."[14] For their part, neo-Malthusians or environmental pessimists are much more cautious. They distinguish between "resource capital" and "resource income." The capital is the resource stock that generates a flow (income) that can be tapped for human consumption and well-being. A sustainable economy is one that leaves the capital intact and undamaged so that future generations can enjoy an undiminished income stream.[15]

Homer-Dixon argues that historically, Cornucopians have been right to criticize the idea that resource scarcity places fixed limits on human activity.

Time and time again, humans have circumvented scarcities and neo-Malthusians have been accused of "crying wolf." Nonetheless, Homer-Dixon notes, in assuming that this experience pertains to the future, Cornucopians overlook a number of factors:

1. Serious scarcities of critical resources in the past usually occurred singly, now we face multiple scarcities that exhibit powerful, interactive feedback and threshold effects making the future highly uncertain for policy makers and economic actors.
2. In the past, the scarcity of a given resource usually increased slowly, allowing time for social, economic, and technological adjustment. However, now populations are much larger and human activities much more resource-intensive hence debilitating scarcities develop much more quickly.
3. Today's consumption has far greater momentum than in the past because of the size of the consuming population, quantity of the material consumed, and density of consumption activities.
4. The free market price mechanism is a bad gauge of scarcity especially for resources held in common like a benign climate and productive seas.
5. Market-driven adaptation to resource scarcity is most likely to succeed in wealthy societies where abundant reserves of capital, knowledge, and talent help economic actors invent new technologies, identify conservation possibilities, and make the transition to new production and consumption patterns. In spite of efficient markets, however, poor countries lack capital and knowledge to deal with these problems.
6. Cornucopians have an anachronistic faith in humankind's ability to unravel and manage the myriad processes of nature. There is no a priori reason to expect that human scientific and technical ingenuity can always surmount all types of scarcity. Humans may lack the mental capacity to adequately understand the complexities of environmental-social systems. Rather than inspire the wave of ingenuity predicted by Cornucopians, therefore, future environmental problems may instead reduce the supply of ingenuity available in society.[16]

Arguably, Cornucopians are geographical possibilists who hold that the natural environment can be modified; that human free will ultimately determines the options available. The environment, especially geography, is but only one of many forces governing the development of human activity. On the other hand, neo-Malthusians are environmental determinists who hold that the environment determines human behavior. Whereas the two approaches appear to be opposites of one another, in our view, the reality lies in between them. The environment may not be the ultimate determinant of human behavior; humankind may possess some limited capacity to change their environment. Nonetheless, we remain circumscribed in our behavior by environmental circumstances. As Dougherty and Pfaltzgraff[17] point out, if the environment does not determine the boundaries of human conduct, it nevertheless provides an important, if not crucial conditioning influence.

In this regard, Homer-Dixon argues that poor countries are more vulnerable to environmental change than rich ones, hence environmentally-induced conflicts are more likely to arise in developing countries where a range of atmospheric, terrestrial, and aquatic environmental pressures will in time probably produce, either singly or in combination, four main, causally interrelated effects: reduced agricultural production, economic decline, population displacement, and disruption of regular and legitimized social relations. "These social effects, in turn, may cause several specific types of acute conflict, including scarcity disputes between countries, clashes between ethnic groups, and civil strife and insurgency..."[18] In the final analysis, we argue herein that essentially, the causes of most social conflicts can be divided into three — underlying, intermediate, and precipitating causes. Ecological factors are, for the most part, the underlying causes of most social conflicts.

Ecological Roots of Social Conflict in Kenya

Kenya occupies a unique position within East Africa. The country's historical background, its geography, and ethnic composition make it a microcosm of the region. Though poor by absolute standards, in relative terms, Kenya is the wealthiest state of the region. In spite of the social tensions and recurrent conflict between its constituent groups, Kenya is regarded as a stable country in the midst of states that are, or were until recently, ravaged by civil wars. Kenya is located between latitudes 4N and 4S of the equator (the equator cuts right a cross the middle of the country) and covers an area of 582,646 square kilometers. The highest point of the country is the peak of Mount Kenya at 5,199 meters above sea level. In the East, the land borders on the coast of the Indian Ocean. In the West, Lake Victoria, the source of River Nile, forms part of Kenya's border with Uganda. Just across the southern borderline, on Tanzanian ground, rises the highest mountain in Africa, Mount Kilimanjaro. The population density is very high in some parts of the country, where more than 300 people live on one square kilometer, while in other regions, such as the Northeast, this number drops to three persons per square kilometer. The urban population tripled between 1960 and 1990, reaching 23 percent of the total, a tremendous growth compared to the less than double overall continental growth from 18.3 percent in 1960 to 33.9 percent in 1990.

There is an enormous variety of ecological systems and climatic zones in Kenya. It ranges from very rich farming areas in the west and central highlands and around Lake Victoria to arid and semi-arid lands in the northern region, where huge and almost empty plains stretch in the north and northwest, bordering Sudan, Ethiopia, and Somalia. The Rift Valley, a branch of the East-African Escarpment, trenches the country from North to South. Overall, 80 percent of Kenya's total land area is semi-arid, and supports a quarter of the country's population and half its livestock.[19] The first major social conflict in modern Kenya was a result of the contest over the best ecological zones of the country, curved out and expropriated by colonial imperialists and renamed "White Highlands" over which they sought exclusive control. The contestation over these most productive parts of the country at this particular moment in time laid the basis for future social conflicts in the country.

Conflict over the "White Highlands"

When the colonialists arrived in Kenya towards the end of the nineteenth century, they began a process of alienation and acquisition of land in the best ecological and climatic zones of the central highlands and parts of the Rift Valley as a prelude to the establishment of the colonial state. Through a series of dubious agreements and sheer force, the local populations were forced out of these areas and herded into less productive areas that the colonialists dubbed "Native Reserves", while they renamed their own areas the "White Highlands." The so-called Native Reserves were created along ethnic lines and placed under the Commissioner for Native Affairs who administered them under separate laws and regulations based on customary law. Local communities that resisted this colonial venture including the Bukusu in Western Kenya, the Nandi in the Rift Valley, and the Kikuyu in Central, among others, were forcefully clobbered into submission through military expeditions that historians have come to euphemistically refer to as "wars of pacification." The sequel to this was imposition of English property law and its acclamation of title and private property rights in the alienated areas. Land tenure reform in the "Native Reserves" under the Swynnerton Plan[20] completed the question by both deepening and diversifying its structure. Since individuals were not allowed out of their reserves, over-population set in, creating all kinds of environmental problems due to scarcity of land for expansion of settlement.

Each of these processes (land alienation and privatization) gave rise to unique but related sets of problems regarding access and control of land thereby laying the basis for what Kanyinga[21] calls a rather complex land question whose solution continues to plague the Kenyan polity. The process had two major consequences with serious implications for social conflicts in the country. In the first place, the alienation of land by the colonialists resulted in mass displacement in alienated areas and especially among the Kikuyu of Central Kenya giving rise to the squatter problem manifested in a mass of people without land rights. The squatter problem later became the basis for the organization of the Mau Mau, an armed rebellion that took place in the 1950s against the colonial state. The rebellion was organized for purposes of fighting for political independence of the country, which was expected to yield the return of the "stolen" land.

Unfortunately, however, the new political elite had no intention of dismantling the colonial political structure to which they succeeded. Jomo Kenyatta, the new president, went so far as to publicly declare his "forgiveness" of the former colonizers and to guarantee them their right to the land they had fraudulently acquired. The irony of this forgiveness was that those being forgiven neither showed any remorse for their past misdeeds, nor did they seek to repent their colonial sins for purposes of genuine reconciliation with the majority they had tormented. Indeed, the "willing-seller-willing-buyer" policy that was settled upon as the basis for the disposal of land in the "White Highlands" meant that only the moneyed few could afford to acquire the land. These were mainly cabinet ministers and top civil servants. As Oginga Odinga observed at the time:

> If Kenya started uhuru [independence] without an African elite class she is now rapidly acquiring one. Ministers and top civil servants compete with one another to buy more farms, acquire more directorships and own bigger cars and grander houses.[22]

At the end of the day, those who had been dispossessed of their land by the colonialists remained landless, the attainment of independence notwithstanding. The new land laws and regulations together with the introduction of the market system to provide an effective exchange mechanism for new cash crops such as coffee and tea eroded local communal societies and the rules, which regulated and monitored land ownership, use, and exchange in the past. Most land redistribution programs were skewed and largely favored the elite. In many cases, more equitable programs gave way to privilege and the elite bought out the poor landowners. Many of the former landowners either moved to cities in search of employment or into marginal agricultural areas where the increased pressure on the land precipitated a process of land degradation. The dual process of social exclusion from land ownership and the concomitant environmental degradation resulting from settlement in marginal areas has culminated in increasing levels of absolute and relative poverty,[23] an eventuality that makes for a politically unstable social setting.

Secondly, as Kanyinga rightly notes, land alienation contributed to the 'ethnicization' of the land question. Most of the squatters moved into the Rift Valley where they sold their labor to the colonial settlers and their numbers added to those of other displaced ethnic groups such as the Kalenjin and the Maasai. Consequently, ethnic tension over the control of land in the area deepened in tandem with the consolidation of the struggle for independence. Some of these other groups feared that the squatters would accede to land rights in the former settler areas and deny them control of what they considered to be their God-given territorial spheres. They consequently began to fence off the squatters from the region and in some cases violently evicted them.[24] The seeds of discord between the various communities in the Rift Valley were thus sowed at this particular historical juncture.

The Shifta War and Insecurity in Northeastern

The Shifta war was the second major social conflagration in Kenya that threatened the integrity of the country in the immediate years following the attainment of independence in 1963. This scenario was a result of irredentism on the part of Somalis in the then Northern Frontier District (NFD—now Northeastern Province). As independence approached, the Somalis sought to disengage from Kenya and join their kith and kin in Somalia. The irredentists explained that their desire to unite with Somalia was not influenced by considerations of fear or enmity for their neighbors in Kenya "with whom they wished to have the friendliest of relations."[25] They asserted, nonetheless, that they wanted an administration that could match their social and political institutions and argued: "The only country that can help us in this respect, is the Somali Republic, which will lift us from our colonial status and give us peace and prosperity, pride and purpose, because it is founded on democratic principles and has a constitution and administration consistent with our way of life and common beliefs."[26]

By "way of life and social and political institutions," the Somalis were tacitly making reference to their pastoralist and nomadic way of life, which was consistent with the ecological set up of the terrain they occupied. Indeed, when Jomo Kenyatta, then Kenya's prime minister, asserted, in the face of heightened Somali demands for secession, that they could pack their camels and leave, they could not countenance loss of the territory they occupied which implied loss of vast tracts of pasture land for their livestock. The secessionists resorted to violence to advance their cause. Armed with mortar bombs, Somali Shiftas (bandits) stepped up attacks on Kenya from mid 1963 that culminated in the murder of Wabera, the Isiolo District Commissioner, and a senior chief in January 1964. It was within this conjuncture that Prime Minister Jomo Kenyatta declared a state of emergency in the North-Eastern Province (formerly NFD) in December 1963 and had the Special Districts Administration Act enacted for the affected districts of Garissa, Mandera, and Wajir. The Shifta war raged on for the better part of the decade of the 1960s, abetting only after the military coup of 1969 that brought Siad Barre to power in Somalia.

Overall, the North-Eastern Province, comprising the three districts of Garissa, Mandera, and Wajir, together with the three neighboring Eastern Province districts of Isiolo, Marsabit, and Moyale is a neglected region, geographically, economically, and socially. It has no roads, has few schools, ill equipped hospitals, and has limited economic significance. The main economic activity is livestock rearing, with communities in the region keeping large herds of cattle, goats, camels, and donkeys. However, because of droughts and high levels of banditry, livestock rearing is increasingly becoming less and less dependable as a means of livelihood. The banditry, skirmishes, and fighting occur between Somali clans and between Somali clans on one hand and the Borana, Burji, and Garre communities on the other hand, each of which enlist support and help from their kith and kin across the borders in Somalia and Ethiopia.

A number of factors explain the conflict between communities in the region, among them the vicious circle of cattle rustling and the proliferation of firearms in the region from Somalia and Ethiopia that feeds the making and remaking of ethnic militias. The key factor that accounts for the deadly skirmishes, however, is the frequency of droughts and the resulting inadequacy of water and pastures. The persistent droughts result into the drying up of dams, boreholes, springs, and rivers leaving only a few that endure the severity of such weather vagaries. Communities around watering points and grazing areas that survive droughts often exact tough conditions on others in exchange for their use of these vital resources. The conditions include relatively exorbitant cash payments or payments in kind in the form of livestock. This eventuality generates hostility and the tendency to resort to survival for the fittest strategies.[27] Furthermore, the fervent inter-communal support of territorial hegemony for the purpose of securing water and grazing land and, of late, political representation in the national parliament and in local councils, is compounded by ignorance and parochialism. This results in distrust between neighboring communities that escalates the vicious circle of conflict, making the region the hardship area it has remained since the colonial times.

Pastoralist Conflicts in the Ilemi Triangle

The third major social conflict in Kenya has been in the pastoralist ecology of the Ilemi Triangle. The Ilemi Triangle[28] is composed of the area covering southern Ethiopia, Southern Sudan, Northern Uganda, and Northern Kenya.

Overall, the Triangle is an oasis in the Horn of Africa, a region whose 70 percent of the total land area is arid and semi-arid, and provides 20 to 30 percent of the region's gross domestic product. With regard to the Kenyan portion of the Horn, it is part of the 80 percent of the country's total landmass that is arid or semi-arid and which supports a quarter of the country's population and half its livestock.[29] Between 1898 and 1900, Emperor Menelik II of Ethiopia conquered and annexed the lower Omo River Valley; the British conquered the Triangle areas of Kenya, Sudan, and Uganda. Due to its permanent pastures as a result of waters from Rivers Tarach, Akobo, and Ajibur, among others that pour into the Lotagipi Swamp, colonial authorities keen to minimize conflict between pastoralists carved the Triangle out as a neutral zone. Free grazing in the Triangle was reserved in April 1924 after a conference at Kitgum, northern Uganda, attended by colonial officials from Kenya, Uganda, and Sudan. Accordingly, the Anyuak, Nyangatom, and Toposa of Sudan; the Merile, Murji, Shangilla, Oromo, and Hamar of Ethiopia; the Borana, Gabra, Pokot, Rendille, and Turkana of Kenya, and the Karamanjong of Uganda were to graze and water their livestock under the supervision of British authorities in Kenya.[30]

However, on the departure of the British colonialists at the end of colonialism in the early 1960s, jurisdiction over the Ilemi Triangle was left indeterminate. Control and rights over the waters and pastures of the Triangle became a free for all. Consequently, fierce fighting among the ethnic groups with claims over the Triangle's natural resources became a common feature. It is on account of this that Amutabi describes the Ilemi Triangle as a death trap, a security nightmare for all governing authorities surrounding it and argues: "Livestock is the main denominator in all the skirmishes and is likely to remain so until peace or alternative economic avenues and ways of livelihood come to the region."[31] Nonetheless, in addition to raiding for livestock, fighting in the Triangle is usually carried out to establish rights over pastureland and water. Raiding purely for cattle purposes occurs mainly during periods of acute resource scarcity on account of droughts and famine. These are the times when raiding for livestock intensifies in the pastoralist domain.

Given the harsh environment in which they operate, pastoralists depend almost entirely on livestock for their subsistence. Livestock ownership defines their social status; it is their main food supplier, means of transportation, and exchange. While for the subsistence farmer it is wise to raise some cattle, to supplement farm products, for the pastoralist it is mandatory since he is defined by it. Cattle are considered to have a value of their own without considering their productivity. It is less important if they supply a lot of milk or meat. What counts is the pure number of heads. This pastoralist predicament has resulted in creation of large, unproductive herds in parts of Kenya where in some cases they have devastating effects on vegetation. To enhance the development of these semi-arid areas, wells were constructed for purposes of engendering a safer water supply and a better distribution of herds throughout the country. Paradoxically, however, this presumed act of development enhancement led to an even further increase in cattle and then, inevitably, to complete destruction of vegetation around these wells, where, ever more herds of cattle gathered.[32]

The carrying capacity of land has thus reached its limit not only in the peasant farming areas, but also in those of different pastoralist ethnic groups, like the Pokot, Samburu, and Turkana. The process of soil erosion and environmental degradation is magnified by the behavior of the affected people. They react to the worsening conditions and the higher risks of loss of cattle by increasing their herds in order to try to compensate for the risk. For the pastoralists involved, raids on the cattle of their neighbors form a part of their survival strategies. For instance, one morning in March 2001, hundreds of young men from the Pokot ethnic group in Kenya, armed with small arms, violently attacked neighboring Marakwet villages. The raiders left 47 people dead, stole livestock, and burned an estimated 300 homes. Around the same time in 1999, close to 1,000 men (also presumably Pokot) killed dozens, including 28 women and 15 children under 12, at a Kenyan market in Turkana. The raiders fled with cattle, goats, and donkeys. The brazen and deadly nature of these raids has been compounded by the introduction of firearms among some pastoralist communities, which first occurred during the 1970s. The situation has been accentuated by the civil wars in Ethiopia, Sudan and Uganda, and the Somali state collapse, which have led to the ready availability of small arms and light weapons in the region. Consequently, cattle raids escalated and became more numerous and bloodier. Because of all this, the readiness to simply steal cattle grew even more, thus worsening the security situation among the communities.[33]

Amutabi argues that the problems affecting this part of Eastern Africa are closely linked to globalization due to the fact that all the combatants have a certain connection to global forces in history. He points out, for instance, that the collapse of the Somali state was a direct consequence of the collapse of the Soviet Union; that the rebel Sudanese Peoples Liberation Movement/Army (SPLM/A)'s support from the international community permeates local groups through arms; that violence among pastoralist groups in the Ilemi Triangle has been exacerbated by availability of the AK-47 rifles, which are procured easily from wars in the region and which are connected to global forces; that the Uganda rebel group, Lord's Resistance Army (LRA), is supported by Khartoum as a counter measure against Kampala's support for the SPLM/A; that the powerful radio communication equipment used by the cattle raiders and bandits, and the modern weapons are all factors of globalization: "...in as much as the problems of the Ilemi Triangle maybe about livestock, they are also about international ideological scheming and counter-scheming."[34]

For instance, in northern Kenya, the Borana have formed alliances with rebel factions from neighboring areas in southern Ethiopia to increase their influence. In recent years, the Oromo Liberation Front (OLF), a faction opposing the government in Ethiopia, and fighting for regional autonomy, has supported Borana claims to key grazing areas. These alliance formations and the introduction of firearms have complicated a hitherto more peaceful coexistence. Competition to control grazing areas wasn't intense in the past. Conflicts over key grazing areas were mediated through reciprocal arrangements between communities. Somali and Borana communities had such an agreement. Communities would be invited to use each other's lands for grazing, allowing for the growth of pasture. But small weapons changed the dynamic, enabling communities to take grazing lands and even livestock by force. Similarly, farmers in South Turkana historically formed reciprocal arrangements to share natural resources with their Pokot neighbors as a way of managing environmental risks. This arrangement was, however, destabilized by colonialism and its concomitant forces of globalization.

On the whole, the raids have also interfered with seasonal migration. The usable area has diminished further in proportion to escalating warfare and has been even more exposed to degradation. A major consequence has been the circumscription of migration and under-utilization of the resources. Grazing areas that are insecure due to violent livestock theft through the use of small arms and light weapons are avoided and more secure key resource areas are badly overgrazed.[35] The ecological poverty in the pastoralist areas is thus perpetuated by the vicious circle of insecurity, warfare, easy availability of firearms, alliance and counter-alliance formation, and environmental degradation. It is a nasty *cul de sac*, extrication from which seems unimaginable even on the most optimistic assumptions, at least from the standpoint of the affected communities.

The Ethnic Clashes of the 1990s

In the run-up to the first multiparty elections in Kenya in 1992, ethnic clashes erupted in Western Kenya between the Bukusu and the Sabaot; and between the Luo, Kisii, and Kuria on the one hand, and the Kalenjin and Maasai on the other in Nyanza. In the Rift Valley, the conflict was between the Maasai and Kalenjin on the one hand, and the Kikuyu, Luhyia, Luo, Kisii, and Kamba on the other. In the run-up to the second multiparty elections in 1997, similar clashes recurred with the addition of Coast Province on the roster of clashes, where the indigenous Mijikenda were pitted against migrant ethnic groups including the Luo, Luhyia, Kamba, and Kikuyu, especially in the Likoni area of Mombasa District and in Kwale District. Whereas the precipitating causes of these conflicts lay in the new competitive multiparty politics[36] given the ethnicized nature of Kenyan politics and the politicized nature of ethnicity in Kenya, the conflicts had ecological roots as their underlying cause, particularly the complex issue of rights of access to, and use of land.

In the case of Bungoma District of Western Province, the clashes began on December 24, 1991 with the burning of houses of the non-Sabaot in Kapsokwony Division of Mount Elgon sub-District. By April 1992, the clashes had spread to Cheptais, Sirisia, and Kimilili Divisions of Bungoma District. In evicting the non-Sabaot, mostly the Bukusu from Mount Elgon, the Sabaot saw this as a step in the right direction toward the recovery of their ancestral lands, which they had lost to the White settlers. Indeed, they got to keep the lands from which they had forcefully evicted the Bukusu and, in 1994 their demand for a separate district was finally granted. Second, in the wider Rift Valley Province, the clashes began on October 29, 1991 at Miteitei Farm, Nandi District, and quickly spread to other farms and into Kericho District. By early 1992, the clashes had spread to Molo, Olenguruone, Londiani, Trans Nzoia, Uasin Gishu, Naivasha, Narok, and Gucha. Ibrahim Injaga Livanze, a witness to the Akiwumi Commission[2] succinctly summed up the underlying cause of the clashes:

> The problem was tribal land. That the Kalenjin did not want other people to live on their land while they were landless... [they said,] 'you will not get land. What you have paid here, you have earned. You have milked this land enough time. Now, you must go.'[37]

In the same vein, clashes at the Coast were undergirded by similar grievances on the part of the indigenous Mijikenda communities. As Akiwumi notes, the coastal inhabitants of Likoni and Kwale have an inferior status in the land of their forebears. Most small-scale and large-scale businesses are in the hands of the non-coastal people. It is reckoned that the coastal people control less than twenty percent of the commercial and business sector including exploitation of the key natural resources of the area such as the sandy beaches, historical sites, magnificent islands and titanium mining, among others. Very little of the revenue generated from the lucrative tourism industry in the area is ever used to uplift the social standing and economic conditions of the locals. Multiparty politics and political instigation of ethnic clashes on the part of political entrepreneurs to disorganize the opposition thus provided an opportune moment for the Mijikenda to demand a rightful share in the exploitation of their natural resources and the return of their grabbed lands and desecrated sacred spaces.

Duraiappah et al[38] are thus correct in contending that the origins of many conflicts in parts of the developing world can be traced to disputes over land ownership, rights of access, and degradation. They argue that a lack of the understanding of new institutions for land tenure, land use, and market exchange by some groups often lead to their exploitation and marginalization by the rational choice of other groups who are more informed. "Conflicts over ownership, access, and rights of use to land are symptomatic of the social unrest which has plagued countries such as Guatemala... Nicaragua... Kenya... South Africa... and India."[39] Moreover, they posit, these conflicts are also a response to the increasing control of economic and political institutions by a small elite and privileged class, the degradation of natural resources, and the widening inequality and poverty among certain communities, which usually results from these disagreements. This position is vindicated by the foregoing analysis and is illustrated with particular clarity by the conflict between the Kikuyu and Maasai in Narok District, which is emblematic of the 1990s ethnic clashes in Kenya.

The Kikuyu-Maasai Conflict in Narok

The Narok clashes between the Maasai and the Kikuyu, which preceded the December 1992 multiparty elections escalated again during the week of October 13 1993 when Maasai Morans in Enoosopukia attacked Kikuyu settlers and killed at least 17 of them. The Enoosopukia clashes in Narok began after Kikuyu settlers allegedly mutilated Maasai cattle for trespassing on their farms. The Maasai accord great respect to cattle, the source of their livelihood, which they consider a gift from God. As a result of this insult, Maasai Morans (warriors) numbering 500 attacked and killed up to 33 Kikuyu as the clashes spread throughout the Narok district and forced 30,000 Kikuyu out of the area.[40] None of those driven from their homes was made aware of their destination. The displaced were forced to gather at the Maela Refugee Camp, about ten kilometers from Enoosopukia, while the Maasai took over their farms. Although the clashes in Narok followed calls by senior government officials for the complete expulsion of immigrant ethnic groups from the Rift Valley Province, the violence was preceded by years of distrust between the indigenous Maasai and the immigrant Kikuyu.

The crisis can be traced to the land reform initiative by the Kenya government aimed at bolstering agricultural productivity. The land reform aimed at displacing indigenous land tenure systems and imposing private property rights along the lines of English land law. The logic of this land tenure reform was that private property in land is a necessary condition for the development of capital accumulation in agrarian societies. It was assumed that this would provide security for investment, which would, in turn, boost agricultural production on which the country's economy depends. This thinking reflects the World Bank's sectoral work and Structural Adjustment Programs. The Bank has been supporting titling efforts on the assumption that this will ensure secure land rights, activate markets, and increase agricultural production.[41] The process of privatization of land tenure originally focused on high potential areas but was extended at the national level to include the arid and semi-arid rangeland areas, including Narok, where major wildlife and tourism activities are located.[42]

On account of this process of land tenure reform, communal land in Narok District was privatized and individual land titles were distributed to the Maasai pastoralists. Within this conjuncture, the landless Kikuyu, products of the land alienation by the colonialists and population growth in Central Province, were able to purchase land in Narok. As a relatively wet and fertile oasis, a water catchment area in an extremely arid region, the Enoosopukia area naturally attracted the Kikuyu, who are traditionally agriculturalists. By 1990 the Kikuyu, who tended to be better educated and skilled, replaced the Maasai, a pastoralist people, as the majority in the area. The Kikuyu and Maasai had traditionally been partners in trade. Perhaps because of this tradition of cooperation, both sides avoided open conflict, and the Kikuyu continued to obtain permission to settle in Enoosopukia from the local authorities, including Maasai hardliner and government minister William ole Ntimama.

The new land tenure system had a major impact on the Maasai. First, for it to succeed, it required a radical change to the pastoralist way of life, particularly from nomadic to sedentary agricultural activities, from pastoralism to farming. Yet behavior cannot be legislated for, especially in the absence of incentives, encouragement, and government institutional support. Without the support of government for livestock management, as demonstrated by the dismantling of the Kenya Meat Commission, pastoralists lacked the wherewithal to follow this route. Second, the situation was compounded by the asymmetry of information, which existed among the different communities and groups. The predominantly Kikuyu commercial and mixed farmers as well as a small group of Maasai elite were more keenly aware of the importance of individual land ownership within the market system and the value of the land itself and really took advantage of land privatization to acquire land tracts of arable land. It was for this reason that Ntimama argued that the Kikuyu had acquired their land by dubious means, cheating the illiterate Maasai out of their ancestral property.[43] Indeed, the Maasai, having survived for generations on the basis of communal land ownership, and having effectively resisted the allure of modernization including schooling, did not fully comprehend the implications of private land title.

Second, it was the agricultural activities of the commercial and mixed farmers, in contrast to the pastoralist activities of the Maasai that were better suited to the new land tenure system. The spatial and inter-temporal dynamics of Maasai pastoralism on the other hand thrived and depended very much on communal land ownership.[44] The resultant inequity generated increasing animosity on the part of Maasai pastoralists towards the mixed farmers. The anger was fueled largely by frustration on the part of the pastoralists at exclusion from traditional lands, particularly for livestock and the accompanying lower income levels, and a widening equity gap between them and the other communities in the district, especially the Kikuyu.[45] The fencing off of wildlife conservation areas further depriving the Maasai of grazing lands further compounded their Maasai predicament and sense of loss. The costs of conserving wildlife have thus been borne largely by the Maasai pastoral community without due compensation.[46]

Thirdly, as Pander[47] posits, the change from communal grazing to individualized land plots had serious ecological impact on account of the concomitant change in land use. Individual tenure increased settlement, fencing, and intensified cultivation. The intensification of crop production in arid and semiarid lands (ASAL) greatly impacted on natural resources especially water and carrying capacity of the soil to a larger extent than other areas. With regard to Narok, the Enoosopukia area suffered the consequences of bad agricultural policy for decades. As a relatively wet and fertile oasis in an extremely arid region, the increased settlement in the area, and constant use of the soil without replenishment transformed the area into a near wasteland. "Only blackened tree stumps, the legacy of unbridled charcoal burning and slash-and-burn agriculture, interspersed with scattered trees and other sparse vegetation, testify to the fact that the area was formerly a lush forest."[48] The consequence in Narok was a downward spiral to poverty, inequality, and environmental degradation.

In view of the foregoing, the degradation of the environment had been a concern for over a decade. In April 1993, six months before the conflict broke out, the Narok County Council declared Enoosopukia, the site of the conflict, a water catchment area and decreed that all inhabitants, mostly immigrant Kikuyu, had to leave. This was in consonant with the demand by a German team that had built a dam in the area that the catchment had to be protected if the construction was to be successful.[49] Indeed most observers agree that the Narok County Council had every right to declare Enoosopukia a water catchment area. The area is truly a catchment zone for streams flowing toward the Maasai dry lands, and it had been devastated in recent years. However, the government, which ordered everyone out, and local belligerents, who massacred innocent people, enforced the local governing body's decision badly. Economics partly explain the government's decision; the expectation that showing firm commitment to environmental protection would convince Western non-governmental organizations and other donors to continue their financial assistance. For the local belligerents, the emergent multiparty politics provided an opportune moment for expelling the non-indigenous groups and thus expanding the reach of their grazing fields.

The Multiparty Political Angle

From the foregoing exposition, it is evident that by the time of the reintroduction of multiparty politics in Kenya in 1991, the relations between the Kikuyu and Maasai were far from cordial. The animosity between the two communities was so intense that it required just a small spark to explode into open hostilities. This came in the form of political incitement on the part of incumbent political entrepreneurs who appropriated the land question and the related complex environmental concerns as a tool to fight the opposition and scatter their supporters. They projected the new multiparty politics as a mere plot, a conspiracy by other ethnic groups to remove President Moi from power. With the re-introduction of multiparty politics, therefore, tempers began to flare between the Maasai elites, who tended to support the incumbent regime, and the Kikuyu, many of whom favored the opposition. Pro-establishment political leaders among the Maasai and Kalenjin communities took advantage of the new politics of ethnicity to unflinchingly defend the perceived interests of their ethnic communities against all others. William Ole Ntimama in particular, an ethnic Maasai and Narok Member of Parliament, demanded that Kikuyus residing in his Narok District support him in the 1992 elections. As noted above, Ntimama contended that the Kikuyu had acquired their land in Narok by dubious means. They had, he asserted, cheated the illiterate Maasai out of their ancestral property (Dietz 1997:8). Ntimama fanned the flames of ethnic hatred by making "blatantly inciting utterances at a public meeting, by saying that the non-Maasai living in Maasai land should respect the Maasai, and further warned that the title deeds owned and cherished by such non-Maasai were mere papers that could be disregarded at any time."[50]

According to Ntimama, the Maasai have historically been the victims of oppression, first, by the British colonialists, and second by the Kikuyu, who allegedly marginalized his people by encroaching on their ancestral homeland. At a conference in Vienna on indigenous peoples, he claimed that the Maasai were becoming an "endangered species" and would have to fight to protect themselves. Ironically, it was Ntimama, as Chairman of the Narok County Council at the time, who allowed the Kikuyu access to the land around Enoosopukia in the first place. For this he expected political support.[51] He espoused the view that settlers must vote with the local Maasai or move out of his district, even though the Kenyan constitution guarantees the right of all Kenyans irrespective of their ethnicity to move, settle, and own property wherever they choose.

It was within this context that Maasai herdsmen began to brazenly herd their livestock onto Kikuyu private land in the Enoosopukia area for purposes of grazing. Feeling justifiably wronged, some Kikuyus reacted by mutilating Maasai cattle for trespassing on their farms during the week of October 23 1993. For the Maasai who consider cattle a sacred gift from God, this was seen as an unimaginable act of desecration. Consequently, 500 Maasai Morans (warriors) mobilized for attack and killed 17 Kikuyus in Enoosopukia. As the clashes spread throughout the Narok District, 16 other Kikuyus were killed bringing the death toll to 33. With the escalation of attacks against the Kikuyu, 30,000 of them were forced out of the area.[52] None of those driven from their homes was made aware of their destination. The displaced were thus forced to gather at a makeshift Refugee Camp at Maela Township, about ten kilometers from Enoosopukia. In the interim, the Maasai took over their farms.

The Narok clashes thawed at the end of October 1993. Nonetheless, the problem of internally displaced people (IDPs) continued much longer. Government attempts to address the plight of IDPs largely failed. At the end of 1994, as many as 8,000 IDPs were still living at the Maela Camp when it was razed down by government troops (Dietz 1997:8). Government forces transferred 2,000 displaced persons to Central Province, the "traditional home" of the Kikuyu, according to the incumbent KANU regime. Families, unaware of where they were being sent, became separated. By the end of the KANU regime in December 2002, the situation had still not been satisfactorily resolved. Most of the displaced persons from the clashes of early and late 1990s were still crying out for resettlement and alleviation of their plight.

Even the post-election violence of early 2008 following the disputed December 2007 elections can be traced to the land question. Part of the driving force of electoral politics in Kenya, especially in the vast Rift Valley Province are historical grievances associated with access to land and its use, and marginalization from the same. In mobilizing to vote for the opposition in 2007, the Kalenjin expected to have their historical grievances addressed. President Kibaki's brazen rigging the electoral outcome meant perpetuation of the status quo and thus defense of entrenched interests. It was within this context that the Kalenjin turned on the Kikuyu, Kibaki's ethnic group, with vengeance, sparking off violence that saw the death of over 3,000 and displacement of 300,000.[53] The import of this and the Enoosopukia conflict is that land is a central dynamic to the broad Kenyan social and political context as it has a major impact on the patterns of social relations. How access to land is regulated, and any changes to the structure of land tenure have significant implications not just for agrarian development and economic organization, but also for environmental concerns and socio-political dynamics.

Conclusion

This chapter has demonstrated the ecological roots of social conflicts in Kenya. The ecological dynamic, it has argued, is the underlying cause of most conflicts in Kenya and in Eastern Africa generally. Conflicts in the arid areas of the country are a function of the harsh ecological conditions and the resulting contest between groups and communities over the scarce natural resources including water, pastures as well as livestock. In the more arable areas, the politics of the rights of access to land, its use, and degradation are at the core of social tensions that frequently escalate into warfare under opportune conditions that act as a spark such as electioneering periods. The prevalence of these conflicts is ample testimony to the fact that diplomacy and policy to manage them has largely been ineffective mainly because of their inherent fixation on resolving competing political viewpoints and inadequate attention to the environmental embededness of social conflicts.

There thus is imperative need for a shift in conflict management approaches away from resolution-oriented ones toward prevention-oriented strategies that rely on extensive information, take into account the multiplicity of social conflict causal variables, and recognize the multiple linkages between multiple actors in given conflict situations. Since these social conflicts are a function of contests over natural resources, as has been demonstrated in this chapter, it is crucial that governments in the region design broadly defined and clearly articulated rights of equitous access to and use of natural resources, be they land, water, pasture, or whatever. This would obviate the need for vulnerable groups and those that feel excluded from the resource dispensation of particular moments to seek access to resources via force of arms. At the end of the day, therefore, neo-Malthusians trump Cornucopians. The policy complacency implicit in Cornucopian theorizing cannot be afforded by developing countries that lack the requisite capital and technical capacity to hedge their communities against the vagaries of environmental scarcities. For them, the most viable option is the neo-Malthusian prescription: a national resource use strategy embedded in a national environmental policy that ensures renewable use of the natural resource capital and that, ipso facto, engenders sustainable development. Failure to take into account this ecological dimension will render efforts at social conflict prevention and management overly intractable at least for the foreseeable future.

Notes

1. P. G. Okoth and B.A. Ogot, eds., *Conflict in Contemporary Africa* (Nairobi: Jomo Kenyatta Foundation, 2000), 257-268.
2. H. Sprout and M. Sprout, *The Ecological Perspective on Human Affairs with Special Reference to International Politics* (Princeton: Princeton University Press, 1965).
3. G. Anyumba, "Diversity and Assortment of Environmental Conflicts in Africa," in P. G. Okoth and B.A. Ogot, eds., *Conflict in Contemporary Africa* (Nairobi: Jomo Kenyatta Foundation), 226-242.
4. Ibid., 226.
5. Ibid., 228.
6. T.F. Homer-Dixon, "Environmental Changes as Causes of Acute Conflict," in R. K. Betts, ed., *Conflict after the Cold War: Arguments on Causes of War and Peace* (New York: Longman, 2002), 495.

7. G. D. Dabelko and D.D. Dabelko, "Environmental Security: Issues of Concept and Redefinition," *Environment and Security*, Vol. 1, No. 1, Spring, 1996, 23-49.
8. See M. Weiner, "Security, Stability, and Migration," in R.K. Betts, ed., *Conflict after the Cold War: Arguments on Causes of War and Peace* (New York: Longman, 2002), 467-482; B. Mwasi, "Land Use Conflicts Resolution in a Fragile Ecosystem Using Multi-Criteria Evaluation (MCE) and a GIS-Based Decision Support System (DSS)," in International Conference on Spatial Information for Sustainable Development (Nairobi, Kenya, October 2-5, 2001); D. Schwarz & A. Singh, *Environmental Conditions, Resources, and Conflicts: An Introductory Overview and data Collection* (Nairobi: UNEP, 1999); B.A. Ogot, ed., *Ecology and History in East Africa* (Nairobi: Kenya Literature Bureau, 1979).
9. T. F. Homer-Dixon, "Environmental Changes as Causes of Acute Conflict," 493-507.
10. J. K. Cooley, "The War over Water," in R. K. Betts, ed., *Conflict after the Cold War: Arguments on Causes of War and Peace* (New York: Longman, 2002), 483-492.
11. Ibid., 483.
12. Ibid.
13. See IRN, "The Nile: Water Conflicts," *Science in Africa* (May, 2003).
14. J. Simon, *The Ultimate Resource* (Princeton: Princeton University Press, 1981), 345.
15. Homer-Dixon, "Environmental Changes as Causes of Acute Conflict," p. 502; Dabelko and Dabelko, "Environmental Security: Issues of Concept and Redefinition," 23-49.
16. Homer-Dixon, "Environmental Changes as Causes of Acute Conflict," 502-503.
17. J. E. Dougherty and R.L. Pfaltzgraff, *Contending Theories of International Relations: A Comprehensive Survey* (New York: Longman, 1997).
18. Homer-Dixon, "Environmental Changes as Causes of Acute Conflict," 494.
19. S. Mbogo, "Pastoral Shoot outs: Breaking the Link between Small Arms and Environmental Conflicts," *Toward Freedom, Online Magazine* (winter, 2003); J. M. A. Orodho, "Development and the Environment," in J.E. Otiende *et al.* eds., *An Introduction to Environmental Education* (Nairobi: Nairobi University Press, 1997), 144-168; J.W. Kabeberi-Macharia, "Policies and Laws Affecting the Environment," in J.E. Otiende et al. eds., An Introduction to Environmental Education, 224-263; H.O. Oruka, ed., *Philosophy, Humanity, Ecology* (Nairobi: ACTS Press, 1993).
20. See J.M. Swynnerton, *A Plan to Intensify the Development of African Agriculture in Kenya Colony and Protectorate of Kenya* (Nairobi: Government Printers, 1954).
21. K. Kanyinga, "Struggles of Access to Land: The 'Squatter Question' in Coastal Kenya," CDR Working Paper 98.7 (Copenhagen: Danish Institute for International Studies, 1998).
22. W. R. Ochieng, "Structural and Political Changes" in B. A. Ogot and W. R. Ochieng, eds., *Decolonization and Independence in Kenya, 1940-93* (London: James Currey, 1995), 100.

23. H.M. Amman and A.K. Duraiappah, "Land Tenure Conflict Resolution: A Game Theoretic Approach in the Narok District in Kenya," Working Paper No. 37 (London: International Institute for Environment and Development and Amsterdam: Institute for Environmental Studies, 2001); A. Dejanvry, "Land as a Source of Conflict," in USAID Workshop on Conflict Prevention Management, (Washington, DC, June 7, 2000); A.K. Duraiappah *et al.*, "Land Tenure, Land Use, Environmental Degradation and Conflict Resolution: A PASIR Analysis for the Narok District, Kenya." Working Paper No. 3 (London: International Institute for environment and Development and Amsterdam: Institute for Environmental Studies, 2000).
24. K. Kanyinga, "Struggles of Access to Land: The 'Squatter Question' in Coastal Kenya," 2; J. Plateau, "The Evolutionary Theory of Land Rights as Applied to sub-Saharan Africa: A Critical Assessment," *Development and Change* (Vol. 27, 1996), 29-86.
25. S.W. Nasong'o, "Ethnonationalism and State Integrity in Africa: Cultural Objectification and the Rwandan Genocide," *Canadian Review of Studies in Nationalism* (Vol. 30, No. 1-2, 2003), 53-63.
26. Ibid.
27. See A.M. Akiwumi, *Report of the Judicial Commission of inquiry Appointed to Inquire into Tribal Clashes in Kenya* (Nairobi, Government Printer, 2001), 2.
28. The Triangle was named after Chief Ilemi (Ilembi) of the Anyuak whose village was located on the Sudanese bank of the Akobo River near its juncture with River Ajibur.
29. Mbogo, "Pastoral Shoot outs: Breaking the Link between Small Arms and Environmental Conflicts."
30. See M. N. Amutabi, "Beasts of War, Pastoralism, Banditry, and Globalization: The Case of Livestock and Conflict in the Ilemi Triangle of Eastern Africa," in Association of Third World Studies 19th Annual Meeting (Savannah Georgia, October 11-13, 2001).
31. Ibid.
32. C. I. Lang, "Environmental Degradation in Kenya as a Cause of Political Conflicts, Social Stress, and Ethnic Tensions," ENCOP Occasional Paper No. 12 (Zurich: Center for Security Policy and Conflict Research and Berne: Swiss Peace Foundation), 1995.
33. See J.A.N. Kamenju *et al.*, *Terrorized Citizens: Profiling Small Arms and Insecurity in the North Rift Region of Kenya* (Nairobi: Security Research and Information Center, 2003); R. Carver, "Kenya: Update to End of July 1995," Writenet Country Papers (1995).
34. Amutabi, "Beasts of War, Pastoralism, Banditry, and Globalization: The Case of Livestock and Conflict in the Ilemi Triangle of Eastern Africa," 1.
35. H. Busingye, "Land Planning and Sustainability of the Natural Resource base: Land Reform Processes and Policies in Uganda," *International Conference on Agri-*

culture Beyond Trade (Paris: Centre de Conferences Internationales, January 8-10, 2002).

36. E.O. Mogire, "The State and Internal Political Conflicts in Africa: The Case of Kenya," in Okoth, P. G. and Ogot, B. A., eds., *Conflict in Contemporary Africa* (Nairobi: Jomo Kenyatta Foundation, 2000), 128-140; F.O. Wanyama, "The Role of the Presidency in African Conflicts," in Okoth and Ogot, eds., Conflict in Contemporary Africa, 30-43; G.R. Murunga, "Urban Violence in Kenya's Transition to Pluralist Politics, 1982-1992," *Africa Development* (Vol. XXIII, Nos. 1 & 2, 1999).
37. Akiwumi, *Report of the Judicial Commission of inquiry Appointed to Inquire into Tribal Clashes in Kenya*, 6, 8.
38. A.K. Duraiappah et al. "Land Tenure, Land Use, Environmental Degradation and Conflict Resolution."
39. Ibid. 7.
40. T. Dietz, *Entitlements to Natural Resources* (Utrecht: Antenna Books, 1997), 7; Weekly Review (Nairobi: October 29, 1993), 4-7.
41. See World Bank, *Sub-Saharan Africa: From Crisis to Sustainable Growth* (Washington, DC: The World Bank, 1989).
42. H. Pander, *Land Tenure and Land Policy in the Transmara District, Kenya: Situations and Conflicts* (Nairobi: German Technical Cooperation Agency, 1995).
43. Dietz, *Entitlements to Natural Resources*, 8.
44. Duraiappah *et al.*, "Land Tenure, Land Use, Environmental Degradation and Conflict Resolution," 26.
45. Ibid., 27.
46. Ibid., see also, M. Ole Dapash, "Coexisting in Kenya: The Human-Elephant Conflict," *Animal Welfare Institute Quarterly* (Vol. 51, No. 1, winter 2002).
47. Pander, *Land Tenure and Land Policy in the Transmara District, Kenya: Situations and Conflicts.*
48. See "The Environmental Factor," *The Weekly Review* (Nairobi: October 29, 1993).
49. Dietz, *Entitlements to Natural Resources*, 10.
50. Republic of Kenya, *Report of the Parliamentary Committee to Investigate Ethnic Clashes in Western and Other Parts of Kenya* [Kiliku Report] (Nairobi: Government Printer, 1992), 60.
51. *Weekly Review* (Nairobi, October 29, 1993), 8-9.
52. Dietz, *Entitlements to Natural Resources*, 7; *Weekly Review* (Nairobi, October 29, 1993), 4-7.
53. V. O. Ong'ayo, "The Post-election Violence in Kenya: An Overview of Underlying Factors," http://www.pambazuka.org/en/category/comment/46094 (2008).

• Chapter 13 •

FAILED PROMISES: A STUDY OF THE PROMOTION OF SCIENCE AND TECHNOLOGY IN NIGERIA

Bukola Adeyemi Oyeniyi

Introduction

The primary aim of this chapter is to assess Nigeria's efforts at ensuring the promotion of science and technology since independence in 1960. The chapter asserts that since 1960, efforts at ensuring the promotion of science and technology dominated national dialogue, but rather than ensuring development, evidence abound to assert that much of the energy is dissipated through non or half-hearted implementation. Going by the number of commissions of inquiry, seminars, symposia, lectures, etc which we have had on how best to promote science and technology, it goes without saying that the failure of the mission is not lost on Nigerians.

It is incontrovertible that science and technology is a key to the development of any nation; hence, the chapter seeks to understand what factors militate against the development and implementation of sustainable science and technology policy in Nigeria. As researches have shown, Nigerians not only have the requisite knowledge, but also put the right policies in place to attain the goal; but the nation lacks the will and the leadership to do what it ought to do. Above all, Nigeria, as this study shows, specifically lacks the culture that is capable of ensuring the development of science and technology.

The study is divided into six sections. The first section, which is this introductory one, sets the basic parameters of the chapter. In the second, attempts are made to explain the basic concepts used in the study and how each of them integrates into one another. The third section reviews Nigeria's journey towards the promotion of science and technology as an end from the colonial period to the contemporary period. In the fourth section, attention is devoted to identifying what went wrong in the nation's efforts at promoting science and technology, while in the fifth section; the study focuses on how to combat the problem and get the nation out of the cauldron. In the sixth, which is the concluding section, the study brings out the basic lessons while at the same time asserting that in spite of the recognition of the need for a culture of science and technology among the citizenry, official efforts, so far, have been concentrated on centralized planning, financing and control under predominantly authoritarian regimes. To ensure that scientific development attains sustainable growth, it is necessary to link it more closely with indigenous technological development and traditional science.[1]

Conceptual and Theoretical Frameworks

Culture, culture of science and technology, and sustainable development are three major concepts that deserve some clarifications, no matter how brief. Culture is derived from the Latin word, *cultus,* which originally describes the practice of nurturing and cultivating domesticated plants and animals. In its modern sense, culture can be understood from different perspectives. Linguistically, it is that distinctive factor which differentiates one people from another. From a geographical viewpoint, it is the delineation of areas having the same dominant and significant traits, complexes, and pattern. From its ethnographic and anthropological connotations, it is a complex whole, a superstructure which includes knowledge, beliefs, arts, morals, laws, customs and any other capabilities and habits acquired by humans as members of a community. It describes the cumulative knowledge of humankind in the production, management, use of artefacts, building, implements, weapons, and other material objects and the meanings assigned to them.[2] The Inter-governmental Conference on Cultural Policies in Africa defines culture as consisting of all material and spiritual assets, works of art, knowledge and know-how, languages, ways of thought, forms of behavior and store of experience accumulated with a view to building a better society, and meeting the needs of a constantly changing world.[3]

Although transmitted from one generation to another, culture is neither inherited nor acquired by instinct. When two people share a similar culture with one another, culture helps to define the group or society to which they individually belong. Having a common culture therefore simplifies many day-to-day interactions. Language, amongst other things, is a critical element of culture that sets humans apart from other species. Members of a society generally share a common language, which facilitates day-to-day exchanges with others.

Nigeria's National Policy on Education and the National Policy on Science and Technology conceive of science and technology as the introduction of western science through formal schooling and educational institutions to goods and production; the offering of labor-saving and mechanical ways of producing goods.[4] By culture of science and technology, we refer to the socio-economic and political environments under which science and technology can best be promoted, flourished and grown in a self-sustaining manner. Development means change in an organism's state, which could be upward or downward. Generally, development is conceived as upward or progressive change in an organism's physical state. In relation to a society, change must be from one bad state to a good state. For any change to be described as sustainable, such a change must be that it is controllable and could be ignited at will. Sustainable development can be described as an upward change in a society's fortune, which can be ignited, controlled at will and sustained for as long as the society pleases.

As far as this study is concerned, three basic theories of culture that are crucial for examination are the functionalist theory, conflict theory and interactionist theory. These theories help in understanding the dynamics involved in how culture interfaces with development in any given society. Essentially, the theories agree that culture and society are mutually supportive, but for different reasons. Functionalism, which describes a view of society as a system made up of interrelated parts, all interacting on the basis of a common value system or consensus about basic values and common goals. Every social custom and institution, as functionalists have argued, is seen as having a function in ensuring that society works efficiently and effectively. In relation to culture, functionalists have argued that for social stability, a society requires a consensus and the support of all its members. Besides, such a society also needs strong central values and common norms capable of providing supports for it.[5]

From the functionalist's perspective, a cultural trait or cultural practice will persist if it performs functions that a society needs or that contributes to a society's overall social stability and consensus. Functionalists often describe society as an organism with a life of its own, above and beyond the sum of its parts. French sociologists Auguste Comte and Emile Durkheim and American Talcott Parsons assumed functionalist approaches in their studies.

Conflict theorists, on the other hand, agree that a (common) culture may exist, but it serves to maintain the privileges of, usually, a small group. Moreover, while protecting their own self-interests, powerful and privileged groups may keep others in a subservient position.[6]

Therefore, from a conflict perspective, the dominant ideology or culture has a major significance. A society's most powerful and privileged groups or institutions control wealth and prosperity, they also control the means of producing beliefs about reality through religion, education, and the media. For instance, a dominant culture in relation to gender in most African societies is that women should submit to their husbands. This dominant culture, as argued, helps to control women and keep women in check.

Interactionism is a sociological framework in which human beings are viewed as living in a world of meaningful objects. These objects may include material things, action, other people, relationship, symbols, beliefs, ideas, etc.[7] While both the functionalist and conflict theory perspectives analyze large-scale, society-wide patterns of behavior; interactionists are concerned with everyday forms of social interaction in order to explain the society as a whole. As far as development and the promotion of science and technology or social changes in general are concerned, the interactionists argue that they are maintained by shared understanding of everyday behavior (culture) of the people in the society and reflected in people's social positions and their communications with others in the society.

From both the functionalist and conflict theorists' perspectives, culture serves important functions in the development of a society. To the interactionists, development of any kind can only take place within a particular cultural milieu. While development implies a healthy growth of the society, culture is the life of the society. One cannot be divorced from the other. Viewed in this way, any developmental effort, to be successful, must agree with the basic culture of the society or, at the least, be ingrained and grafted into the society's culture. The promotion of science and technology in any society therefore involves a healthy relationship between culture and the science and technology. In essence, sustainable development in science and technology, or anything at all, cannot be achieved when it is pursued as an end rather than a means to an end.

Towards the Development of a Culture of Science and Technology: The Journey So Far

Two major issues that are of concern in this section are: (i) establishing a linkage between culture and sustainable development whether in science, technology or medicine; and (ii) tracing Nigeria's attempt to attain sustainable development in science and technology. In doing these, the section is divided into two sub-sections. The first examines the linkages between different aspects of Nigerian cultures and how they help to promote sustainable development. The second sub-section briefly chronicles the various efforts made by the nation, over the years, to ensure the promotion of science and technology.

As earlier noted, culture is a functional, multi-dimensional phenomenon whose use revolves around people's perception. As argued in sociology, the process of introducing a new thing or idea to a society, called innovation, could either be through discovery or invention. While discovery relates to the introduction of or sharing of the existence of an aspect of reality; invention occurs when existing cultural items are combined together to form another thing that did not exist before. The finding of the DNA molecules and the identification of a new moon on Saturn are discoveries while television, automobile, democracy, and Catholicism, etc., are inventions.[8]

From the foregoing, it is incontrovertible that science and technology, in as far as they result from existing cultural items, which are combined together to form another thing that did not exist before, are inventions. As cultural items or elements, science and technology can be transferred from one culture area to another. To be effective and efficient, science and technology must exist either within an existing cultural milieu or within an entirely new culture.

Development of science and technology is culture-bound. To be successful, science and technology depends entirely on its culture or, if ingrained or grafted within another culture, its ability to integrate into the new culture. Democracy, to consider a common example, benefits and develops efficiently and effectively from an existing culture of openness and collective efforts in a society. In precolonial Yorubaland, the processes of making an Oba were known to the generality of the Yoruba people. In Oyo, the kingmakers, the *Oyo Mesi*, with the approval of the oracles, confirmed the new Alaafin who must be chosen from the royal families before such an Alaafin could begin his reign. While the processes of making an Alaafin were generally known, the actual requirement for making the stool was closed to only the male members of the royal family.[9] In non-centralized societies of Igboland, the Aro to be precise, where oracular priests rule, qualification to office was not only limited to the initiates, the procedure for election to office is generally unknown and the generality of the people had little or no say in the making of their rulers. Given the level of openness in precolonial Oyo or any other centralized states, democracy would thrive more in Oyo than in Igboland or any other non-centralized closed society. For any innovation, whether a discovery or an invention, to be successful, it must relate with the society or be grafted into it. Unless this is done, innovations may not succeed.

Although Nigeria's attempt at promoting science and technology began before the nation's independence in 1960, much of the energy dissipated was wasted through non-implementation or half-hearted implementation. Although colonial administration devoted little or no attention to the development of manufacturing and good production in Nigeria, the need for scientific and technological development was not lost on them. Efforts were limited to the provision of essential services, especially in health, post and telegraph and road construction. Engineers, doctors and technicians required for the opening up and effective administration of Nigeria were lacking. The then on-going World War II starved the administration of capable hands from Europe and most of them had become preoccupied with the war efforts. After the war, the task of reconstruction directed the attention of government away from the colony. This, coupled with increasing agitation for independence, led to a situation whereby the needed manpower had to be sourced within the colony. A beneficial result of this development was the establishment of the University College, Ibadan which was established in 1948, consequent upon the recommendation of the Elliot Commission on Higher Education in West Africa.

The primary aim of setting up of what later became the University of Ibadan was the training of the needed manpower for the development of not only Nigeria, but for the manpower requirements of British colonies in Africa.[10] Since inception, the university never made the pretension of focusing on science and technology. To focus specifically on science and technology, the Yaba College of Technology was founded also in 1948 in Lagos.

By 1954, in the process of decolonization, Nigeria adopted a Federal constitution initially with three regions: East, West and North. Higher education was placed on the concurrent list, that is, leaving it open for the Federal and Regional governments alike to establish and finance universities of their own. The Eastern Region in 1955 passed the University of Nigeria Act and was seeking external assistance to finance it.[11] Other regions initiated their own plans to establish universities in their respective regions. To rein in and coordinate efforts of both the Central Federal Government and the regional governments, the Federal Government set up the Ashby Commission in 1959 to review the manpower needs of the whole country and make recommendations for a structure of postsecondary and higher education to meet those needs in the 1960s and 1970s.[12]

Besides the University College, the Ashby Commission recommended the establishment of a second Federal University to be located in Lagos to cater for commerce, business studies and law, not available in Ibadan, as well as for regional universities in the east and north where no universities then existed.

It also recommended that all the universities should be national in outlook and subject to some central planning and direction, with the bulk of capital grants and a proportion of recurrent funds coming from the Federal Government. However this attempt at a fair distribution in the location of new universities failed to take account of political realities. The West regional government, led by the party in opposition at the Federal level, regarded the proposals as discriminatory and pressed for a west regional university. Hence between 1960 and 1962, the University of Nigeria (Nsukka), the Ahmadu Bello University (Zaria), the University of Ife and the University of Lagos were established and the University College, Ibadan, became an autonomous university. In 1963 the Mid-west region was carved out of the West and it also began to demand its own university. It was the national crisis of 1966- 70 that delayed the establishment of the University of Benin till 1972.[13]

Pursuant to the implementation of the Ashby Commission report, the Central Federal Government, in a White Paper, proposed the establishment of (i) an inter-regional manpower board; (ii) an all-Nigeria academic council;

and (iii) a National Universities Commission. While the National Manpower Board was established in 1962, the academic council was never established. The National Universities Commission, on the other hand, was constituted in 1962, but its twelve members, appointed in their personal capacities, were largely administrative officials with no single Nigerian academic on it.[14]

In 1974, Sir Eric Ashby who was then appointed by the Nigerian government to review the Nigerian educational system, noted that "the habit of apprehending a technology in its completeness: this is the essence of technological humanism, and this is what we should expect education in higher technology to achieve...the path to culture should be through a man's specialism, not bypassing it."[15] Ashby, noting the primacy of university education to development, went further to say that:

> Universities have become absolutely essential to the economy and to the very survival of nations. Under the patronage of princes and bishops, they were cultivated as garden flowers of no more significance to the economy than the court musician. Under the patronage of modern governments, they are cultivated as intensive crops, heavily manured and expected to give a high yield essential to the nourishment of the state.[16]

Besides Ashby Commission, numerous other commissions, committees, and panels have been raised to examine, reappraise, and evaluate Nigerian educational policy, especially as it involves the promotion of science and technology. Some of these commissions were centrally planned. Some were established by the states as well as the non-governmental agencies. Notable examples include the Cookey Commission, Longe Commission, the Ndayako Committee, the Awolowo Foundation's National Dialogue on Education, and other NGO studies and reports, etc.

As Ade-Ajayi had noted, our system of education, notwithstanding these commissions, committees, reviews and recommendations, is collapsing or, perhaps, has already collapsed.[17] Despite this, no one can fail to applaud our National Policy on Education (1976). Similarly, a recent survey, describes the state of science and technology education in Nigeria as "pathetic", in spite of our National Policy on Science and Technology (1986) and, at least on paper, several national agencies and commissions for the sciences, besides creating seven or more separate Federal Universities of Technology and two of Agriculture established to implement these policies and promote science and technology.

From colonial to the post-colonial, and independence periods; Nigerian educational policy aims primarily at the attainment of the following three basic things:

i. prepare the personnel needed to fill high level scientific, technical, professional and managerial jobs... (to be) leaders in Africa's battle against intellectual colonization with which the world-wide explosion of knowledge threatens the region;
ii. generate the knowledge and innovation needed for development, through indigenous scientific research and as agents for the acquisition, adaptation and dissemination of scientific and technical knowledge developed elsewhere; and
iii. as institutions, and their faculties as individuals... provide necessary (consultancy) services needed for development in both the public and private sectors.[18]

Besides the Ministry of Education, the other visible institution controlling the nation's university education is the National Universities Commission. Soon after its inauguration, the commission embarked on its major assignment of preparing a quinquennial program for university development in the country, as part of the first National Development Plan 1963-68. The program was eventually embodied in the Commission's report entitled *University Development in Nigeria: Report of the National Universities Commission.* The main recommendations of the Report can be summarized as follows:

i. that the universities should aim at an enrolment target of 10, 000 students by 1967-68;
ii. that priority should be given to the development of the scientific and technological departments of the universities and that of the total student population of 10, 000 recommended, 7,580 should be taking courses in the sciences, pure and applied;
iii. That for the present, only three engineering faculties should be developed, i.e., the engineering faculties at Ahmadu Bello University, Zaria, the University of Nigeria, Nsukka, and the University of Lagos;
iv. that veterinary science should be developed jointly by the University of Ibadan and Ahmadu Bello University, the pre-clinical course being given at Ibadan and the clinical at Zaria;
v. that there should be only two faculties of medicine, during that quinquennium at Ibadan and Lagos;
vi. that all Nigerian universities except Lagos should develop strong faculties of agriculture and the University of Ibadan should build up a vigorous post-graduate school in agriculture and the three regional universities should each develop an agricultural extension department in cooperation with the appropriate Regional Ministry of Agriculture;

vii. that all the Governments of the Federation should make most of their undergraduate scholarships tenable in Nigerian universities in order to increase the number of scholarships available to Nigerians and in order that the income of the universities may be increased;

viii. that the University of Lagos Act should be amended to enable the Senate of the University to have an overall responsibility for the academic affairs of the University of Lagos Medical School;

ix. that the University of Ife should move to its permanent site at the earliest practicable date and that it should be compensated for the loss of its Ibadan campus buildings which should be taken over by the University of Ibadan;

x. that the universities should be assured of a capital grant of £17.63 million and a recurrent grant of £30 million during the quinquennium 1963 to 1968;

xi. that the Federal Government should provide the entire financial needs of the universities of lbadan and Lagos together with 50 per cent of both the capital and recurrent needs of the three regional universities;

xii. that there should be a National Universities Fund into which all grants and subventions to the Universities should be paid;

xiii. that the National Universities Commission should be converted into a statutory body with the same terms of reference as those set out at page 2 of the report and with responsibility for the management of the National Universities Fund;

xiv. that each of the laws establishing the Regional Universities be amended so as to provide for nominees of the Federal Government.

It must be noted that most of these recommendations, with the exception of recommendations (xiii) and (xiv), were approved by the Federal Government. Recommendation (xiii) was not implemented until much later in 1974.

The take-over of Government by the Armed Forces in January, 1966, the transition from four Regions to twelve States in 1967 (and nineteen in 1976) and the civil war of 1967-70 had untoward consequences on higher education in Nigeria. The NUC and the various government ministries concerned with educational planning were not fully functional as all efforts were concentrated on prosecuting the war. In fact, the secretary of the NUC had to double as the Commissioner for External Affairs leaving only a skeleton staff to carry on the work of the NUC during the war years.[19]

Despite the fact that as early as in 1963, the NUC had advocated for the promotion of science and technology, the Federal Government did not establish the National Policy on Science and Technology (NPST) until 1986. The

NPST was aimed at coordinating the development of science and technology through academic planning, teaching, research, and training. As enunciated in the NPST White Paper; science and technology cannot be promoted except through learning and formal education. The policy therefore advocated for the creation of seven (and later more) separate Federal Universities of Technology and two Federal Universities of Agriculture. In addition, the Mid-West Institute of Technology, Benin-City which was founded in 1972, also became recognized as the University of Benin. In addition, the policy established several other national agencies and commissions for the sciences. The aim of all these were to coordinate the teaching and learning of science and technology as well as providing policy framework to implement them.[20]

As if buttressing the above, the World Bank, in a recent report on Education Policies for Sub-Saharan Africa, asserts that "Higher education is of paramount importance in the process of African development." So, to be effective, the NPST White Paper on science and technology emphasizes training through university education and focuses primarily on the admission of more science and technology students than in other disciplines. Given this official preferences for science and applied sciences in admission into higher institutions, the White Paper advocated a 60 by 40 formula for University admissions.[21] The policy works with the principle of admitting 60% of total qualified students for science and applied science, while other disciplines, the humanities and the social sciences, share the remaining 40%. Among other incentives, scholarships were also awarded in a way to favor the "useful" disciplines such as Science and Technology, and discriminate against the so-called "use-less" ones, such as Sociology, Classics, Theatre Arts, History and Philosophy.

Obstacles to the Development of a Culture of Science and Technology in Nigeria

From the consideration of government and institutional efforts on science and technology, it could be guessed that institutional conception of science and technology was such that science and technology differ from other disciplines. Ironically, this conception undermines the NPST White Paper which states clearly that, in order to foster the development of science and technology, it is necessary to create an awareness leading to the development of science culture in the citizenry. Given the ways and manners policies on science and technology have been implemented, it goes without saying that government attitude and approaches to the promotion of science and technology clearly show that many in government circle could not see the relationship between culture and change in the society. For most, culture is seen solely in terms cultural festivals of traditional dances, music and art. Hence, these individuals fail to see the place of culture in the attainment of science and technology. Culture, as noted above, is the totality of our historical experiences as a people, our values, our traditions, our beliefs, world-view, and our way of doing things. It informs our worldviews and directions. Social change, no matter how small, is intricately linked to this way of doing things.

Although many factors have been identified as militating against the development and promotion of the culture of science and technology in Nigeria, the plural nature of the nation, especially its cultures, has been regarded as the biggest impediment to growth and development. Advocates of this view have argued that given the colonial intrusion, which ultimately fused the disparate groups that make-up what is today called Nigeria together, the various groups have worked in disservice of one another. Rather than been integrated after many years of independence, the various groups have existed in an atmosphere of mutual suspicion, hatred and alienation. They therefore asked; which of the nation's culture would best serve the promotion of culture of science and technology?

Another argument also raised is that the slave trade and the divide and rule tactics of the colonial administration fostered ethnicity so much so that the nation remains in the throes of ethnicity, which was reinforced by inordinate ambition and competitive bidding, first for colonial favors, now, for government contracts and financing. Despite the contentions expressed above, evidence of countries like India, China and many others in Europe who are also as plural in nature as Nigeria negate the view that plurality in Nigeria has been a factor for disunity and underdevelopment. The fact that India, South Africa, etc. went through similar experiences as Nigeria and are better economically, socially and technologically than Nigeria betrays both arguments.

In the main, this study asserts that three different factors: education, language policy and culture; are the underlying factors in the nation's failure in the promotion and development of a culture of science and technology. Of these three factors, education is the most serious problem, as it has succeeded in achieving cultural alienation, *disaggregation* and disorientation rather than cultural integration.[22]

Alienation, as used here, means that the educated elite are alienated from their traditional cultures and values. *Disaggregation* here is used in the sense that western education places strong emphasis on individualism in place of communal values; and individualism in an alienated group promotes selfishness in place of social responsibility. *Disorientation* is used in the sense of being confused and lacking a sense of direction because-of alienation from traditional values, with nothing concrete in its place.

Education: There is the worn-out argument that colonial education fragmented Nigerian (African) culture and separated the educated elite from their cultural roots. Under a system of centralized development planning, as the nation has implemented since 1960, one expects an improvement on this, even if not a total solution. Besides, the National Policy on Education was designed specifically to confront this problem by promoting national unity and reconciling (rather than alienating) the educated elite with their cultural roots as a necessary step in the development process. Thirty-three years after setting up a central coordinating institution for our higher education, Nigeria's educational institutions are still delivering varieties of western, if not colonial, education and no national education. The nation's National Policy on Education and the National Universities Commission places little or no importance on a holistic approach to learning by promoting integrative and collaborative teaching and learning; rather the institution places importance on one course and rates one aspect of a society's life as higher than others.

Resulting from this development, the educated elite not only remain fragmented and alienated from traditional cultures and values, but also have become very eclectic and selective in the values they choose to live by. In Nigerian communities today, Nigerians are losing their sense of moral solidarity, and the capacity to inspire development. As military and civilian rules in postcolonial Nigeria aptly demonstrates, government is less concerned with public welfare and development than the former colonial rulers, and yet government is daily maximizing the area of institutional control over the citizen by destroying the federal structure and assuming central government authority over everything, including education, agriculture, health, even religion to a considerable extent.

As an integral part of the Federal Character Principle (FCP) enshrined in the constitution, the nation's educational policy thrived on jettisoning merit and social justice for equal representation in opportunities, appointment and promotion. For instance, the Joint Admission and Matriculations Board (JAMB), the sole body in charge of entry-level examination into Nigerian universities, admits students on the basis of whether applicants are from educationally-less-developed states (ELDS) or from educationally-developed states (EDS).[23] Different minimum entry scores are set for applicants from ELDS as against EDS. The primary motive is to ensure equal representation in opportunities, appointment and promotion, as required by the FCP. Rather than aspiring towards merit and social justice through competition; the FCP mandated for equal representation. Hence; in some states of the federation, students with an aggregate score of 200 marks out of 400 would be admitted to read Petroleum Engineering while another student from another state who scored 250 would be asked to seek admission elsewhere as the quota for his state has been filled. Out of frustration, many of such students have had to settle for 'use-less' courses in the humanities and the social sciences. The cumulative effects of this development range from alienation, disaggregation and disorientation to underdevelopment.

Language, as integral part of a people's culture, is not only a unit of communication but also a factor of identity. Hence, language is of fundamental importance to a nation's socio-economic, political and cultural development. It is the soul of culture, and the means to the acquisition of knowledge, even of science and technology. It is through language that we know ourselves, relate to others, learn, think and communicate. Language is also loaded with values, history, traditions. Meaningful social change and development involves language intimately because change begins as an idea in the mind of one or more people, and ideas can only be conceived in language and symbols, which can then be articulated in words and signs and later communicated to others. Because of the crucial importance of language to development, our National Policy on Education lays great emphasis on the necessity for each child to master his/her mother tongue and use that mastery to acquire one of more Nigerian languages plus English and one or more foreign languages. We could almost say that we completely staked the success of the educational system on this somewhat ambitious language policy. What, then, has been the result in our schools and training colleges?

Today, majority of the pupils that leave primary schools could hardly read in either the mother tongue or English. If they are able to speak the mother tongue at all, it will be because of what their parents, rather that the teachers, had taught them. Undoubtedly, the level of mastery of such languages would not likely be with the effectiveness that the National Policy envisages. In most schools, the use of mother tongue even in inter-personal communication is forbidden. Mother tongues have been re-christened as vernacular. There is no gainsaying the fact that the formulators of the language policy did not have a proper grasp of what obtains in most schools. Besides, there is the dearth of manpower competent enough to teach the different Nigerian languages, and English and/or French, not only in the areas, which may be culturally homogeneous, but also in the urban areas where children with different mother tongues attend the same school. Undoubtedly, lack of adequate control of language inhibits not only clarity of thought and understanding, but also creativity.

As enunciated in its White Paper, the language policy intends to promote traditional cultures and weans educated elite that appreciates the value of traditional knowledge and processes of education. It is after attaining this that the policy aims at expanding to introduce elements of western science and technology for solving problems in our daily lives, thus laying the basis for culture of science and technology within the society.

An integral part of education is teaching; hence the National Policy on Education states clearly that no educational system can rise above the quality of its teachers. Teachers are the ones to make it possible for a child to be grounded in his mother tongue, acquire one or more Nigerian language, plus English and one or more foreign languages. They are the ones to ensure a proper foundation of scientific education for the child. What then have we done to produce such teachers and improve, or at least sustain, the quality of teachers in this country? Rather than developing a corps of teachers that are well grounded in Nigerian languages, English and French; then science and technology, the 6-3-3-4 system of education imported tools and machineries from different parts of the world for teachers' use in the training and dissemination of knowledge they did not possess.

In a survey in twelve randomly selected primary schools in Lagos, it was found that the major instructional strategies emphasized the content rather than the acquisition of science process skills and scientific attitudes. The teachers gave more of factual information and dictated notes to pupils than encouraging problem-solving and higher level cognitive activities.[24]

Although the survey did not specify in what languages the frustrated and demoralized teachers were trying to communicate, such teachers epitomize the failure of the nation's National Policy on Science and Technological Education under which Integrated Science, Mathematics, Agricultural Science and Health Science were introduced into elementary schools, with curricula prepared by the Nigeria Educational Research and Development Council, to have enough trained teachers to implement such programs.

Culture

Tomlinson, a one time Country Director of the World Bank, noted that:

> Throughout the centuries, we have witnessed countless cases of waste and tragedy that can occur when we take a culturally illiterate approach to development. How many health clinic that have been funded by international donors, have failed to serve the communities that desperately need them because aid workers failed to respect local norms and customs? How many Family Planning initiatives have floundered because they have ignored cultural sensitivities and values placed on women's fertility?[25]

If human development is to be sustainable, it cannot afford to ignore culture, as culture is central to human progress. The promotion of science and technology, like other imported ideas, failed to achieve its ends in Nigeria (Africa) because it was considered as unrelated to culture and that culture is anti-development. As already emphasized, culture is the fertile ground for development. In order for development process to be meaningful and successful, it must be in accordance with people's worldview. The people must understand it for it to help in actualizing their potentials and thus meet the need for social reproduction.

As Olurode and Olusanya had noted, the trouble with Nigeria's policy makers was that they viewed education, science and technology inclusive, as that which is to be gained in the precincts of the four walls of a school. Education is equated with a diploma and a good academic performance which, above all things, seeks to obliterate what was learnt at home. Hence, there is no continuity between the formal and the informal learning environment. No appreciable reinforcement is aimed at. Rather than orientating our pupils and students towards employment and positions that relate to our socio-economic and political needs, our educational system orientates them away from our society. In a society where the major occupation are farming and trading, our educational system inculcates a hatred for social existence on the farm or market. The school calendar does not allow children to be at home during peak farming season; rather, the school curriculum resembles that of a country in Europe, its patent irrelevance is demonstrated by the un-ceased production of graduates that could neither help themselves nor their society. The suggestion that a graduate returns to the farm after his degree is seen as a curse. Thus, after graduation, the products of the school system queue up for many years in expectation of non-existing employment opportunities in government establishment and offices. The educational system, to say the least, is of little or no benefit to both the recipient and the society at large.[26]

Despite the fact that Nigerians, no matter how educated, still retain their strong affinity to traditional medicine, our health policy still remains western in outlook, orientation and delivery. Nigerians believe that ill-health has both the material and immaterial components. Our health personnel, trained in Western medicine, have never hid their abhorrence for non-western medicine. The resilience of Nigeria's cultural heritage in health is testified to by the number of air-time devoted to alternative therapy by radio and television stations. The result is the absolute neglect of our rich-cultural heritage in traditional medicine.

The Nigerian Civil War, which lasted between 1967 and 1970, revealed the extent of advancement in indigenous technology. Both the Federal Troops and the erstwhile Biafra forces testified to the effectiveness of the numerous arms used by the Biafra forces during the war. As noted in Ade-Ajayi, in response to the pressures and deprivations of the civil war, scientists and engineers in Biafra succeeded in building a giant petroleum refining facility out of entirely locally fabricated materials and thereby made the technology so diffuse and more universally understood and applied than anywhere in the world.[27] Local blacksmiths and joiners have demonstrated similar ability to forge implements of daily use. Rather than incorporating this into our school curricula, Nigeria's science and technology policies thrive on importing ideas rather than developing a home-grown, useful technologies.

From the foregoing, one can see that governments, from colonial to post-colonial periods, have operated on the principle that science must precede, so that technological development could follow. In indigenous Nigerian societies, the technology that resulted from thinking went hand-in-gloves with science. Hence, indigenous culture is seen as unrelated to modern science and technology. The result is that there is no continuity between the imported science and technology and the existing culture of science and technology which had been in Nigeria long before the western model of science and technology. Underlying this situation is the perception by policy makers that Nigeria's indigenous culture is anti-development and therefore must be relegated to the background for western ones.

Some Suggested Solutions

The emphasis of the National Policy on Education and the National Policy on Science and Technology is to introduce western science through formal schooling and educational institutions, see it grow and be acculturated through cultural links between the educational institutions and society, thus establishing a culture of science. We have seen the weakness of the formal education sector under authoritarian military control and their civilian counterpart. We cannot gainsay the importance of formal training and research in the schools and training colleges, especially at the tertiary level, for the development of science and technology. However, our emphasis here is on what non-governmental initiatives, and the development of a science culture in the private and informal sectors, can contribute.

A careful reading of the history of the cultivation of science and technology into sustainable growth in other countries would suggest that more lasting success is likely to be achieved when technological development is placed ahead of science. Technology responds to basic needs in society, offering labor-saving and mechanical ways of producing goods, and enhancing productivity. As technology expands, it helps to create a culture of science at the grassroots level, thus encouraging the study and the expansion of science education in the schools and laboratories in a sustainable manner. If that had happened here in Nigeria, the kind of precipitous decline that we have experienced would not have taken place.

As for promoting technological development, although we have had some success in transfer of technology from abroad in such industries as the cement production, the manufacture of soap and detergents, etc., but the road to sustainable growth is to cultivate our traditional crafts and industries, mechanize the processes, and develop the technology whether as small or medium-sized or even large-scale industries. Among our traditional crafts worth developing and where we have recorded notable successes in the past are (i) textiles, including weaving and dyeing; (ii) metallurgy - iron, tin, bronze, etc.; and (iii) brewing, cosmetics, food preservation.

In the process of developing these areas, other possibilities will emerge, meeting new needs for processed material, or exploiting availability of new raw materials to produce new lines of goods. As technology expands within the culture, science will grow in response to the challenge of solving various problems that surface in the development of technology or industrial enterprises. Entrepreneurs and the private sector generally have a good deal to contribute to this process. Our traditional apprenticeship system used in the propagation of arts and crafts can be systematized into science-based training programs. To take the example of one professional body, the Nigerian Society of Engineers, though rather late, has now been drawing attention to the potentialities of locally developed "appropriate technology" which does not rely so heavily on imported sophisticated equipment. It is also advocating the need "to acquire new capabilities for mechanizing the traditional production process in order to improve productivity", and labor intensive "community engineering" to find our own home made solutions to the problems of water supply in the villages and small urban centers; controlling erosion; and using wind mills, small dams and solar energy to generate power.[28]

Another approach would be to study traditional science, a good deal of which remains within various cultures in Nigeria. The aim will be to make a survey of the indigenous scientific knowledge and practices that sustained agriculture, health, crafts and industries within our different cultures and traditional societies. We have so far been paying lip service to this. A few agriculturists are beginning to come to terms with the scientific explanations for many traditional agricultural practices that were initially written off as primitive, but which are now being copied, such as intercropping and other devices to keep up the fertility of soils. We can learn a lot of traditional science from analyses of the practices of, and knowledge necessary to be, good farmers, herbalists, hunters, pastoralists, fishermen, miners, dyers etc. Such scientific knowledge being already available within our cultures will relate closely with practical problems as viewed within the society. The solutions offered by traditional science will form the basis for building up a program of further research and instruction. Such an approach will provide material better adapted than even the NERDC manuals for teaching "process skills and scientific attitudes" in a practical and lively manner. Similarly, educators trying to improve the methodology of teaching mathematics need to seek for fresh insights from deeper studies and understanding of traditional mathematical patterns.

The advantages in doing the above are many. One that must be singled out for mentioning relates to the bottom-up rather than top-down orientation of the above as against the erstwhile approaches. During the colonial period, the emphasis was that science and technology flows from above; and that science precedes technology. The first point that should be made is that historical experience does not support these assumptions. Western science was introduced in the 1920s into selected Government Colleges in Lagos, Ibadan and Umuahia. The Mission schools were not able to contribute much until after World War II, but the Government Colleges were sending their products to the Higher College, Yaba, and this prepared the way for the early flowering of science education at the University of Ibadan where some excellent work was done in a few disciplines and specialized areas such as in Chemistry of Material, Atmospheric Physics, Biology, and particularly Tropical Medicine.[29] The retreat from the high standard achieved in the 1960s to the sorry state we are in now can be blamed on under-funding and the other issues of over-centralized control, which have been discussed. What it amounts to, however, is that it is possible to make a good start in the development of science without its reaching the take-off point for sustainable growth if it is not adequately grounded within society and in response to the demands of technological and industrial development.

From the above, it must be noted that the inter-relationship between science, technology and society, in terms of the economy and politics, is complex. The promotion and development of science at the Government Colleges and the University College, Ibadan came from above, in the context of colonial rule without adequate grounding and acceptance in society. The story would have been different should the drive for science and technology be informed by the needs and desires of the people. Sadly, government has since not.learn from this experience as science and technology still remain a top-down issue even till date. The top-down approach adopted by the colonial administration, the missionaries and later the post colonial leaders have failed to attain the desired goals, especially once the drive (e.g., funding) dried up from the source. This goes to show that the culture in which science and technological development could be sustainable has not yet been created in Nigeria.

Technology, without doubt, is applied science. It is generally believed to be driven by science. More often than not, the reverse is the case. As the example of the Biafra forces earlier mentioned illustrates, it is the demands of technological and industrial change that cultivates and systematizes the growth of scientific knowledge. Thus, it is the response to the social and economic pressures within society that may be said to drive both technology and science. Similarly, it is said that the market in computer hard and software has done more to spread computer science and technology in the last ten years than the university departments of computer science in the last thirty years or more, and that Alaba Market, a market in Nigeria where electronics and computer materials are sold, is doing more to create awareness about computer and other basic electronics than several polytechnics put together.

By Way of Conclusion

From the above, it can be argued that one of the major constraints facing the development and promotion of the culture of science and technology in Nigeria is the fact that many people, at the level of government, fail to see the relationship between culture and the development of science and technology. They fail to understand that while development implies a healthy growth of a society, culture is the life of the society itself. One cannot be achieved without the other. In fact, as has been demonstrated, scientific and technological development is a means; the overall wellbeing of a society is the end itself. One leads inexorably to the other.

In spite of the recognition of the need for a culture of science and technology among the citizenry, efforts so far made to attain this have been concentrated on centralized planning, financing and control under predominantly authoritarian regimes. This has led to undue politicization and inefficiency in government, and lack of social responsibility among the intellectual and political elite. It has led to a collapse of the educational system that was over-dependent on government. As the case of Nigeria suggests, when science and technology is imported from outside, it remains a problem until it is integrated or grafted into the society's indigenous technological development and traditional science.

Notes

1. J. F. Ade-Ajayi, *The Cultural Factor in Technological Development, Eleventh Annual Convocation Lecture*, Akure: Federal University of Technology, 1994.
2. R.T. Schaefer, *Sociology, 10th Edition,* Boston, USA: McGraw-Hill, 2005, 53.
3. Dare Oguntomisin and Edo Osaro, *Culture and Civilization,* Ibadan, Nigeria: Bookcraft, 2005, 23.
4. Federal Republic of Nigeria, *National Policy on Science and Technology*, Lagos, Nigeria: Federal Ministry of Science and Technology, 1986, 35.
5. Schaefer, *Sociology,* 57.
6. Ibid.
7. Ibid., 67.
8. Ibid., 56.
9. Obaro Ikime, *Groundwork of Nigerian History,* Ibadan, Nigeria: Heinemann, 2005, 56.
10. J. F. Ade-Ajayi and Tekena Tamuno, eds., *The University of lbadan, 1948-73: The First Twenty-Five Years,* Ibadan: Ibadan University Press, 1973, 123.

11. J. S. Coleman, *Nigeria: Background to Nationalism*, Benin-City, Nigeria: Broburg and Wistrom, 1986, 34.
12. J. F. Ade-Ajayi and Tekena Tamuno, *The University of lbadan, 1948-73: The First Twenty-Five Years,* 75.
13. O. O. Akinkugbe, ed., *Nigeria and Education: The Challenges Ahead, The Second Obafemi Awolowo Foundation Dialogue,* Ibadan, Nigeria: Spectrum Books, 1994, 203.
14. J. F. Ade-Ajayi and L. K. H. Goma, "Scientific and Cultural Development in Anglophone Africa in the 20th Century", in J. F. Ade-Ajayi, *et al, UNESCO's History of the Scientific and Cultural Development of Humanity*, vol. vii, 1996.
15. J. F. Ade-Ajayi and L. K. H. Goma, "Scientific and Cultural Development," 201.
16. Ibid, 342.
17. J. F. Ade-Ajayi, *The Cultural Factor in Technological Development.*
18. Ibid.
19. J. F. Ade-Ajayi and Tekena Tamuno, eds., *The University of lbadan, 1948-73: the First Twenty-Five Years,* 109.
20. Federal Republic of Nigeria, *National Policy on Science and Technology*. Lagos, Nigeria: Federal Ministry of Science and Technology, 1986, 5.
21. J. F. Ade-Ajayi, *The Cultural Factor in Technological Development.*
22. Ibid.
23. Akinkugbe, (ed.), *Nigeria and Education: The Challenges Ahead,* 234.
24. Ibid., 290.
25. Cited in Akinkugbe, ed., *Nigeria and Education*, 234.
26. Lai Olurode and P. O. Olusanya, *Nigerian Heritage: The Yoruba Example*, Lagos, Nigeria: Rebonik Publications Ltd, 2005, 4.
27. J. F. Ade-Ajayi, *History of the Nigerian Society of Engineers*, Lagos, Nigeria: N E, 1995, 69.
28. Ibid., 151-160.
29. Ade-Ajayi and Tamuno, eds., *The University of lbadan, 1948-73: The First Twenty-Five Years,* 25.

• Chapter 14 •

SUSTAINABLE DESIGN: A DEMOCRATIC TOOL FOR SOCIAL UPLIFTMENT

Jolanda Morkel and Maria Leus

Introduction

> "Human nature exists and operates in an environment... as a plant... in the sunlight and soil. It is of them, continuous with their energies, dependent upon their support, capable of increase only as it utilizes them, and as it gradually rebuilds from their crude indifference an environment genially civilised."
>
> – John Dewey

The UNESCO agreement on diversity of cultural expression acknowledges the importance of culture for the world economy. South Africa, a country synonymous with racial and cultural contrast and often conflict, is aiming to instead transform into a symbol of cultural pluralism, the proverbial "rainbow nation." In so doing, this African nation has become an example to the rest of the world. Over the last decade and a half the young democracy has battled with the effect of policy and legislation. The relationship between social justice, education and ecological sustainability is transpiring. By linking human rights and social justice, and addressing both into the constitution, a message of national concern for environmental protection is conveyed. In this sense environmental education proves to be invaluable. Education is crucial to ensure an awareness of democracy, the environment and ethics for sustainability. Since 1994 the spatial dimensions of education had to be rethought to be aligned with democratic ideals.[1]

To provide an architectural answer to the problematic issue of "post apartheid" and redress of the imbalances of the past is not simple. The question is how place can be defined so that the emotional value of the past can be positively reinterpreted from a future-focused perspective. In which way does architecture, considering the meaning of inter cultural exchange, serve to contribute to cultural diversity? Furthermore, how can architecture make a contribution towards empowering and transforming communities by presenting sustainable and democratic learning environments? This chapter aims to address this question by presenting a cross-continental collaborative community project.

Architecture as facilitator for dialogue

Intercultural dialogue and interaction are indispensable facets of a multi cultural society. These are the keys to share knowledge and culture. The architectural dimension of culture may be viewed in the framework of the large issues of the world such as ethics, sustainability and accessibility (to opportunities).

The architectural conditions for inter cultural dialogue are evident on various levels. In addition to physical accessibility there are a number of functional and sustainable demands. Physical accessibility is the *conditio sine qua non* or the necessary condition for dialogue. A building must be accessible to as many people as possible. This accessibility assumes not only access in the general sense, but also access to opportunities in order to achieve betterment.

The next factor as requirement for establishing dialogue is functional in nature. A building that promotes interaction between people must provide those functions related to such activities, such as meeting places, open spaces and circulation spaces. In this way architecture obtains a social dimension, providing a sense of belonging.

The third aspect investigates the possibilities of architecture in relation to dialogue. A building is in fact carrier of meaning and must convey a message to its users. It should not only be legible to the broader public, but also relate to its users. Design clarity and architectural legibility can improve the ease and comfort of use (of such building) and hence reduce user stress. The quality of the spaces contributes to the quality of the human experience.

It is within this vision that fits the concept of the school family room of Peter Petersen.[2] A school family room is a home base, a living room, a space where one feels safe and a sense of belonging. According to Petersen the primary role of education is to form the person as a whole. Therefore, if such educational spaces succeed in providing the users with a feeling of belonging and self-worth, their potential will be more easily realized.

Sustainable design in action

The main brief for the architectural concept of the educational hub in Philadelphia, South Africa, was that it should be a community facility. The project is considered an important social instrument for the empowerment of the poor, a democratic concept in which the needs of the users enjoy top priority. In this architectural project based on sustainable design and capacity building principles, issues around public participation, social responsibility and democracy in architecture is addressed. The building organisation and design promotes user engagement, encourages human interaction and the opportunity to learn.

The building design was developed from the cross-continental co-operative service learning project, academic collaboration, undertaken by the two departments of architecture, which, together with a process of community participation and professional input, resulted in the successful implementation thereof. All these aspects are considered within the context of diverse values that different groups of the population attribute to the democratic process.

Identifying the special needs of this community proved the first step to developing a sustainable concept for a school building that effects the social upliftment thereof. Joseph Ki-Zerbo, one of the most prominent African philosophers, suggests three conditions for making true development possible, namely minimal economic development, a generalized educational system and a democratic system at all levels, inside societies as well as on political level.[3]

Learning by doing

There is an important relationship between education, health and the success of a society. Because the built environment plays such an important role in social welfare, sustainable buildings can make a valuable contribution to the social aspect of a community.

Fundamental questions asked are "What is the social role of a school in a community and how can a design concept be generated from this?" A child is a product of the relationships it has with fellow humans. The educational process takes place within a society that upholds certain norms and values. The school is such a society. The school must not be isolated from the society, it must form part thereof. Against this background the interaction and multi functional use of the school for the community is of key importance.

Education and training are interrelated; the learning activities must be integrated into the daily lives of the child. The self activity of the child provides the only real learning results. Each member of the school community has the responsibility to contribute to the fundamental cooperative nature of the school.

According to Dewey learning is about adapting one to the social living environment.[4] Knowledge is, according to him, only useful if it is evident from the experience that it contributes to the development of the living environment. Applying the 'learning by doing' concept is an experienced-focused approach. 'The school may not be the preparation of life, it is life itself.'[5] Experience is the key concept.

"Learning by doing" is a central principle in economical growth thinking. One must learn in order to progress. That is also relevant to government, especially in developing countries. This asset-based community development is a new approach to dealing with slum areas, and for development aid in general. Here we refer to the home economy, or the principle of mentoring where the coach empowers the community, for sustained growth and development

Sustainability in project design

"Human development" can be considered a democratic process through which people are afforded more choice. The most important choices of people are leading a healthy life, education, and the enjoyment of a basic standard of life. Additional choices are political freedom and guaranteed human rights and integrity.[6]

In September 2008 the World Green Building Council of South Africa became the 13th full member of the World GBC. They define a green building as a building that is energy efficient, resource efficient and environmentally responsible, incorporating design, construction and operational practices that significantly reduce or eliminate its negative impact on the environment and its occupants.

The Brundlandt report 'Our Common Future' of the 'World Commission on Environment and Development' of the United Nations' development provides the following definition: "Sustainable development is development that meets the needs of the present without compromising the ability of future generations to meet their own needs. It contains within it two key concepts: the concept of 'needs', in particular the essential needs of the world's poor, to which overriding priority should be given; and the idea of limitations imposed by the state of technology and social organization on the environment's ability to meet present and future needs."[7]

Sustainable development requires a multi-player environment. The design process may be participatory or a form of trans-disciplinary learning. At the same time, a policy of shared knowledge may generate shared understanding. The emphasis in both policy and process is on the dynamic nature of sustainable development where goals and means are moving constantly.

Sustainability is a complex given and encompasses, in addition to ecological aspects, also economic, social and spatial ones. Sustainable building is implicitly a constant balancing act between a number of factors, and not only the energy issue. Schools could play an important role in this respect. Finally, it is much easier and convincing to teach learners the principles of sustainable building if those ideas behind sustainable development are in fact being applied by the school itself.

With this project we aim to combat poverty in a sustainable manner. Sustainability in this context has multiple meaning. In the first place the project had to be long lasting, ensuring that the community continues to benefit for generations to come. On the other hand sustainability also has to do with the environment. Cognizance should be given to the available natural resources to ensure a healthy living environment. This project therefore is strongly directed at 'capacity building ' and a process of 'learning by doing', by involving the local community as much as possible. In so doing, a feeling of self worth and pride is established, and a need for the community to maintain the building.

Sustainable building has a social component due to the fact that the built environment plays an important role in social welfare. The life span of a building is dependent on the appreciation of the building. In this respect the term 'affectionate building' is appropriate. If we want to achieve a peaceful co-existence between a democratic society and sustainable building, we should not seek standard solutions, but rather seek to understand the characteristics of that society.

Democracy and architecture

Democracy and architecture are connected. Architecture is the three dimensional expression of the values of a society. Architects cannot change a community. However, they can strive towards making spaces that express the important democratic values, warm human spaces.

John Dewey is an American philosopher who, with his concept of the creative democracy attempts to contextualize a contemporary notion of democracy.[8] Dewey's democracy concept serves a society that is able to solve complex social problems in a manner that results in educational experiences and consequent individual and social growth. The authority creates the material as well as the immaterial conditions to ensure democracy. Education and training are important contributors in the forming of citizens who, on the basis of effective freedom, may choose for a voluntary participation in society.[9] Dewey stresses the importance of the social environment.

On the basis of an environmental psychological interpretation of human behavior, Dewey concludes that we need to; in the first place, adjust the environmental factors if we want to change habits. "We cannot change habit directly, that notion is magic. But we can change it indirectly by modifying conditions".[10] There is an important relationship/ link between education, health and the success of a society.

Architectural concept

The idea of the school as an organic community is translated in a central space, the public space of the school building, a general teaching space for multi functional use. Key design concepts include flexibility, openness, and overlapping and multi functional use.

School buildings can be considered public space, as space that respects the difference between people and creates opportunity for development. According to Moore, the public space is another form of "inside". A space is a 'place' that emits optimism and assists the community to acquire a central position in the world.[11]

Hannah Arendt argues that there's no place for plurality in our contemporary society. Plurality is necessary for acting in a situation.[12] The absence of plurality where everybody is the same, results in a totally undemocratic system. Acting or 'vita activa', conversation and education are closely related. Acting creates new narratives, something new is created, and freedom manifests itself. We are only free if we can act in freedom. The concept of plurality, however, has a complex meaning. It contains equality as well as inequality. We are equal if we all can act in freedom in the public space, where we meet our equals. We are not identical; everyone has his own unique identity. The public space is for Hannah Arendt the only space in which human qualities surface. Stouthuysen states that "freedom is only guaranteed if we feel safe".[13] This, according to Arendt is the responsibility of the government. They should ensure that the citizens do not fear each other.[13]

Ronald Beiner sees architecture as generator for citizenship.[15] Also Kenneth Frampton and George Baird have attempted to use Arendt's 'vita activa' as guideline for understanding and design of buildings and public space.[16] Ronald Beiner sees no use in politics to provide people with citizenship. He aims to prove that the concept of public space that Arends deals with is architectural in nature. He considers sustainable architecture as 'ensembles of buildings that need not so much be aesthetically responsible than rather to establish the relationship between the citizen and his or her place'. We should not underestimate the role of architecture. It can focus on the individual and facilitate this visibility.

Architecture can support active pluralism. Architecture that takes on this important task may not at all be oppressive. It must be democratic and respect and challenge the individual. The need for a school that would initiate this thought process and participate in the domain of public life was identified. These ideas provided the backbone for the conceptual ideas for the new school building for the Meulenhof Primary School in Philadelphia.

Transforming design

The school, located between the formerly separated white and black areas, was ideally situated to act as an integrating cultural centre for the previously disadvantaged coloured and black communities, serving to provide general health and HIV and aids education, adult literacy programmes, skills development and training.

When the school governing body of the Meulenhof Primary School approached our University in Cape Town for assistance in 2004, the small but rapidly expanding village of Philadelphia, South Africa, was in need of job opportunities, housing, a clinic, outdoor sports facilities, adult learning facilities, outdoor sports and community facilities. The most pressing need, however, was to provide safe and healthy spaces conducive to teaching and learning, that would replace the then more than 30 years' old temporary prefabricated structures.

However, their application for funds from the Western Cape Education Department would place them at the bottom of a long waiting list, for them a certain dead end. A different approach had to be taken. So the two departments of architecture from two different continents pooled resources to develop a brief for the project, by means of student research and design projects. This material was used to find private sector donations and sponsorships, and soon the Western Cape Education Department was convinced to make their contribution, outside of the standard process of providing financial aid.

The single most important requirement for the project as expressed by the school governing body was that it should be sustainable. The primary school of Philadelphia is situated in an almost entirely temporary and dilapidated prefabricated structures dating from the 1970s. These structures no longer comply with safety standards and do not provide a healthy environment for teaching and learning. The project posed a challenge for the creation of an adequate and appropriate learning centre that would reflect sustainable and democratic principles. The concept of the school is informed by the broad and critical vision whereby aspects such as sustainability and pedagogical identity and democracy play an important role.

In this proposal, the architectural concept of the buildings represents prototypes to stimulate creative solutions. Through the educational process of the local community they can, over time, build the required "pavilions". Each pavilion represents one school living room. A frame provides the structural support. In fact, the frames might be built with the assistance of a contractor, after which the non-structural infill panels could be added as required. A variety of affordable and re-used materials, even waste that is readily available, such as timber and PVC, can be used. The structural elements offer a high potential of recycling and reuse, reminding one of the meccano building system.

The overall sustainable layout concept was realized by creating courtyards covered with pergolas that aid in thermal control. They allow heat gain in the winter and provide shade in summer. Sustainable solutions are found in bottom-up strategies, studying local indigenous vernacular architecture and applying low-tech design.

The first phase included four classrooms for 40 learners each, of which three could be opened up into a multi-purpose hall to accommodate large school or community gatherings. These, together with a serving kitchen and store room, were housed in a wing that would partly enclose a courtyard. It was decided to gradually replace the old redundant buildings, by providing new wings on their foundations. The result would be a series of buildings defining useable outdoor spaces, thereby allowing the teaching spaces to expand to the exterior. The second phase will include more class rooms, workshops, an administrative section and computer and resource centre.

According to the principal their wish to establish a centre of excellence through the project, is being addressed: "It is a way of getting our children ready for the world out there. The project shows in which direction education should move: forming partnerships between local community, universities, parents, provincial education departments and business. By being involved in this project, people can experience how different levels of education come together: primary school, tertiary institution and the professional world. In short: sustainable facilities for continued learning." By reinforcing best practises, the South African-Belgian co-operation proved to enrich students and learners on both continents.

Conclusion

> The human being is always and everywhere essentially the same. He possesses the same mental equipment, although he may use it differently, depending on the cultural or social background, depending on the distinctive way of life of which he coincidentally forms part. [Aldo Van Eyck][17]

In this project the wider definition of sustainability and democracy was employed, including the involvement of its users in the design process. Both learners and teachers were asked for their inputs, the children using drawing as a means to express their needs and dreams for the project.

Research on current trends in education showed a move from teacher-based education to learner-cantered education. In addition to the need to ensure environmental and economic sustainability, lessons learnt include that the whole school be designed as a learning environment, that the school be part of a community facility that educational facilities provide for flexible and adaptable multi-functional spaces.

The conclusions formulated by the student work, relate to the themes of context, technology and function. These assisted in formulating the brief which was developed by a local firm of architects who provided their services at cost. A local contractor, in turn, agreed to make his contribution, and so construction was soon underway.

The design concept for this educational hub goes beyond simply planning and construction. It has to refer to social issues that are relevant for everyone who is living in this area. The inhabitants of this district have to redefine their identity, which creates tension between integrating in the modern globalised world and maintaining their sense of belonging in the roots of the traditional community.

By allowing the community to participate and formulate their vision of how a society should be and how such a facility might be shaped and used, it was possible for people to construct these spaces together. The collective process which resulted in this untypical educational centre creates social links that express the new identity and social construct based on democratic principles.

Notes

1. I. Low, "Spaces and Transformation: Reflections on the Cape Schools Programme," in *Counter Currents, Experiments in Sustainability in the Cape Town Region*, ed. E. Pieterse, Jacana Media (Pty) Ltd in association with the African Centre for Cities, University of Cape Town, 2010.
2. P. Peersen, *Der Kleine Jena-plan*, 62th Edition, Weinheim /Basel: Beltz, 2001. Peter Petersen was German professor in pedagogy who developed the Jenaplan education in the 1920s.
3. Y. Kluyskens, *De ontwikkelingssamenwerking: zin en onzin*, 3 February 2009, www.ugent.be/ge/nl/raden/cifg/os/ontwsamenw_kluyskens.
4. J. Dewey, *The school and the society*, Chicago: University of Chicago Press, 1925.
5. Ibid, p. 125.
6. United Nations, *Process of preparation of the Environmental Perspective to the Year 2000 and Beyond.* General Assembly Resolution 38/161, 19 December 1983. Retrieved: 2007-04-11.
7. World Commission on Environment and Development, *Our Common Future, Report of the World Commission on Environment and Development*, Published as Annex to General Assembly document A/42/427, Development and International Co-operation: Environment August 2, 1987.
8. Dewey considers it the task of philosophy to provide theoretical solutions for the approach to concrete social problems.
9. L. Logister, *Creatieve democratie, John Dewey's pragmatisme als grondslag voor een democratische samenleving*, Budel: Damon, 2004, p. 328.
10. MW, *The Middle Works of John Dewey*, 1899 - 1924, Carbondale/Edwardsville: Southern Illinois University Press, pp. 18-19.
11. K. Keim, *You have to Pay for the Public Life: Selected Essays of Charles W. Moore*, Cambridge, MA: The MIT Press, 2001, pp. 23-27.

12. H. Arendt, *Vita Activa, De mens: bestaan en bestemming*, Amsterdam: Uitgeverij Boom, 1994.
13. P. Stouthuysen, *Kerstessay , het is niet wat je zegt maar hoe je het zegt', Ik denk dus ik zeg,* De Standaard, 28 December 2006, p. 24.
14. Ibid., p. 22.
15. Ronald Beiner referred to the lectures of Hannah Arendt. R. Beiner, *Our Relationship to Architecture as a Mode of Shared Citizenship. Some Arendtian Thoughts*, 3 February 2009, http://scholar.lib.vt.edu.
16. K. Frampton, "Labour, Work and Architecture," In *Meaning in Architecture*, eds. George Baird & Charles Jencks, New York: NY George Braziller, 1970.
17. K. Frampton, *Moderne Architectuur: Een Kritische geschiedenis*, Roeslaere: Roularta Books, 2001, p. 340.

• Chapter 15 •

TRADITIONAL ARCHITECTURE AND ENVIRONMENTAL CONSERVATION AMONG THE ABALUYIA OF WESTERN KENYA: LESSONS FOR CONTEMPORARY SOCIETY

Maurice N. Amutabi

Introduction

After the Stockholm Conference on the Environment in 1972, many societies realized that they could not take environment for granted. Connections started to be made between climate and use of resources. The Earth Summit at Rio de Janeiro in 1992 and the promulgation of Agenda 21 on Global Environment, and the 2005 Kyoto Protocol made environment a collective global agenda. These international agreements have changed public opinion about the environment, and it is against this background that I seek to use architecture in interrogating environmental ethos among the Abaluyia. Indigenous architecture among the Abaluyia has not attracted a lot of scholarly attention until recently.[1] Oral history, archaeology and physical anthropology have started to unearth new information about Abaluyia traditional (what I prefer to call indigenous) architecture as well as their environmental ethos.

The selection of construction sites, the choice of construction materials, the design of buildings, the position of houses in the homestead, among others, are aspects that were guided by rich, elaborate and respected cultural norms of the Abaluyia. Many of the norms, it has been established, were consciously environmentally friendly. Through these norms, many of which were systematized as taboos, a lot of environmental conservation efforts were practiced. Water catchment areas and rare plant species, among others, were preserved. This chapter utilizes construction-related taboos and guidelines among the Abaluyia to show how they assisted in environmental conservation. The chapter uses them to draw parallels with modern construction industry in its extravagant, carefree attitude towards the environment. The chapter comes up with lessons that the contemporary construction or building pursuits can learn from indigenous practices of the Abaluyia architecture.

In this chapter, my intention is to make a retrospective analysis of indigenous architecture among the Abaluyia people of western Kenya, and its implications on contemporary architectural practices and relationship to environment. It is fitting to point out here that it is not easy to isolate architecture per se in the indigenous set up because it is very much intimately and intricately intertwined with the contemporary social fabric. For, it should be noted that even the simplest settlement as we see them today are the product of a whole multitude of economic, social, political, physical and technological forces. Some of the features require decoding, analysis and interpretation. For example, a collection of pottery remains at a hillside rock might appear meaningless to an uninformed observer and yet laden in such innocent-looking ceramics remains might be all manner of facts about what took place, who participated and why. The earthenware might be indication of an old settlement site or a sacrificial site. Therefore, appearances can easily be missed or even be misinterpreted. Not even detailed analyses of manifest spatial layout are sufficient for a thorough appreciation of how indigenous architecture functioned or operated. Each indigenous settlement among the Abaluyia is a balanced solution to a problem of habitation and needs to be viewed as a physical reflection of the way of life of its inhabitants, replete with its own attendant attributes.

There is overwhelming architectural evidence which suggests that the Abaluyia have been conscious about their environment. Chitambe 'Fort' near Webuye town and numerous earth fortifications found in many parts of Western Kenya testify to the fact that the Abaluyia have produced monumental architectural and construction engineering masterpieces which we can use to make certain conclusions about their past activities. Looking at the burial sites of the Nabongos (kings) of Wanga kingdom at Elureko, and shrines of rainmakers in Ebusiekwe, one comes out with a sense of deep knowledge of environmental ethos and careful planning. Archaeological data is ever more adding to what we previously knew—that the Abaluyia built enormous houses and produced complicated homestead planning guided by social values and culture. These archaeological artifacts are providing useful information to insights into how diverse architectural fundamentals evolved among the Abaluyia.

Examination of worship sites (olusambwa), sacrificial platforms, irrigation terraces and canals, clay structures, pottery, huts made of wood and grass, and mud mounds, and a mixture of numerous building materials, we are able to learn a lot about the relationship between architecture and environment. By looking at how these features were conceived and produced, and examining the architectural motifs, we can get a sense of the history of the people. It is against this background that this essay will endeavor to seek answers to the following question: How did indigenous architecture function in regard to the environment? Was it user friendly? How does modern architecture affect the environment? Can we draw a comparison between indigenous and modern architecture with the view of acquiring compromise approaches?

Architecture and Environment

The 2009 Summit on climate held in Copenhagen, Denmark recognized the importance of developing countries in fighting against environmental degradation. There is now realization that environment is a wide phenomenon that encompasses all disciplines, generating a universal attention. There has been more concern on inter- and multi-disciplinary approach to any issue affecting the environment. Indigenous architecture is therefore starting to be appreciated after the realization that it is environmentally sound and friendly. When architecture is examined vis-à-vis the environment, it attracts enormous attention. This erases previous ambivalence that has attended this relationship. It is therefore gratifying to see attention from disciplines such as history, sociology, anthropology, engineering, planning, with the realization that such interdisciplinary approach is useful in proving a more holistic study of the relationship.

Does a historian have a role to play in the work and general understanding of the art and science of building with its attendant attributes of planning, design, decoration, etc? The very root of the word architecture - arch, the first or principle; and techne, art or craft - connotes its early and central place among man's civilizing activities, of serious interest to the historian. Architecture is place, time, and culture specific and operates at all levels of society.[2] Place, time and culture are pertinent parameters for the historian for they help in understanding historical facts. The architect and the historian are bedfellows, for they operate in society. Architecture requires serious investment of all kinds (effort, skill, materials, wealth, risk) and provides a powerful trace of human enterprise and social structure, all subjects of history.

Architecture as "mistress of the arts"[3] has always doubled as eyewitness to culture, because temples and burial places are generally accepted as the first true works of architecture, and they constitute very significant elements in cultural history. As Fletcher puts it, "the study of ancient architecture opens up the enjoyment of contemplating buildings with an appreciation of their purpose, meaning, charm, and every structure conjures up the conditions of past ages."[4] One should perhaps add that ancient architecture in Africa is more than buildings. There are many architectural artifacts that have not been included in what is examined in indigenous architecture, such ancient river crossing sites, where there were tree trunks or huge boulders or stones used as bridges. These were creative inventions and yet they need careful examination in order to understand how decisions were made to construct them at particular points.

Therefore, the study of indigenous African architecture demands an interdisciplinary approach just like that of any universal aspect touching the whole society. One needs to pay attention to archaeological and historical and sources from other disciplines. This makes it exciting, but one of its inherent drawbacks is that it is inevitable that things will be said with which specialists in individual disciplines will take issue with any claims that are made by another discipline. Take for example the case of pre-colonial Chitambe fort among the Abaluyia on which there are many oral testimonies but yet it has not attracted scholarly attention because some scholars had thought that it is a more recent creation than it is has been claimed to be.[5] Shilaro's reference to Chitambe is due to the selection of Webuye (Broderick Falls) as a European outpost. Although this silencing of African historical sites is tolerable for production of neutral or impartial compromises, it has tended to undermine scholarship in indigenous knowledge. Moreover, since this particular chapter covers over ten different Abaluyia sub-ethnic groups, those with special knowledge of some of these will inevitably find generalizations which they might consider imprecise or even totally misleading. That is expected when dealing with diverse groups such as those of the Abaluyia but there are obvious commonalties that can be utilized to arrive at acceptable generalization so as to obliterate the minor differences. Thus, we can use Chitambe to produce information on how the Abaluyia have managed their environment.

There are many scholars who have written and tried to recover.knowledge about indigenous African architecture beginning from 1970s to the present.[6] Unfortunately, they have not made a connection between architecture and environment. Most commentators (both from the South and North) on architecture and on environment define these concepts pretty much as they please. There is almost a permanent divide hence the notion of traditional (what I prefer to call indigenous) and modern architecture. Consequently, there has arisen over time, a plethora of different meanings and connotations of the two terminologies. It is open to anyone to define and employ these two phenomena in some other sense or to show that their use of these concepts is more perceptive and dynamic than other definitions. I define indigenous architecture as ways, means and forms through which pre-colonial or pre-contact societies have prepared their dwelling places and related infrastructure. I see indigenous architecture as involving houses, forts, all kinds of shelters, infrastructure such as roads, bridges, worship paces and burial sites. To me, indigenous architecture is represented by pre-colonial structures as well as more contemporary ones but which use past designs and plans in their schemes. Indigenous African architecture is closely linked to what we know today as western architecture, largely as a result of diffusion and cross borrowing, to the point that today, the two can be said to exist side by side, or as combination of both.

Is indigenous architecture still relevant today? There has been a lot of concern over the issue of transplanting western architecture in African environments. This has come about as a result of inconsistencies in geological or seismological, meteorological, cultural, political or even social dynamics, where some scholars have attributed problems in many urban areas to western architecture. Some scholars regard western architecture as dangerous in some parts of Africa, such as Kenya's Rift Valley which sits on a major fault line and is dangerous for high rise buildings. Many people in Kenya are killed or permanently maimed for architecturally unsound structures or suffer consequential damage accruing from architectural malpractice. In Kenya, many storey buildings have collapsed killing hundreds annually. Looking at beautiful buildings that have come up in Nairobi recently, imitating African huts such as Safari Park Hotel, one can concur that African architecture is still relevant.

The problem has been that modern architecture as practiced in Kenya has not drawn on indigenous knowledge and models. There has been a lot of emphasis on the 'Skyscrapers architecture' at the expense of the 'Manyata architecture'. My argument is that whereas the 'Manyatta (Masaai house) architecture' has been here for centuries with little or no damage at all to the environment, the less than a century existence of the "Skyscrapers architecture" has caused untold environmental devastation and human suffering in Kenya. There have been concerted efforts to produce mini-Londons, New Yorks, etc. and none for major reenactment of our own Great Zimbabwe's, Meroes, Axums, Timbuktus, which are some of the most acclaimed, serene past African cities. Buildings fashioned after African architecture look beautiful and are more stable. Societies need to adopt to change, very carefully; otherwise they will end up mimicking what others are doing. As Homi Bhabha notes, whatever is mimicked can never be the same as the real or original one. He says,

> It is from this area between mimicry and mockery, where the reforming, civilizing mission is threatened by the displacing gaze of its disciplinary double, that my instances of colonial imitation come. What they all share is a discursive process by which the excess or slippage produced by the *ambivalence* of mimicry (almost the same, *but not quite*) does not merely 'rupture' the discourse, but becomes transformed into an uncertainty which fixes the colonial subject as a 'partial' presence.[7]

This is the challenge that historian Arnold Toynbee's model of analysis, of 'challenge and Response' talks about. Bhabha's position is a rejection of Toynbee's cyclic theory of social change that seeks to allow change as normal and on the assumption that they will be replaced by newer ones.[8] Inherent in this theory is the assumption that nothing is permanent. The modern architectural designs and impressions were introduced among the Abaluyia with the assumption that the people were responding to the need of an improved habitation to escape the ravages of nature and yet that is not the case.

The assumption by Toynbee theory would be that the people had the in-felt conviction that the modern way of architecture was an improvement on their state. Now, it is apparent that modern architecture has brought with it even more complex problems in their existence, ranging from deaths caused by unstable buildings to indirect effects such as those emanating from water harvested from asbestos roofs to mercury and other chemicals entering the water cycle through modern sanitation via our modern bathroom soaps, among others. Of course no one is calling for a return to everything old, but there is evidence that modern architecture has come up with all kinds of problems, especially in urban areas.

Perhaps there is need for what Homi Bhabha has called the 'hybrid moment' of change where a society takes what works well, from the indigenous and modern and using it.[9] The environmental degradation caused by modern architectural developments is devastating, but there are certainly positive attributes that people have acquired from it. They need to make a balance as they respond to everyday demands of now. This response to the challenges posed by modern architecture by the people is in no way permanent according to Toynbee. Change is one phase among the many phases of the cycle of change which govern societal evolution. As Bhabha has suggested, there is need for dialogue and negation to find out what works well and best. There is now emphasis on sustainable development and use of environmentally friendly materials, especially recycling and sparing use of non-renewal resources. Modern architecture can borrow from indigenous architecture and vice versa so as to ensure that people benefit from great experiences of both worlds. The attraction to indigenous architecture which has been initiated by environmentalists is not without its merits.

Indigenous architecture

Indigenous architecture forms part of the more elaborate material culture of the African societies to which the historian has access through his tools such as oral sources. Archaeologists and physical anthropologists have also illuminated the historians' horizons on this important African heritage. Today, there is a rich literature on the advanced state of the position of indigenous architecture in Africa from the ruins of the Great Zimbabwe in the south to the sphinx and Great Tombs in the north. The fortifications left behind by pre-colonial Kenyan societies like the Miji Kenda and their Kayas, or the Babukusu with their forts exemplified by the Chitambe ruins near Webuye town, show high levels of sophistication and environmental awareness.[10]

Ironically, despite these obvious architectural wonders, some early European adventurers saw Africa as a dark continent with nothing worth of mention. Many of their feelings were echoed by historian Trevor-Roper when he said that before the white man, Africa had no history, as darkness is not part of history! This position has dominated the minds of many Eurocentric Scholars who talk of European innovations as "universal". Anything good in Europe is easily termed as "the best in the world" as if the world is just Europe and North-American. In presenting what he calls a systematic universal history of architectural development, Fletcher admits that there are geographical, geological, climatic, religious and social influences. However, in his illustrations through what he calls the 'Tree of Architecture', he completely fails to mention Black Africa but as usual has Greece at the center followed by the Roman architecture. Others in the tree are Egyptian, Assyrian, Indian, Chinese, Italian, English, French, German, Spanish, and Belgian-Dutch, with American at the apex representing revival of modern styles.[11] Homi Bhabha talks about this silencing of all that does not conform to the imperial set up. He says,

> There is a conspiracy of silence around the colonial truth, whatever that might be. Around the turn of the century there emerges a myth, masterful silence in the narratives of empire, what Sir Alfred Lyall called 'doing our Imperialism quietly…Around the same time, from those dark corners of the earth, there comes another, more ominous silence that utters an archaic colonial 'otherness' that speaks in riddles, obliterating proper names and proper places'.[12]

To the list of silencing, I would like to add indigenous architecture and its relationship with the environment. In modern architecture, there is still this misconception of looking at everything in Africa as untested and unworkable. There is also tendency to look at everything in Africa under European gaze or eyes. It is common to find such statements as "African houses had generally no Chimney."[13]

They also made claims that African houses were inadequate. Many of these scholars "find it difficult to accept that indigenous buildings of the continent merit more than passing consideration," as Denyer rightly adds that:

> One only has to consider for a moment the vocabulary used to refer to them (traditional African buildings) including such basic words as "mud" and "hut" (as opposed to house) which in English have such derogatory overtones, to realize that even for those who know

> and respect other aspects of African culture it is hard to avoid being drawn into a web of selective and distorted perception.[14]

When reviewing literature for this chapter, I came across such derogatory statements as 'primitive architecture', etc. and I have intentionally edited out reference to such literature and such abuses to the African past. Emotions aside, sensational modern accounts of Africa's underdeveloped architecture only serve to reinforce earlier stereotypes of 'Dark Africa' and its backward or primitive material culture, as represented by Joseph Conrad in his book *Heart of Darkness* and some of his contemporaries. It is for us scholars to correct the wrong impressions created by such imperial and neo-colonial stereotypes.

Incidentally, not all early adventurers to Africa came with open eyes and closed minds on African culture. Some of them gave true accounts of what they saw. Some even showed the admiration and great surprise for what they saw in Africa. Others even served as advocates for the African glory and wonders. One such recent explorer of the 19th century, to Kenya had this to say about what he saw:

> ...it seemed a perfect Acardia, not just the fertile and pleasant countryside but the whole landscape of well planned neat farms interspersed with immense shady trees and the charmingly neat circular huts with conical roofs and walls hanging out all round with clay worked prettily into rounded bricks and daubed symmetrically with spots.[15]

That is no doubt a near-paradise compared to the rubble and mix of the present concrete jungles that we call cities in Kenya and elsewhere in Africa. Many are poor imitations, or what Bhabha calls 'mimicry' of western cities. But the picture of prosperity painted by Thomas above, especially the serenity of the Africa he describes and planned neatness was not just confined to Kenya or East Africa for that matter. Commenting on "similar splendor found in pre-colonial West Africa", Denyer had this to say:

> ...the walled towns of Hausaland were the foci of Park-like farmland, shaded with mature cotton trees. The countryside was landscaped and planned around the towns and an air of studied permanence, commending itself to many visitors.[16]

The account of one of the early missionary travelers in Southern Africa in present day South Africa Rev. Robert Moffat is also very edifying for lovers and admirers of gloria-Africana alike. It brings out the reality that Africans were aware of a whole range of architectural methods and styles. Writing his true impressions on the architectural settings in the Transvaal in 1829 just before the Great Trek to the interior by the Afrikaners (Boers) and which disrupted the African serenity, Rev. Moffat wrote thus:

> ...the walls (were)...so well polished that they had the appearance of being varnished. The walls and doorways were neatly ornamented (decorated) with architrave (moulded frames surrounding a door or window) and cornices, the pillars supporting the roof in the form of pilasters projecting from the wall and fluted showing much taste.[17]

The beauty of indigenous architecture

There are many obstacles in reconstructing indigenous architecture, as what remains today suffers from acute influence from the North (especially Europe). The past architectural pieces are only in museums, the archives and some artistic impressions based on oral transmissions from the past. The information is therefore patchy. This is partly a reflection of the official policies of colonial government that did not see the essence of preserving pertinent aspects of this material culture. Also, colonial researchers, mainly physical anthropologists pursed their limited research interests with precision beneficial to their objectives and left out the "irrelevant details." Majority of the present day researchers do not have access to some of the original sources many of which are in European metropolis of the former colonial powers.

Another major problem is that very little of the published material is specifically in material culture especially architecture most of which comes either as anthropological, archaeological or anecdotal accounts. As scholars, we must be prepared to bear with these inadequacies and employ other fitting ways of creating or generating knowledge as this present chapter has endeavored to do.

The historical tools of research and documentation utilize methods similar to other humanities and social sciences. Therefore in cases discussed, this was the method applied. It must however be pointed out that in all but a few cases, the material aspects of culture that touch on architecture, were assessed without the aid of professional architects or even specialized builders. Again, one can seek consolation in the words of Olaudah Equiano, an Igbo man in slavery in the Europe and Americas who said that "Every man is a sufficient architect for the purpose. The whole neighborhood [can] afford their unanimous assistance in building... and in return receive and expect no other recompense than a feast."[18]He said this is reference to his Igbo homeland but his words could equally be relevant for any society in tropical Africa then as of now.

Abaluyia Architecture

In discussing the Abaluyia architecture the aim is not to describe the finished product alone but also forms and groupings of buildings and their social, economic and environmental factors and features associated with their construction, production and use. The Abaluyia who consist of over twenty sub ethnic groups (Bukusu, Tachoni, Maragoli, Banyore, Nyang'ori, Tiriki, Batsotso, Marama, Kisa, Banyala, Wanga, Kabras, Idakho, Isukha, Marachi, Bakhayo, Samia) inhabit many parts of Western, and some parts of Rift Valley Provinces of Kenya.[19] Indigenous architecture among the Abaluyia was a personal adaptation of a group solution. The houses were erected by use of wood, grass, ropes, mud/clay found locally. This was after an elaborate ritual of setting up a homestead.

Among the Banyore, such aspects as source of water, nearest homestead, and more significantly the population of fire flies at night were considered. Thick population of fire flies at night was evidence of dangerous animals especially of the cat-family in the vicinity. Also, the presence of many anti-hills was detested because of the damage ants could do to a house in a short period and therefore, it also mattered. The houses were generally round and this style, because of its preponderance among the Abaluyia must have been communally worked out over several generations and consequently closely tailored to their needs. Construction was preceded by planning which was reflected in the arrangement of buildings in the homestead. A whisper was no supposed to be heard in the next building. Landscaping was part of homestead-making.[20]

The Abaluyia, except for a few parts of Busia District, inhabit an area that is generally wet throughout the year. In indigenous architectural terms just like in modern architecture, the annual amount of rainfall was not as significant as how it fell. This therefore influenced the few variations in roofing styles or method and the type of grass used in the geographical locations. The Abaluyia inhabiting heavy rainfall areas made houses with high roofs for easy flow of water downwards, a specification that was not very much adhered to by those inhabiting areas with less rain.[21]

Walls were made of different material depending on purpose of the building. Residential houses had generally mud walls. Walls for granaries were of either reeds or thin long sticks or millet stems. This was to allow more and free circulations of air in the food stored in the granaries. The Walls were at times treated with repellents to keep dangerous and unwanted adversaries at bay. The walls of animal dwellings especially in leopard-infested places were thicker and were supported by thick wood and stones to make burrowing difficult, if not impossible. There was a special soil used for walls, especially the very gummy or sticky red soil. An ordinary wall benefited from at least three layers and coatings. The final coating was very fine and smooth, usually white, red, brown, or grey. Some parts of the walls were ornamental.

Openings in many buildings were for exit, and ventilation. The doors were usually large enough to allow exit of the biggest adult in the home or neighborhood. The shapes of doors differed from one building to another but ranged from rectangular, to near oval. Doorways were covered or closed with doors made from mat-like materials from reeds, animal skins, fibers, wood, etc. There was a provision for windows which were usually round or oval openings put at strategic places. Windows were covered by disused materials that would be stuffed in the openings when not in use. Stones shaped to fit, or pottery works, were also used to cover windows. The openings that acted as windows were used for ventilation and also to study or view happenings outside especially at night. The openings both doorways and windows were also decorated using special materials constituting the architrave. This was done by specialists or professionals, either hired from outside or members of the same family. The character and shape of an opening differed from one locality to another and depended on local factors. Even the type of material used for covering openings was determined by many factors. The type of weather, common wild animals and their aggressiveness on human settlements, were some of the factors considered.

Roofs were mainly made from grass, although at times certain shrubs would be used together with grass or alone. Reeds were also used for roofing certain type of houses. The design of the roofs was mainly conical to allow water to flow down easily. There were no level roofs among the Abaluyia, as such would not have survived the vagaries of heavy rain of a double maxima found in Abaluyia. The grass for roofing was usually left to dry after being cut or uprooted. It would then be tied in small bundles or bunches for easy transfer on top of the house.

The thatching was usually done by men. The grass used was called *ebembe.* There were specialists for each type of thatching style, and there were three types. First, was *Okhutuya (*compact/close) thatching which used a lot of grass and lasted longer, and was mainly for the well to do in society. Second, was *ebileche* (steps/ascending) thatching was the most popular and therefore the most common. Grass used for the same size of house was half that used for *okhutuya*. This style was very suitable for the Abaluyia country because it allowed water to descend fast and such thatches lasted longer than any other. The third was *okhumasia* which was quickly done and where very little grass was used. This style was used for very temporary buildings, like shelter for ceremonial occasions such as weddings, funerals, etc during rainy seasons. Such thatching was also used in emergency house-building where for instance a man had died without a house and one was required before his burial.[22]

Columns (pillars used in houses as support, ornament or monument) were common in indigenous Abaluyia architecture. Every hut was built upon a central pillar, usually a big, strong poll at the centre of the hut on which rafters and other support structures were linked. There were also pillars around big huts especially the main hut for the head of a homestead. These pillars were useful for holding out wider roofs on the verandah. The pillars also held back on the wall useful things placed outside the house but under the shelter of the house like firewood, etc. Some pillars outside the house were ornamental because on them, skulls usually of major bulls and cows slaughtered in the homestead were strategically placed. Some pillars especially those near the door were elaborately decorated. The decorations were either professionally and artistically produced or just ordinarily produced by one of the members of the family.

Molding of certain figures outside or inside the house was also a common feature of Abaluyia architecture. Cooking stoves (usually made of three "stones") were specially molded. Cupboards and even beds were at times also molded. Raised platforms were built in huts for different uses, like for water posts. The seats for pots were also molded. Oil lamps were also placed at strategically molded places within the house so as to shed light everywhere. It should be noted here that Abaluyia huts were usually not partitioned to the roof, leaving leeway for one source of light to be used in the whole hut. The doorways and windows also enjoyed special moldings by use of mud or clay and done by women.

Ornaments were also very common in the Abaluyia architectural pursuits. Ornaments, especially animal parts, man-made tools and items, plants, unique stones and other features were common features of Abaluyia homesteads. Skulls of bulls and buffalo were common features at gates and main houses. At the apex of the hut, unique features would be permanently displayed. At times totems were dried and placed in the lounge of the main house.

Among the Abaluyia buildings were renewed after every few years, through either major overhauls or complete destruction and setting up of a new building. Among some Abaluyia, especially the Bukusu, Tachoni and Kabras if an elder advanced in years died, he was buried in his house in a sitting position and the surviving members of the family would move to a new site but within range of the previous homestead. The former homestead provided a rich farming area. In such ritualistic practices, permanent buildings would have been embarrassment. Also, if on settling at a particular spot was occasioned by unreasonably high deaths on the members of the family that was enough reason to move to another safer spot. It was assumed that the place was inhabited by either bad spirits or that it was not favored by ancestors.

Further, separation of spouses (and not divorce unless dowry had been refunded in full) was frequent and the adoption of children common. Therefore families changed often in size and composition. Houses therefore changed to suit new needs. Also, when a head of a family died and a new generation was started, houses within the homestead or in a new site were re-oriented as they were rebuilt to reflect a new social arrangement, all of them built with the bearing from the house of the head of the family.

It would be wrong to see the temporary nature of many Abaluyia buildings as epitomizing an unstable and unsure society. On the contrary, the Abaluyia exhibited a high level of permanence than may Western societies in the same period. Feelings of permanence among the Abaluyia were strengthened through religion and social order. The planting of trees at grave sites was meant to be a permanent reminder to the society of somebody's resting place.

The importance of these temporary buildings to the Abaluyia environment is worth noting here. First, former homesteads, because of the high deposits of human and animal waste especially urea, were cultivated and yielded enormous outputs. Also, graves were never cemented like is the practice with modern architecture. Graves were represented by sacred trees grown at the time of burial but cultivation could take place anywhere on space previously occupied by a grave. It is sad to note that today, cemented graves and cemeteries combined occupy a very significant proportion of the Abaluyia especially in Vihiga district. This practice of cementing graves threatens to engulf the entire settled areas in the coming years going by the present rate.

The taboos surround building, from choosing construction site to the type of material used also played an important role to the environment, factors that have been obliterated by modern architecture as mere suspicion. For instance there was no compromise on building a house where an *Omukhuyu* tree grew, for the sacred tree unlike its cousin *Omuriba* chose where to grow and could never be planted. However much you tried, you could not grow *Omukhuyu*. Because of this and the fact that it was sacrificial, compounded by its potential to attract all sorts of animals, birds and insects underneath it, perhaps attracted by its fruits, it could not be cut by ordinary beings. Taboo decreed it for any domestic use apart from its shade. There were many other species of trees that were prescribed in house construction or domestic use, like *omurembe, lifutu, esikhoni*, etc. Reasons advanced for their being banned ranged from attraction of lightening, bareness, attraction of snakes, to some other obscure explanations. What is clear here however, is that these taboos were instrumental in safeguarding some of the plant species that would otherwise be extinct now. Some of these plants like the *Omukhuyu* are very important for an ecosystem because of their dominant role.

Besides the discomfort that would obviously be caused by water emerging from beneath ones house, it was seen as bad omen, a sign that ancestors disapproved of the site. It called for immediate vacation of such place. Of course this was important for preservation of water catchment areas unlike modern architecture that will easily circumvent such indications through draining or strong foundations.

Among the Abaluyia there was also a high degree of building specialization. As many as five different areas of construction - i.e. logging, thatching, muddying (smearing), door/window making, landscaping, etc. were well developed. Experts were utilized to minimize wastage of resources. Areas surrounding the home were well taken care of, and in the village, the centre was the most secure rather than the vulnerable periphery which neighborhood uninhabited bush and place of malevolent forces. The village symbolized the strength of the community. Nowadays, modernity makes isolation to be the best and serene accomplishment epitomized by the suburbs.

There was a good deal of careful planning often based on sound practical principles but sometimes related to religious beliefs and taboos. It was quite appreciated especially among the Babukusu that some relationships were better kept separate, especially in-laws. Many of these boundaries were invisible and no physical features were put. This was a plus for the environment compared to fortifications in the name of high walls that surround homes today, and which have compromised the ecosystem.

Many of the village configurations among the Abaluyia were semi-circular or horseshoe, arranged with all houses facing one direction. Burial of adults was done in front of the house whereas aborted births and people with questionable standing in society, the mentally-sick, epileptic, etc were all buried in the backyard. These formalities implied conscious planning, although the mere plan means nothing without knowledge of the intricate relations and invisible boundaries recognized by the inhabitants.

For instance, within a homestead married sons were not given their own granary or cooking place until they established their own homes. Even if they did not move out, their cooking was done in their mothers' kitchen. This served many local purposes. Environmentally, it meant less use of firewood, and therefore less stress on the ecosystem.

The Abaluyia also organized communal ceremonies and rites of passage such as circumcision. There were special isolated houses put up by the community for recuperation of inmates after circumcision. This communal approach minimized environmental dangers for example firewood used for warmth in such houses, and the material that would have gone in the construction of each house for each inmate.

However, each homestead exhibited its own internal plan. Requirements of the family were simple. They included: a place for cooking, storage of food, for sleeping, for sitting and talking and use for craft work, basket making, spinning, weaving, etc. Smaller domestic animals like goats and sheep were generally housed inside the house because of animals of prey, but the bigger ones like cattle and donkeys spent in the cattle boma (shade) within the compound.

It should also be pointed out that in Luyialand the greater part of the days are warm especially during relatively dry spells. Thus, many activities can be performed outside the house. So, a demarcated space was often sufficient for some activities and this must be considered as much a part of the homestead as the buildings. Men preferred to sit under shady trees to talk and eat their food. There was an evening fireplace for elders and circumcised young men. Women also cooked from outside during dry spells, partly surrounded by low walls for shelter from the wind.

In their farming, as the Abaluyia were and remain mixed farmers, many of their methods were environmentally friendly. In the plains, they practiced a sophisticated system of land husbandry using field or crop rotation. As already noted, the Abaluyia also had a tree planting and management system especially for their herbal supply. Composting and manuring was practiced from livestock cleared everyday by women and girls. In hill farms, stone terracing was used and these were favorable areas for growing finger millet, sorghum, etc.

Agro-forestry was also very common and fruit trees were not removed even when they were on farms. External fencing was only useful against any ethnic groups, but unlike other Bantu groups that used trenches, the Abaluyia used fortifications (like the Bukusu), thorn or euphorbia. Fortifications like those unearthed by archaeologist, Wandiba Simiyu and exemplified by Chitambe near Webuye town in Western Kenya, deterred enemies. The euphorbia was also effective because of its white latex which is carcinogenic and will blind eyes and bun skin.

The decoration of the Abaluyia houses in natural colors is one of the aspects that should also be highlighted here. Most of them were sometimes decorated on walls inside as well as outside houses with geometric patterns in many colors such as red using soil, ochre or flowers), yellow (ochre, flowers), white (chalk, bananas) black (suit, carbon), and green (leaves).

These are only a few aspects touching on indigenous architecture among the Abaluyia. There are others but because of time and space, we cannot discuss all. The few mentioned are a good pointer of environmental consciousness in indigenous societies.

Conclusion

Although we today know a lot about African indigenous architecture from researchers and writings availed to us by scholars, we have, in this chapter limited our discussion to what has happened among the Abaluyia before and during colonialism, and even today. The idea is to chronicle the authentic Abaluyia thinking and view vis-à-vis indigenous architecture - and not the rehearsed opinion that has been shaped by hindsight.

Presently, the world experiences calamity after calamity emanating from modern-day architectural innovations. Kenya's own deaths at collapsed building at Sunbeam Supermarket, Nairobi, and Nakumatt Downtown are still fresh in the minds of many Kenyans. Most observers and analysts attribute these architectural problems to poorly adopted technologies many of which are not suitable to Africa. Since the last massacre at Sunbeam, there have been reports at Kenya's Daily Newspapers ranging from the sinking picturesque Kenyatta International Conference Centre (KICC), through the crumbling View-Park Towers on Uhuru Highway, Nairobi. At the same time the professional bodies, like the moribund Architectural Association of Kenya is blaming quacks, it is passing the buck.

It is instructive to note that such meager projects like the American MacDonald's food chain, or another American chain Pizza Hut, remain storey-less yet very popular. It is only in areas like Europe where space in urban areas is very scarce that these chains have taken their businesses in storey buildings. Yet Africa, particularly Kenya is on an endemic rush to become a Manhattan-Long Island in Africa. The architects, as the experts, have sat back as observers. Kenya's problem is not and at the moment. Many vast areas of Nairobi with high potential remain unexploited to warrant the kind or rush we are witnessing to an already congested and overcrowded city centre. Geologists/seismologists are repeatedly warning builders not to concentrate buildings in one spot because of the sinking of Los Angeles and what is happening to Tokyo today. Indigenous Abaluyia architecture offers a solution.

The colonial-created green zone between the Wetlands and Eastlands in Nairobi has almost disappeared save for a small stretch, thanks to environmentalist, pragmatic and flamboyant Nobel Peace Prize winner Wangari Maathai of the world renowned Greenbelt Movement. How long the patches will remain amidst the present level of satanic grabbing mania in Nairobi, time will tell.

This chapter has one appeal; that there should be collaborative approach in solving environmental problems, where no knowledge is ignored because of who has produced it. The architects, planners, engineers , etc. have obviously failed us in Kenya, at least going by the dead Nairobi River, the constantly bursting sewerage systems, the congestion in a small city of three million such as Nairobi, the collapsing buildings, the poor road network, the many accidents, just to mention a few. The argument of this chapter has been that there is a lot to draw from indigenous past. That Kenya and other African countries need to set up a center of excellence for indigenous practices that can be positively adopted for our development. Further delay will only lead to more degradation. We need to appreciate the importance of indigenous architecture and environmental ethos now. The Abaluyia architecture has endured for generations and will still be relevant in the future.

Notes

1. Maurice N. Amutabi, "A History of Traditional Architecture among the Luhya and Environmental Conservation: Lessons for Contemporary Practitioners," Conference Paper presented at the Second All Africa Conference on "Green Architecture and Environmentally Sensitive Sustainable Development," Panafric Hotel, Nairobi, Kenya, 13 - 18 July 1996.
2. B. Farmer and H. Louw, *Companion to Contemporary Architectural Thought* (London and New York: Routledge, 1993), 3.
3. B. Farmer and H. Louw, *Companion to Contemporary Architectural Thought* (London and New York: Routledge, 1993).
4. B. Fletcher, *A History of Architecture on Comparative Method* (London: B.T. Batsford, 1948).
5. Priscilla Shilaro, *A Failed Eldorado: Colonial Capitalism, Rural Industrialization, African Land Rights in Kenya, and the Kakamega Gold Rush, 1930-1952* (Lanham, MD: University Press of America, 2008).
6. Amutabi, "A History of Traditional Architecture among the Luhya and Environmental Conservation." See also Thomas Barrie, "The Context of Sacred Architecture," *Journal on Religion, Art & Architecture* 3(1997): 9; S. Denyer, *African Traditional Architecture* (Nairobi, Ibadan, London: Heinemann, 1978); Nnamdi Elleh, *African Architecture Evolution and Transformation* (New York: McGraw Hill, 1997); Richard Hull, *African Cities and Towns before the European Conquest* (New York: W.W. Norton & Co., 1976); Louise E. Jefferson, *In the Decorative Arts of Africa* (New York: Viking Press, 1970); and Amos Rapport, "Vernacular Architecture and the Cultural Determinants of Form." In *Buildings and Society*, 281-305 (London: Routledge and Kegan Paul, 1980).
7. Homi Bhabha, *The Location of Culture* (New York: Routledge, 1994), 23.
8. A. J. Toynbee, *A Study of History, Volume II* (London: Oxford University Press, 1957), 5-7.
9. Bhabha, *The Location of Culture*, 41.
10. Amutabi, "A History of Traditional Architecture among the Luhya and Environmental Conservation."
11. Fletcher, *A History of Architecture on Comparative Method*, 11.
12. Bhabha, *The Location of Culture*, 175-6.
13. S. Denyer, *African Traditional Architecture* (Nairobi, Ibadan, London: Heinemann, 1978), 19.
14. Denyer, *African Traditional Architecture*, 1.
15. Joseph Thompson, *To the Central African Lakes and Back* (2 Volumes, 1881), 14. London.
16. Denyer, *African Traditional Architecture,* 1.
17. R. Moffat, *Missionary Labours and Scenes in South Africa*. London, 1842.
18. Quoted in Philip Curtin, (ed.), *African Remembered Narratives by West Africans from the Era of Slave Trade* (Madison: University of Wisconsin Press, 1967).

19. See Gideon S. Were, *A History of the Abaluyia of Western Kenya, 1500-1930* (Nairobi, Kenya: East African Publishing House, *1967*); and John Osogo, *A History of Baluhya* (Nairobi: East African Literature Bureau, 1967).
20. Amutabi, "A History of Traditional Architecture among the Luhya and Environmental Conservation."
21. Maurice N. Amutabi, "Environmental Dialogue From Below: Retrospection of the Role of Regular People, CBOs, NGOs and Governments as Stakeholders in Environmental Questions in Africa," Paper presented at the conference on the Environmental Justice and Governance: African Perspectives in the Neo-Liberal Era, University of Washington, Seattle, April 15 -16, 2009a.
22. Maurice N. Amutabi, "The role of Indigenous Knowledge in Environmental Conservation in Africa: The Case of the Abaluyia of Western Kenya," conference paper presented at the 2009 Africa Conference held at the University of Texas at Austin on the theme: Science, Technology and the Environment in Africa, March 28-30, 2009b.

• Chapter 16 •

CLIMATE CHANGE AND SUSTAINABLE DEVELOPMENT IN NIGERIA

Soji Oyeranmi

Introduction

> Climate change is a grave threat to the developing world and a major obstacle to continued poverty reduction across its many dimensions....further warming will bring poor countries high costs and few benefits....[1]

> Everyone in the world is affected by climate change, but we are not all affected equally. It is the billions of the world's poor who will suffer the most...[2]

The dangers posed by climate change are nearly as dire as those posed by nuclear weapons.[3] This seems like an alarmist projection or speculation but with the evidence around us, many will be worried. The evidence is alarming: cyclone nargis in Burma, April 2008 (over 200,000 deaths and 1.5 million displaced); earthquake in China, May 12, 2008 (50,000 deaths, almost 30,000 missing and nearly 300,000 injured), tsunami in 2004 claimed almost a million lives in more than five nations; floods and drought had continued to ravage many African countries (with attendant loss of lives and properties); and the current global food crisis is inflicting hardships on millions of people.[4] Many scholars and scientists have attributed this seemingly irreversible human tragedy to climate change occasioned by the catastrophic impact of human activities on nature and the environment. And according to recent report,[5] the worst is yet to happen. Therefore, humanity cannot afford to waste any further time in saving "our collective future."[6]

While climate change actually threatens our future,[7] the threat appears to differ from place to place. In reality, it is the billion of world's poor (especially in Sub-Saharan Africa) that will bear the greatest loss. The industrialized countries have been taking some steps to implement sustainable development strategies and the poor ones are yet to have specific programs either to solve the existing environmental crises or avert the impending ecological cataclysm. Due to this fact, the international community needs to fashion more adequate ways of responding to the complex global challenge of climate change (especially in saving the poorest segments of humanity from this climatic doom).

With the current speed of unstable climatic conditions, poor countries in Africa (Nigeria inclusive) may suffer more due to geographical locations, fragile economies, low income, heavy reliance on climate sensitive sectors such as extractive industries and agriculture,[8] and lack of adequate modern conservation skills and facilities. This definitely will have catastrophic consequences on their already battered economies. Obviously, climate change has the potential of reversing any progress (if any) made by these countries towards achieving the Millennium Development Goals, especially the goals of eradicating extreme poverty, combating communicable diseases and securing environmental sustainability.[9] This chapter argues that the poorest of the poor Nigerians, as in other parts of sub-Saharan Africa, are mostly affected and may still suffer more if urgent steps are not taken to abate the climatic fluctuations.

But what can we make out of the hues and cry over climate change? Is there any nexus between climate and economy? Is the cynicism about climate change real? What are the causes? Can the impact be positively reversed? Is there any link between poverty and climate change? Do they reinforce each other? What are the ways out of the impending climate doom for poor countries in Africa (especially Nigeria)? What are the implications of climate change on the Nigerian economy? This chapter shall be guided by the above questions and many more.

Climate Change

It has been established that human beings and the natural world are on a collision course.[10] This is largely because man's activities had contrived to destroy the natural equilibrium that had facilitated for centuries the beneficial co-existence of man and nature.[11] Consequently, our world continues to witness hard and often irreversible damages on the environment and the critical natural resources. Evidently, the problems facing our environment are vast and diverse, but most scholars and scientists agree that climate change occasioned by global warming resulting from pollution, ozone layer depletion, emission of green house gases etc which currently threaten the bare existence of humans and other living organisms is by far the gravest.[12] For the purpose of this discourse, climate change is seen as ceaseless fluctuations in both atmospheric and biospheric conditions leading to climatic catastrophes such as flooding, drought, desertification, earthquakes, tsunami, hurricanes, food crisis and other natural disasters. On the surface, it appears that Africa is naturally shielded against the climate change calamities. This is unlike Asia which has lost millions of lives and valuable properties to natural disasters (tsunami, cyclone, earthquake, flooding etc.) ostensibly due to climate change. But as many recently conducted researches have shown, no nation in the world is spared from the cataclysm of climate change but the impact varies from place to place. In some places, the worst has happened; in some, the effects are mild, while in others the real tragedies are just waiting to happen. However,most climate related researchers generally agree that "it is the billions of the world poor that will suffer the most."[13] The case of sub-Saharan Africa is particularly pathetic in this regard.

The conditions of 21st century sub-Saharan Africa remain serious[14] primarily because it consists of the world's poorest countries with absolutely fragile economies. With increasing climate fluctuations, these countries will bear most consequences due to the grave poverty, deepening corruption, geographical location, lack of sound conservation strategy and strong dependence on climate sensitive economic sectors such as oil exploration, agriculture, mining, and gas flaring.. According to a recent report by the World Bank, if not properly handled, climate change is capable of reversing any possible progress made (especially in Africa) towards achieving the Millennium Development Goals including the goals of eradicating poverty, combating communicable diseases, and achieving environmental sustainability.[15] It is most disturbing that while industrialized countries are already combating climate change through implementation of sustainable development policies, most countries in Africa are yet to realize that confronting climate change is fundamental to economic development and their poverty alleviation agenda.

It is, therefore, not surprising that most African countries never seriously consider environment or climate change in their developmental agenda. Yet, most of them depend mainly on environmental resources for raw materials such as ores, oil and other minerals and agricultural like timber, cotton, sugar, and coffee. Recent researches have shown that these natural resources are decreasing, thereby forcing the people to exert greater pressure on the environment and over drawing their natural resources which is causing further environmental degradation and climate change. As the Brundtland Commission clearly puts it:

> The recent crisis in Africa best and most tragically illustrates the ways in which economics and ecology can interact destructively and trip into disaster. Triggered by drought, its real causes lie deeper. They are to be found, in part, in national policies that gave too little

> attention too late to the needs of small holder agriculture and to the threats posed by rapidly rising population.[16]

The challenges are enormous. As Walter Rodney once observed "man has always exploited his natural environment in order to make a living."[17] This explains why humanity has long appreciated the forest for the energy, food, medicine, construction and other purposes. According to the UN estimates, about 1.6 billion around the world, including many of world's poorest, derived at least part of the food, income or medical needs directly from the forest. Africa's rural poor are primarily forest dependents. Although forest products, mostly unfinished logs, account for only about 2 percent of sub-Saharan African exports, the forest generates an average of 6 percent of the region's gross domestic product. Eighteen African countries, including Cameroon and Ghana, are among twenty-four countries that rely on forest for 10 percent or more of their economies.[18]

Ordinarily, this forest dependence ought not to be harmful to the economies or the environment because of the presence of abundant forest resources. Africa contains about 15 percent of the world's remaining forest and is second only to South America in the amount of dense tropical forests "that are the most effective in removing carbon from the atmosphere."[19] But the problems lie in the absence of adequate/modern knowledge about the role of forest in supporting agriculture, preserving biodiversity, protecting water supplies and most importantly moderating the impact of climate change are not well understood. The conversion of forestland to agriculture, both subsistence and commercial, is by far the most destructive cause of deforestation in Africa and other tropical regions.[20] This is mostly due to the response to population pressure resulting to more demand for arable land. Many experts are blaming the poorest of the poor for the continuous exhaustion of forest reserves. Thus they argue that:

> It is the rural poor themselves who are primarily agents of destruction as they clear forest for agriculture, firewood and other necessities. Lacking other means to meet their daily survival needs, rural people are forced to steadily erode the capacity of their natural environment to support them.[21]

Some, however, are reasonably sympathetic to assisting the African poor. For example, Christian Lambrechts believes that poverty is forcing people to depend on their forest to gain access to essential products such as medicine which they don't have money to buy. "Such subsistence exploitation causes no damage when done sustainably," he concludes. In the same manner, many African environmentalists have also accused some developed countries for promoting activities that are damaging the African environment and its natural resources. The Brundtland Commission specifically blamed the economic backwardness of Africa on the unjust international economic relations which develop the West almost at the same level that it is underdeveloping the rest of the world through gross exploitation of their natural resources, debt burden and trade barriers. The commission comments that:

> ... Their roots (poverty and ecological tragedy) extend also to a global economic system that takes more out of a poor continent (Africa) than it puts in. Debts that they cannot pay force African nations relying on commodity sales to overuse their fragile soil, thus turning good land into desert...Trade barriers in wealthy nations and in many developing ones make it hard for Africans to sell their goods for reasonable returns putting yet more pressure on ecological system.[22]

Using Senegal as a basic illustration, the commission portrayed how the crazy urge for foreign exchange through exports has led to despoliation of soil and forest. Rather than blaming the environmental disasters (including climate change) on poverty, the Brundtland Commission accused the endless exploitation of resources for exports to the Western world for this tragedy.

In Africa, as in other parts of the world, economic growth has been recognized as the greatest weapon to fight poverty. But often the policy makers fail to match their efforts towards economic development with ensuring safe and healthy environment. A world in which poverty is endemic will always be prone to ecological and other catastrophes and also a world that is overwhelmed by environmental hazards will be perpetually in poverty.

Nigeria presents a paradox: a rich country with poor people.[23] By far the most populous country in Africa (with about 140,003,542 people, 2006 census and United Nations, 2007); with mega oil wealth; rich agricultural and mineral resources; and yet rated 159th out 177 countries in UNDP's 2004 Human development Index. The failure of Nigeria to channel its wealth to the development of its people and to the productive use of land and other resources has continued to impact negatively on the people and the economy. Indeed, the recent ECOWAS poverty profile captures the ugly trend that:

> The Nigerian economy has declined, especially since the mid-1980s. The real GDP per capita from 425 in 1980 to 358 in 1990 remained slightly over 350 up to 2002 then increased to 387 in 2003 and 402 in 2004. The average GDP growth between 1996 and 2005 was four per cent.[24]

It is therefore not surprising that poverty remains endemic in the midst of plenty as more than half of the Nigerian population still wallows in abject poverty.[25] As the figure of those into poverty appreciates, so does the fight against poverty appears to increase, though with no concrete reduction in its growth. Central to the reasons for the failure of poverty reduction strategies is overdependence on non-renewable resources (crude oil, coal and gas) which are said to be accelerating global warming leading to climate change. This is coupled with the fact that Nigeria has no adequate technology or expertise to abate the resultant climate malfunctions. Writing about the real danger of this unsustainable development, Jekwu Ikeme says that:

> Few countries stand more threatened by this development (global climate fluctuations) than the oil producing Nigeria: Nigeria's economy today remains monocultural and heavily dependent on the oil sector, which accounts for around 80% of government revenues, 90-95% of export revenues and over 90% of foreign exchange earnings... Nigeria is believed to have an estimated 124 trillion cubic feet (TCF) of proven natural gas reserves –9th largest in the world. Due to a lack of gas utilization infrastructure, Nigeria currently flares 75% of the gas it produces...Despite the diversification agreement between government and oil companies in 2004, Nigeria's economy remains dependent on fossil fuels. This is particularly worrisome because fossil fuels are chief culprits implicated in the environment issue of climate change phenomenon commonly referred to as global warming.[26]

No where is Nigeria's problem more pronounced than in the Niger Delta. The Niger River enters Nigeria from the northwest, crossing the western part of the country to join the Atlantic ocean in the south. Near the coast, the river forms this delta with mangrove forest, lagoons, and swamps stretching about 100 km (about 60 mi) inland. The Niger Delta is the largest in Africa, covering an area of about 36,000 sq km (about 14,000 sq mi.). Oil pumping stations in the delta of the Niger River operate around the clock. Petroleum deposits were first discovered in Nigeria in 1956. The country is now a major supplier of oil.

Global warming (which refers to an overall increase in temperature in earth's atmosphere and oceans) has been described as "the greatest threat facing 21st century humanity."[27] But many skeptics are questioning this assertion claiming that though the earth is warming but no one is certain of both the causes and consequences as the "physical processes that underlie climate systems are complex and not fully understood."[28] But is it real that the earth temperature is rising? Most recent scientific reports have concluded that global warming is unequivocal and that mankind is largely responsible for it. For example, the UN sponsored intergovernmental panel on climate change (IPCC) specifically blamed human economic activities in the last 100 years for the increase in the concentration of green house gases (carbon dioxide Co_2), methane, nitrous Oxide, N_20 and others.[29] This gives direction to any discerning minds that despite some scientific uncertainties, the balance of opinion suggests the certainty of climate change and favors early actions for precautions. This has informed the international community to take various abatement measures against climate change.

Many developed countries are already expending huge resources into research and development of alternative and renewable energy sources that would lead them away from fossil fuel economy. Governments have also encouraged private sectors to do same by giving them incentives. For instance, the Non Fossil Fuel Obligation (NFFO) in the UK requires electricity companies in England and Wales to secure specified amounts of electricity from renewable energy.[30] Denmark has also put in place tax levies to restructure the power markets, raising a levy on conventional energy and reinvesting it on renewable energy. The Dutch government equally subsidized wind power by up to 40% which increased wind power capacity to 120MW. While the above examples clearly demonstrated the actions of some developed countries curtailing climate change by curbing the use of fossil fuels, many African countries still largely stand akimbo and show little interest in environmental issues.

Many reasons have been given for this non-chalant attitude. One of such reasons was the assumption that since Nigeria does not fall in the category of developed countries required by the Kyoto protocol to urgently reduce their green house gas emissions, then the country needs not take any abatement action now. Some argue that since the present emissions of green house gas from Nigeria and Africa are negligible, there is no cause for alarm. Others submitted that since the industrialized nations are the main cause of emissions, then, they should bear the main responsibility and the cost of combating climate change. While these points seem to be sensible, a more critical evaluation of the present situation and the impending doom traceable to the environment will spawn any reasonable African government to action. But what are the real and possible economic implications of climate change on Nigeria? This will be the next focus of this chapter.

Some Real Economic Impacts of Climate Change in Nigeria

Nigeria has not escaped the negative impact of the horrendous environmental changes. The country is practically under siege as most cities are either engulfed by raging floods or confronting surging desertification. From East to West, North to South erosion and desert have continued to eat the land. All these are crippling the peoples and national economy and it seems that the country is helpless.

Environmental hazards disrupt economic activities in addition to the massive displacement of people. The people of Abeokuta, Ibadan, Ode Remo, to mention but a few, hardly escape the agony of flooding every rainy season. According to one report by the *Tell magazine*, incessant flooding has damaged some parts of Lagos State. This is not limited to Lagos or even south-west. Indeed in many communities across Nigeria "tales of woes and pains usually follow any down pour."[31] While most residents are usually displaced, their economies also got dislocated. In Lagos, industrialized and commercialized areas like Victoria Island and Ikeja usually encounter floods, resulting in loss of valuable goods.[32]

The floods have also resulted in erosion and landslides in many parts of the country. In fact, several communities in South-eastern Nigeria are affected by erosion and landslides. According to the *Tell magazine*, South east has the highest number of erosion sites, with the majority in Anambra state. The landslide in Nanka community literally cut off the community from the rest of Anambra State, halting all the economic activities. Other areas affected include Ozubulu, Onitsha, Ukpor, Agulu, Aguleri, and Amakpo.

If southern Nigeria is under the surging siege of downpours, the northern part is perpetually confronting the ravaging desert. Experts are warning that "the biological potential of land to grow plants in 10 northern states of the country is diminishing at an alarming rate."[33] This phenomenon actually portends hardship and economic doom for a region with over 20% population that depends on farming and animal husbandry. Unfortunately, the doomsday is already here as statistics show that Nigerians are already suffering the economic impact of desertification. The country has already lost about 350,000 hectares of arable land to desertification yearly with a large number of people already afftected. For examples, about 50,000 farmers in 100 villages have been affected in Yobe State, and 40,000 farmers in Borno State have lost their means of livelihood.[34] As a recent survey suggests the impact is likely to spread further as the Sahara desert is advancing southwards at a rate of 0.6 kilometres yearly.

There is a general consensus that desertification is by far the most pressing environmental problem in the dry lands. Nigeria loses over 350,000ha annually to advancing desert leading to deterioration of ecosystems, degradation of various forms of vegetation, destruction of biological potential, diminution of biological potential, decay of a productive ecosystem, and reduction of productivity. Nigeria is equally losing the capacity to feed its population. Thus, experts have attributed the current food crisis that is biting harder on the poor Nigerians to climate change. According to the Federal Ministry of Agriculture and Water Resources, food production in Nigeria dropped by 20% in 2007, partly responsible for soaring prices of foodstuff. This further proves the serious relationship between economics and ecology which may be positive or negative depending on the strategies adopted by the affected people in conserving their environment.[35]

With the above illustrations, it becomes obvious that agriculture has come under direct climatic attacks. Attesting to this, the government claimed that the country loses about 750 million U.S. dollars annually to the depletion of its 350,000 hectares of forest land by direct human activities and climate change.[36] As Kayode Ojo points out, though oil is the most important source of government revenue, agriculture remains the mainstay of the Nigerian economy as "about 70% of the labor force is still employed in agriculture."[37] It is, however, lamentable that most of these employees are still peasant farmers who still largely depend on nature as their work is strongly tied to the seasons and productivity is usually affected by the rain which is climate dependent. Consequently, the variability and unreliability of rainfall has become the most critical trend in agricultural production in Nigeria.[38] This has not only impacted negatively on the crop production but also livestock production in a number of ways. Most importantly, as a result of low crop production the income of livestock farmers would decline resulting in greater financial stress. Indeed, economic survival of many will be threatened due to significant impact on characteristics of labor, employment, demography and so on.[39]

Possible Future Economic Implications of Climate Change in Nigeria

The possible future economic implications of climate change for Nigeria could be multidimensional. In the first instance, projections suggest that Nigeria will experience massive environmental refugee migration. For a 1-m rise, more than 3 million people are at risk, based on the present population. The estimated number of people that would be displaced ranges from 740,000 for a 0.2-m rise to 3.7 million for a 1-m rise and 10 million for a 2-m rise.[40] The most vulnerable is the coastal region of the country. A large percentage of Nigeria's urban population lives in coastal cities. Estimates put the total population living along the coastal zone to about 20 million people which translates into 22.6% of the national population. Similarly, most of the economic activities that form the backbone of the national economies are located within the coastal zone. Coastal areas also form the food basket of the region. Offshore and inshore areas, as well as estuaries and lagoons, support industrial fisheries accounting for more than 75% of fishery landings in the region. This naturally dovetails into another possible negative economic implication of climate change which is that of its effect on food security. As populations are displaced and climatic and vegetational belts migrate away from the traditional geographical location, a general disruption in food production is expected to be the outcome.[41]

Environmentally, Nigeria's climatic regime stands to be severely disrupted leaving its forests and water resources at risk. Studies have shown that biological productivity in Nigeria will decrease in the event of global warming with an additional consequence of severe fuel wood shortages. Already Nigeria has experienced definite shift in the long-term rainfall mean towards more arid conditions. These climatic changes have had adverse implications for water resources availability for power generation and agriculture. Likewise, Nigeria's low-lying lagoon coasts stand threatened by sea-level rise, particularly because most of its major and rapidly expanding cities are on the coast. If sea level rises, inundation could occur along more than 70% of the Nigerian coastline, placing land at risk many kilometres inland. In Nigeria, inundation is the primary threat for at least 96% of the land at risk. With a 1-m rise in sea level, up to 600 km^2 of land would be at risk. This area includes parts of Lagos and other smaller towns along the coast. The periodic overflow of the Atlantic across the Bar beach bank is an indication of a phenomenon that may accelerate as climate change intensifies and the sea level rises even further.

Sea level could have adverse impact on multimillion dollars investments in mining and oil exploration thereby disrupting the economic and some social infrastructure in large cities of Port Harcourt and Warri. However, the most critical concern is the possible impact on the supply and demand of hydroelectric power and fuelwood, which are the most widely used sources in Nigeria (as in other parts of sub-Saharan Africa).[42] For example problems related to the seasonal aspect of river flows as well as unreliability of rainfall as sources of water supply will cut the hydroelectric power production. Expectedly, reduction in hydroelectric power is already imposing grievous economic hardships on many Nigerians, especially on the poor ones. Worse still, the expected elimination of some sources of fuel wood would lead to further economic degeneration among poor Nigerians. The situation would be aggravated by the growing energy demand due to an increase in population.[43]

Earth temperature is rising and the climate is changing, no matter the cause or causes. Though the world is already reaping the consequences but the worst is yet to happen in most African countries. And if we are going to ameliorate the already worsening situations and prevent the impending catastrophe, then Nigeria and Africa must act fast. One of the best ways is to replenish the existing but neglected indigenous conservation techniques and improve the modern ways of conserving the environment and natural resources. Indeed, it is very significant to note that in Nigeria, various communities have many forms of traditional mechanisms usually adopted to conserve their environment. Popular among these methods is shifting cultivation (which allows land to rest and regain its productive powers), seasonal fishing, and erosion control mechanism.[44]

Over the years, the Nigerian government has adopted several methods of conserving the environment, especially by making laws such as the Federal Environmental Protection Act of 1988, The National Park Decree of 1991 and many more. The federal government has also inaugurated Building Nigeria's Response to Climate Change (BNRCC) to discuss matters relating to climate change. At the launching ceremony, the then Minister for Environment and Urban Development, Fatima Alao, admitted that "Nigeria is more vulnerable to impacts of climate change because of its size and population, amongst others... It has led to greater soil erosion and flooding in areas of higher rainfall, heightened drought and desertification in the northern parts, accelerated sea level rise and water salt erosion in coastal areas".[45]

In terms of policies for remedying the situation, the key is the diversification of the economy away from oil production. This will ensure that the global switch away from fossil fuels and the consequent reduction in fossil fuel demand have little impact on the Nigerian economy and foreign exchange earnings. Also research on climate change and the socio-economic implications for Nigeria is necessary for developing adequate response strategies. Furthermore, study of the science of climate change and its potential impact on Nigeria is very important for creating awareness and providing the background information for targeting policies adequately. This point is made in view of the recognition that the major constraint to adequate forecasting and formulation of adaptation policies is the paucity of climate data in Nigeria. Long term studies on national and regional climate change in Nigeria should be embarked upon and vigorously pursued. The findings of such studies will be crucial for the formulation of adequate response and adaptation policies such as adequate resettlement programs for those that may be displaced by climate change.

Furthermore, increased government participation in the global climate change deliberation in order to negotiate a better deal for Nigeria and Africa is necessary. The suggestion that oil-producing countries should be compensated for their projected income losses in the event of the implementation of the Kyoto protocol and assisted in their economic diversification attempt should be vigorously argued and canvassed. Nigeria can only be sure that its interest is protected in the emergent global abatement strategy if it increases its level of participation. Its participatory capacity in turn will be enhanced by findings from studies and research into various ramifications and dimensions of the climate change issue as suggested above. Most times Nigerian government does complain of shortage of fund but we are also aware of the deepening corruption that has attended the management of ecological funds across the country. But this does not foreclose the need for more commitment from international community in resolving fundamental ecological problems in Nigeria. At the 35th meeting of the Ecological Committee in Abuja, which centered on dealing with the threat of climate change, Nigeria's Minister of Environment, John Odey, announced the approval of 1.46% of the national budget for ecological problems (such as desert encroachment puts at 650,000 meters per annum, soil and gully erosion). The breakdown shows that 60% goes for drought and desertification control and afforestation;

25% for soil and gully erosion control and 10% for administration.[46] While this seems laudable, it is not enough. There are a number of mechanisms already deployed with respect to coping with the impact of climate change. Although these adaptive responses have been more at the community level through the use of underground storage tanks, improved rainwater harvesting and cleaning of water bodies, governments, both at the national and sub-national levels, have made some efforts in support of these community based interventions. But there are still a number of ways through which the government can become a more active participant than is currently the case in Nigeria. This can be done via enactment and enforcement of suitable laws, improvement in medical facilities in case of climate change induced epidemics, better response to emergencies and involvement of the private sector as well as Non-Governmental Organizations (NGOs). Most importantly, better advocacy with regard to integrating climate considerations into the broader poverty reduction strategy of the government and the drafting of a National Adaptation Plan of Action (NAPA).

As David Okali, the chairman of the Nigerian Environmental Study / Action Team (NEST) has rightly suggested, in combating the menace of climate change, Nigeria's poverty reduction strategies must be made sustainable. To attain this level, it must strive to meet international climatic standard and adapt to climate change through "improved governance, promoting gender equality and more sustainable natural resource management."[47] All these are necessary because any further efforts at reducing poverty without paying due attention to issues and impact of climate change will remain futile. For example, there is no way any government in Nigeria can effectively tackle poverty and climate change if the current energy crisis is not seriously resolved. Near-zero availability of electricity in the country has not only crippled national economy but has further engendered poverty especially among the poorest of the poor. Consequently, most Nigerians mount too much pressure on the environment and the natural resources because they are forced to rely on alternative energy supply (electricity generating sets) and cheaper means of energy such as coal and fuel wood which are said to be some of the greatest sources of global warming with the attendant climate change. With this seemingly unabated ugly trends, achieving sustainable economic development in Nigeria will remain a mirage.

Conclusion

To conclude, we have demonstrated that Nigeria cannot afford to continue to ignore the potential impact of the global climate change and it is in its interest to begin to introduce measures to reduce its greenhouse gas emissions. The Nigerian economy must be diversified and steered away from fossil fuels both in terms of production and consumption.

The most important step to be taken is to raise environmental consciousness of Nigerians. Every citizen, no matter the status, must know that the earth is fast losing its ability to nurture the natural resources we all depend upon and become more serious about saving the environment. By saving the environment, we are only saving ourselves. Environmentalism should therefore move from being a philosophy only championed and promoted by a passionate minority to a way that governs mainstream behavior of the people and public policy. Collaboration among individuals, NGOs, other groups and government agencies is the key to achieving this.

Notes

1. For full details of real and possible economic costs of climate change, see *The Stern Review on the Economics of Climate Change*, a 700 page report released on October 30, 2006 by economist Lord Stern of Brentford for the British Government which discusses the effect of climate change on the world economy.
2. This was the opinion of Ngozi Okonjo-Iweala, Managing Director, World Bank on why climate change is fundamental to development and fighting poverty. It was extracted from issue 42, 2008 of Develop*ments Magazine* of the DFID, p.14. For more information you may visit www.worldbank.org or www.development.org.uk.
3. See the *Bulletin of the Atomic Scientists* as quoted in August 2008 *Awake,* magazine of the Jehovah witness titled "Global Warming is planet Earth in Peril?" 3.
4. According to the UN office for coordination of humanitarian affair, weather-related disasters are on the increase around the world. In 2007 for example, about 26 countries were affected from developed ones (Britain, US) and underdeveloped, (about 14 countries were said to be affected in West Africa).
5. For instance, at the meeting of the 2008 G8 leaders summit held in Japan the problem of environmental disasters, especially how to save the world from the impending climate doom, was the heart of their discussions.
6. Thijs de la Court, *Beyond Brundtland, Green Development in the 1990s* (London, Zed Books Ltd.1988) 7-11.
7. See Okonjo-Iweala in Develop*ments Magazine* of the DFID. p. 14.
8. See Develop*ments Magazine* of the DFID, p.14.
9. *Ibid*
10. World Commission on Environment and Development (WCED) *Our common future.* (Oxford: Oxford University Press, 1987), 43.
11. European Commission, *Key Facts and Figures about Europe and the Europeans* (Luxembourg, 2007), 53.
12. See G. Asheim, C. B. Froyn, J. Hovi and F. C. Menz, "Regional versus Global Cooperation for Climate Control," *Journal of Environmental Economics and Management* (2006)51: 93-109; S. Barrett, *Self-enforcing Environmental Agreements*. Oxford Economic Papers 46: 1994,878-94; S. Barrett, Heterogeneous *International Environmental Agreements, in Carraro* (Ed.). S. Barrett, *Environment and Statecraft*.

Oxford University Press, Oxford, 2003; S. Barrett, "Climate treaties and "breakthrough" technologies," *American Economic Review*, 2006, 96: 22-25.

13. See *Developments*, 14.
14. See Paul Collier, *The Bottom Billion, Why The Poorest Countries are failing and what can be done about it* (Oxford: Oxford University Press, 2007), 5-12.
15. See Okonjo-Iweala, *Developments,* 14.
16. See *Beyond Brundtland*, 9-20.
17. Walter Rodney, *How Europe Undeveloped Africa,* (London: Bogle L'Ouverture Publications, 1986), 30.
18. *Africa Renewal*, Vol 21, No. 4 J (New York: United Nations, January, 2008) 4.
19. *Ibid.*
20. *Ibid.*
21. See *Beyond Brundtland*, 21.
22. See *Africa Renewal*, 6.
23. Several Reports such as World Bank, 1995, 1996, 2004, 2006 and UN 2006, 2007 have confirmed this. But the latest was ECOWAS, *ECOWAS Poverty Profile,* (New York, UN Statistics Dept. of Economics and Social Affairs, 2007), 100-1006.
24. See ECOWAS, ECOWAS Poverty Profile, (New York, UN Statistics Dept. of Economics and Social Affairs, 2007), 100-1006.
25. Ibid.
26. Jekwu Ikeme, Comment on "Assessing the future of Nigeria's Economy: Ignored Threats from the Global Climate Change. Debate," Comment posted on November 20th 2008 www.AfricaEconomicAnalysis.org (accessed on March, 20th, 2009)
27. See the *Bulletin of the Atomic Scientists,* 10 – 20.
28. See I. M. Mintzer (ed), *Confronting Climate Change: Risks Implications and Responses*, (Cambridge, Cambridge University Press, 1992), 163-170. A Time International Magazine 2000 Special_edition an Earth Day, UNEP 2007 Reports on Climate Change. The 2008 G8 Summit in Japan also focused on global environmental crisis.
29. See *Climate Change 2007: The Physical Science Basis*, Contribution of Working Group 1 to the Fourth Assessment Report of the Intergovernmental Panel on Climate Change (IPCC) (Cambridge: Cambridge University Press, 2007). IPCC, *Climate Change,* (Cambridge: Cambridge University Press, 1996). For further information visit IPCC official web site www.ipcc-org.See also *The African Renewal,* July, 2007 and January 2008 editions.
30. *See Developments Magazine*, 7 -12.
31. See *Tell Magazine* October 2007 Special edition on Environmental disasters in Nigeria.
32. *Tell Magazine* August 11, 2008 edition captioned "Environment: Nigeria's Nightmare," 20-31.

33. *Ibid.*, pp. 26-27.
34. *Ibid.,* pp. 28-29.
35. See Ukoha Ukiwo, "Indigenous and Received knowledge in the conservation of the Niger-Delta Environment" in Kayode Fayemi et al (eds), *Toward an integrated Development of the Niger Delta*, (Ikeja: Centre for Democracy and Development, 2005), 224.
36. John Odey, Minister of Environment, disclosed this at a forum with the House Committee on Climate Change, adding that the Sahara Desert in Nigeria is moving southward at a rate of 600 meters annually. Extracted from http://news.xinhuanet.com /english/2009-07/22/content_11753989.htm.
37. Abraham Olukayode Ojo, *The Climate Dilemma* (Lagos: Lagos State University Press, 2007), 9-29.
38. Ibid.
39. Ibid
40. According to IPCC 2007 projections, in Africa, by 2020, between 75 and 250 million of people are projected to be exposed to increase water stress due to climate change. By 2020, in some countries, yields from rain-fed agriculture could be reduced by up to 50%. Agricultural production, including access to food, in many African countries is projected to be severely compromised. This would further adversely affect food security and exacerbate malnutrition. Towards the end of the 21st century, projected sea level rise will affect low-lying coastal areas with large populations. The cost of adaptation could amount to at least 5 to 10% of Gross Domestic Product (GDP). By 2080, an increase of 5 to 8% of arid and semi-arid land in Africa is projected under a range of climate scenarios (TS). For possible future impacts of climate change in Nigeria, See also Temi E. Ologunorisa, "Global Warming" in W. I. Bell Gam et-al, eds, *Perspectives on the Human Environment* (Port Harcourt: Colleagues Publishers, 2004), 42-43.
41. See Temi E. Ologunorisa, "Global Warming" in W. I. Bell Gam et-al eds: *Perspectives on the Human Environment* (Port Harcourt, Colleagues Publishers, 2004), 42-43.
42. L.G. Awosika et-al, "Impacts of Sea Level Rise on Nigeria" In the Proceedings of the IPCC Symposium on the Rising Challenge of the Sea, 14-16 March 1992 in Venezuela.
43. L.G. Awosika et-al, "Implications of Climate Changes of Sea Level on Niger Delta, Nigeria Phase 1," A Report for the United Nations Environment Program.
44. See Ukoha Ukiwo "Indigenous and Received knowledge in the conservation of the Niger-Delta Environment," mimeo.
45. See *Tell Magazine* August 11, 2008 edition captioned "Environment: Nigeria's Nightmare," 20-31.
46. See *This Day Newspaper* of 21-08-2009.
47. *Tell Magazine* August 11, 2008 edition captioned "Environment: Nigeria's Nightmare," 20-31.

Part IV

AFRICAN ENVIRONMENT AND CONSERVATION METHODS REVISITED

• Chapter 17 •

THE CHALLENGE OF BIOFUELS DEVELOPMENT IN AFRICA: AN INSTITUTIONAL ANALYSIS AND DEVELOPMENT (IAD) PERSPECTIVE

Ogundiran Soumonni

Introduction

Biofuels have been promoted as having the potential to reduce the demand for fossil fuels, promote economic development, and mitigate the impact of global climate change. The two main types of biofuels that are currently consumed are ethanol produced from carbohydrates (starch in corn or sugar in sugarcane) and biodiesel from plant oils such as soybean and oil palm. Some of the countries that have established significant biofuel programs include Brazil (sugarcane), U.S. (corn), Indonesia (oil palm) and South Africa (maize, sugarcane).

The development of biofuels has however been marred by controversies and challenges, which have questioned their viability as a sustainable energy option for the future. One major controversy emerged when media outlets around the world reported riots in several impoverished nations due to grain shortages and soaring food costs that have been blamed to varying degrees on increased demand for ethanol.[1] It has also been reported that the growing demand for ethanol is leading to higher crop prices, which could lead farmers to till more acres, and remove environmentally sensitive land from conservation programs, thereby negatively impacting soil and water resources.[2] Another important debate involves the clearing of forests and savannas to be replaced with plantations of oil palm, soybean, jatropha and other biofuel crops. It has been reported that this results in a "carbon debt", that is, more carbon dioxide is released in this process than the reductions that can be obtained from using the biofuels.[3] It should be noted, though, that cellulosic ethanol from crop wastes (stalks, cobs, leaves) or non-edible plants, do have the potential to reduce the destruction of habitats, reverse carbon debts, and lessen the competition with food crops, but the required technology is still at the developmental stage.[4] Nevertheless, because land and labor costs are low in many developing countries, many of their governments and businesses have been keen to invest in conventional biofuel crops in order to meet the demand for them, and generate national revenue. Some African governments have also begun to pursue biofuels primarily in response to the demand for raw materials by industrialized nations and as an opportunity for economic growth, but with little discussion about domestic use. Community groups have alleged that this push is causing the loss of agricultural land and forests and undermines the livelihood of vulnerable peoples who depend on these resources.[5] However, very few African nations seeking to pursue biofuels have developed a set of comprehensive policies to tackle such a complex and multi-dimensional undertaking.[6] In light of this background, the current chapter seeks to employ the Institutional Analysis and Development (IAD) framework in order to assess existing practices and successful strategies used to manage the common pool resources that are impacted by biofuels.

The first part of the chapter describes the IAD framework, which has been recognized for playing an important role in the study of common-pool resources (CPRs), and emphasizes those aspects that are deemed relevant to the topic of biofuels development in Africa. The framework has been used as a general organizing tool for long-term research on CPRs and on other problems where individuals find themselves in repetitive situations, which are affected by a combination of rules, physical factors and a cultural world.[7] The chapter analyzes four African countries which have been selected from each of the following four sub-Saharan African economic regions: Uganda (EAC),[8] South Africa (SADC),[9] Benin (ECOWAS),[10] and Cameroon (ECCAS).[11] It then highlights some of the ways in which African social groups have successfully resisted the degradation of their resources, and how community groups and transcontinental networks continue to struggle to have biodiversity and livelihoods seriously incorporated into national biofuel policies.[12]

Some of the strengths and weaknesses of the IAD framework, especially with respect to Africa, are presented, and some examples are provided regarding African models that follow the basic outline of the framework, but go outside of it to provide specific approaches that could promote biofuels in a sustainable manner. At the local level, we can cite the Songhai Center in Porto-Novo, Benin, which offers a practical model of sustainable development in both agriculture and bioenergy use.[13] The next level is based on ideas provided toward an integrated African energy policy in order to better identify, monitor and track shared interests, strategies or dangers, given that most of the continent's energy resources are transnational.[14] The third level or the constitutional level focuses on the concept of a consensus-based model of decision making in traditional African life and governance as practiced by the Asante of Ghana.[15]

IAD Framework

The IAD framework provides a common linguistic element or meta-theoretical language for analyzing behavior, and testing hypotheses in a wide range of problems and at multiple levels of analysis.[16] Institutions are defined in terms of the shared rules, norms and strategies that govern behavior and its analysis is made from the local perspective of the 'appropriators', such as the daily users of forests.[17] Because they are essentially shared concepts, the challenge of scholars in general is to learn how to recognize institutions on the ground, given that they are sometimes 'invisible", that is, they are shared as implicit knowledge rather than in a written form.[18] Hence, the IAD framework refers to institutions in terms of these shared concepts that are used in repetitive situations.[19]

The conceptual unit in the framework is the action arena, which is a social space for analyzing, predicting, and explaining behavior within institutions.[20] It is composed of actors and an action situation, where individuals engage in appropriation or provision activities, solve problems or even fight.[21] The action arena has three tiers or levels of social organization, namely, the operational tier where actors interact in light of the incentives they face, the collective-choice tier where decision makers are constrained to make decisions within a set of rules and the constitutional tier where decisions are made about who is eligible to participate in policymaking, and about the rules that can be used to undertake policy making.[22] In general, the variables that need to be identified in order to analyze a shared resource problem such as the overfishing of a lake using IAD include who the set of participants are and their positions, the potential outcomes and level of control over choice, the information available to the appropriators and what kinds of costs and benefits accrue to them.[23] These correspond to the elements of a game theoretic model whereby over repeated interactions, the rational, self-maximizing model of the individual in neoclassical economics gives way to cooperation. [24] However, three main sets of underlying variables can be said to affect the structure of an action arena. The first of these includes rule configurations such as entry and exit rules, position rules, scope rules, authority rules, and aggregation and payoff rules. The second set consists of the attributes of the physical and material world, that is, what is physically possible, what can be produced, how actions can be linked to outcomes, and what kind of information about this world the actor possesses. The third set of variables refers to the attributes of a community that include the norms of behavior broadly recognized by the community, the degree of common understanding that the actors share about the structure of different types of action arenas, the extent of homogeneity in the preferences of the people in the community and how the resources are distributed among them. This bundle of variables, the author notes, is frequently referred to as "culture". Furthermore, whether a vernacular language is written or not is considered to be a crucial variable to institutional analysis because without writing, individuals are thought to find it much more difficult to accumulate their learning in a "usable form" to transmit to future generations.[25]

One widespread thesis that Elinor Ostrom challenged was that of the "zero contribution thesis" which argued that individuals are unable to overcome collective action problems and need external rules to be enforced on them in order to achieve what is in their self-interest.[26] She states that this thesis contradicts observations that can be made in daily life, in which people organize themselves voluntarily to benefit from trade, defend themselves, and implement rules that protect their natural resources. In fact, she argues, all known resource governance systems that have survived over long periods of time, have done so by the participants contributing their resources toward monitoring and sanctioning the actions of others in their group to reduce the opportunity to free ride.[27]

The IAD framework has also been recognized for playing an important role in the study of common-pool resources (CPRs) generally. A common-pool resource can be defined as a natural or manmade resource system, where excluding prospective beneficiaries or limiting the appropriation rights of existing users is expensive and where the yield of the given system is subtractable, that is, one person's use of the resource reduces its availability for others.[28] With respect to renewable resources, it is possible to define a replenishment rate for which the average rate of withdrawal must not exceed the average rate of replenishment in order for the resource use to be sustainable over time.[29] The development of the literature on CPRs in the U.S. can be dated to the foundational article by Garret Hardin entitled "The Tragedy of the Commons." In it, he argues that if the world's population continues to grow, it will use up its resources and the only way to stop this is by "giving up the right to breed."[30] He identified two main factors as being responsible for his prediction, the first of which is the tragedy of freedom in the commons, which means that population growth increases demand for natural resources and environmental services and will ultimately outgrow the resources. The second factor, which is its converse, is that the pollution problem is the consequence of population.[31]

Hardin's analysis has been widely criticized for many reasons such as his assumption that individuals act rationally, and will choose those actions that maximize their welfare in a common pool resource, thereby leading to resource destruction.[32] It was further argued that even though appropriators are actively engaged in creating the dilemmas they face, they can nevertheless become active in resolving those dilemmas if given the opportunity and under certain conditions. In the view of this particular author, the missing link is an empirically grounded model of individual decision-making that can better explain why individuals cooperate, although such work is ongoing.[33] It has also been pointed out that many social groups, which include the herders that Hardin used in his original metaphor, have struggled against resource degradation by developing and maintaining their self-regulating institutions.[34] A critique from the Global South, argues that the uncritical but widespread use of the phrase "tragedy of the commons" has resulted in a confusion which erroneously equates the term 'communal' with the term 'common'.[35] While the tragic overuse of the commons is a consequence of poorly defined property rights, communal land tenure is a form of property rights, which are clearly defined in the case of the indigenous communities of Mexico, for example, even if land is not owned by an individual but by the entire community.[36] This clarification puts us on much firmer ground in helping us think about the application of the linguistic elements of this framework to African situations.

Application to Africa: Case Studies

The development of large scale biofuels projects is becoming more prevalent in Africa due in small measure to the need for energy security but primarily for the export of biofuels for economic growth. [37] Some of the areas being impacted by this development include parts of the African rainforest, which is surpassed in size only by the Amazon of South America. It extends from West Africa to the island of Madagascar in the Indian Ocean and from the Congo basin in Central Africa to the archipelago forest islands off the eastern coast of the continent.[38] Because these projects are in their early stages, very few rigorous, in-depth studies have been done on the impact of biofuels on land-use in Africa. However, environmental organizations and world rainforest bulletins have reported their impact on land use, livelihoods, biodiversity, and the economy.[39] The action situations in these case studies consist of the forests or lands on which biofuels are grown. The actors refer to everyone from those directly affected by biofuels policies, to officials and to the citizenry at large, which elects its leaders. Together, the actors and action situation make up the action arena. The three tiers or levels of social organization in an action arena include: 1.) the operational level, that is, the people who live on the land and depend on its use; 2.) the collective-choice level, which refers to the interactions between the local people and the government officials that are supposed to enforce or coordinate the laws and regulations at that level; 3.) the constitutional level, which in this case refers generally to the macro-level political institutions namely, the national government, the legislative branch and the judiciary, despite the fact that there are few explicit biofuels policies approved by the various legislative branches in Africa and made available for public review with the exception of South Africa.

Uganda (EAC)

In the case of Uganda, the relevant and interlinked action arenas are the local forests as well as the national forest system in the country. It has been reported for instance, that a large oil palm project in Uganda covering an area of 26,500 hectares on the Bugala Island has been awarded to an oil refineries company.[40] However, residents of the island and the members of parliament of the district charge that the project is taking place on one of the most unique ecosystems in the world and one of the last few pristine natural large scale forests in the country.[41] Any additional demand for biofuels is therefore not expected to alleviate the situation but rather exacerbate it. It is for this reason, that it is helpful to identify cases of successful forest management so that those forests which do experience overuse can have examples from which to learn to construct more effective strategies in anticipation of the potential impact of biofuels.

Two of the most important independent variables that have been found to affect the type and degree of consumptive use of forests in different settings include the degree of enforcement of the rules that govern the use of Ugandan forest resources, and the security of tenure that residents of forest areas possess.[42] This is because those who do not have secure rights tend to use up their forest resources for fear of losing them to others. In addition, if the rules that govern access to the use of forest resources are not sufficiently enforced, then that common pool resource in fact becomes one of open access rather than one of secure tenure. In the study of all the Ugandan forests, the researchers found that deforestation was not extensive only in those situations where institutional arrangements were viewed by residents as legitimate, which then led to the extensive monitoring of forests. [43]

South Africa (SADC)

The Republic of South Africa is the only country that has so far been found to have made an extensive outreach effort on its national biofuels strategy. A draft strategy published in June 2006 was released for public comment by all the stakeholders. As a result of these comments, the originally planned biofuel penetration level of 4.5% into the national liquid fuel supply was reduced to 2% due to questions about land, water and food security, environmental concerns, crop selection and choice, incentives and many others.[44] At this level, which is part of the constitutional tier where decisions are made about who is eligible to participate in policymaking, we can say that rules or at least a rule-making process is in place that can be used to undertake policy making in the area of biofuels in the country. The response to the initial Strategy from rural communities, NGOs and farmers organizations alleged land grabs of communal land and very little remuneration for industrial plantations of oilseed rape, maize and soya crops.[45] They further alleged that this was due to deals that had already been struck to meet the demand of the European Union. [46] These reports show that the local institutions at the operational-tier were not robust enough to withstand the push for biofuels and similarly, institutions in terms of repetitive actions of shared norms were not operationalized at the collective-tier. The groups called for a new strategy that would integrate poverty into any energy strategy and would foster genuine participation by often marginalized segments of the society, in particular, women.[47]

Benin (ECOWAS)

The relevant action situation in Benin consists of the prime agricultural lands of the South of the country, which represent about 7.7% of the national area and on which about 50% of the population lives. It is also on this land that the government of Benin was seeking 300,000 to 400,000 hectares of land in order to establish oil palm plantations for the development of biofuels.[48] The push continued to occur despite the fact that several hydrological studies have reported a climatic anomaly which has been observed due to the disruption of rainfall patterns in the form of low rainfall in Togo, Benin and Cote d'Ivoire since the 1980s.[49] Other studies have also shown that the oil palm crop requires abundant rainfall to be successful, which is consistent with the low yields that have been collected in recent years.[50]

The social organization at the operational and the collective-choice tiers had been fraught with a history of corruption and disagreement between the communities that produce and sell palm oil and the government, until private companies were brought in to trade. The latter would present an offer to buy palm oil at higher prices but fail to pay, thereby forcing the communities to bear the financial losses without government redress.[51] Similarly, in the quest for arable lands for oil palm for biofuels, the government has not explicitly stated where these lands will come from, and it is unclear whether the residents of potentially affected communities have been consulted.[52] Given the brief background of the recent history of the local development of this crop in Benin, it appears that the interconnections between the constitutional, the collective-choice and the operational tiers of action arenas do not follow a stable or effective institutional paradigm.

Cameroon (ECCAS)

The biofuel boom is reported to have fueled the growth of the oil palm industry in Cameroon. This expansion has been carried out primarily by SOCAPALM, the largest of five companies which is reported to have colluded with corrupt government officials to forcibly seize land without compensating local populations, and tear down the forests that have been traditionally used by these peoples.[53] These industrial plantations have mainly been supported by government initiatives and external capital, and seek to continue to expand.[54] Based on these reports, it can be deduced that the local institutions at the operational tier have been overrun by the more powerful government and its allies among the multinational companies. Therefore, neither the collective-choice nor constitutional tiers can be said to be functioning based on any shared concepts with the local people in repetitive situations.

Strengths and Limitations of the IAD Framework for Africa

Some of the main strengths of the IAD framework in general and that are also relevant to the case studies presented above are outlined below. First, it provides a common set of linguistic elements that can address land use destined for biofuels, for any other consumptive uses, and for a general diversity of situations in a way that no single discipline can. Secondly, it provides concepts and categories that allow the researcher to break down the problem into different component parts that can be analyzed separately and later synthesized into a coherent explanation. It is therefore one of the most explicit theories of the policy process and one with a high degree of explanatory power. Thirdly, it helps to address the problems of nested structures of rules within rules of an individual level of social organization (tier) of an action arena and across action arenas by linking the arenas thereby allowing for multiple levels of analysis. Finally, by focusing on the link between people, working rules and biophysical contexts, IAD provides a framework for analysis of the resulting patterns of order, thereby showing that no one model of order can be appropriate for all situations.[55]

The IAD framework is intended to be compatible with the most explicit theories of the individual namely, neoclassical economics and game theory, which involve fundamental assumptions such as the full maximization of net benefits or the capability for virtually unlimited computation.[56] However, several objections can be raised in particular to the model of the individual as rational and self-maximizing as postulated by neoclassical economics but only three of them will be outlined here. First of all, it is worthwhile to point out that this type of rationality has evolved over time. Before the twentieth century, some theorists believed that all rational individuals are purely motivated by self-interest while others thought that they were motivated by pleasure.[57] The difficulty of these postulates lay in the fact that it was pointed out that some individuals do act altruistically and based on moral or idealistic objectives. In an attempt to resolve this problem, the rational individual is now said to be one who seeks to maximize his or her expected preference satisfaction, where preference is defined broadly enough to include idealistic and moral goals.[58] It can be deduced then, that generalizations about rational individual behavior are agentistic, that is, based on preferences and rooted in a broader ideology rather than legitimate scientific generalizations, which are deterministic and follow natural laws.[59] These agentistic generalizations assumed by neoclassical economics are based on the ideology of the functioning of a market economy and can be said to represent a description, explanation and justification of the economic relations in the society that gave rise to the theory.[60] Normative expectations of this type of behavior have been set through policy adoption, maintenance and enforcement through the action of governments such as antitrust or fair trade laws.[61] It is clear then, that the IAD framework bases itself on a model of the individual that is neither scientific nor universal but is instead specific to a particular type of society and culture. Therefore, to the extent that modern African governments are organized based on Western models, this representation of the individual could in principle apply to the members of the largely urban and elite civil society to which their officials belong.

However, this leads us to the second point, which is that even in the case of those nations that gave rise to and/or foster the use of neoclassical economic theory in policy-making, its predictive power has long been intensely debated and the arguments for its success by some of its strongest proponents such as Milton Friedman, have been demonstrated to be less vigorous than they appear by other eminent economists such as Wassily Leontief.[62] The latter concluded that year after year, economic theorists continued to produce several complex mathematical models, which had been unable to appreciably advance the structure of economic theory dating back to Adam Smith more than two hundred years earlier.[63] With respect to biofuels development, this limitation is much more apparent because forest resources are complex and uncertain and cannot match the kind of settings that closely approximate such competitive and constrained models of choice.

A third objection can be raised to the characterization of culture in the IAD framework as referring to a set or bundle of variables which represent some of the attributes of a community and that as a set, represents one of the characteristics of an action arena.[64] The apparent subordination of culture to a framework that is based on an economic model that has been demonstrated to primarily be an ideological construct and a cultural and historical invention makes it difficult to apply epistemologically to African situations. As has been stated in a critique of the supposed primacy of another economic model as the appropriate lens through which to view African realities but equally applicable here, "Such a frame of analysis precludes any understanding and appreciation of the primacy of culture and of its subsequent ever-presence in all matters of life."[65] Furthermore, a study of the role of wood in the development of various civilizations showed that societies have been making choices regarding their relationships with their forests for millennia and in the last 5,000 years, the dominant choice across Asia, Europe and more recently North and South America, has been cutting down trees for fuel and wood and replacing them with crops or urban centers.[66] In contrast, ecologically-oriented cultures in Africa and South and Central America, which had developed informal norms that sustained their forests are quickly disappearing due to pressure by market forces, government policies and changes in the values of forest peoples.[67] This point strongly suggests that a deeper understanding of different cultures and the roles of individuals within them may be more beneficial than developing a sweeping model of 'the individual' whose behavior is necessarily conditioned by culture. This is the avenue that is pursued in the next section.

Toward an African Model of Sustainable Biofuels Development

By taking advantage of the linguistic terminology in the IAD framework, we can highlight some specific African examples of sustainable development that have existed, currently exist, or have been proposed for the future, but also go outside of it to explore model ideas and ideals that the people of the continent can aspire to.

At the local or operational level, the Songhai Center in Porto-Novo, Benin is a pilot center for the training, production, research and development of sustainable agricultural practices.[68] The center has put in place a biogas methane system where all the sewage, animal excrements and plant wastes are used as a source of energy to run parts of the center such as its restaurants and the student cafeterias. Its main objective is to improve the living standard of African peoples through the use of local resources, the hybridization of traditional and modern farming techniques, the encouragement of communal and individual initiatives, and the inclusion of diverse opinions.[69]

In a keynote address on Africa and its future energy issues to a scientific symposium in Kinsasha, Congo in 1985,[70] the late Professor Cheikh Anta Diop cited the early evidence of anthropogenic global climatic disruption and the drastic reduction of fossil fuel resources by 2020 as two of the reasons for Africa to develop its many sources of energy. He briefly mentioned Brazil's sugarcane methanol program as evidence of the impending energy crisis, but excluded this source of energy for Africa due to uncertain climatic consequences and pollution. However, he did call for bioenergy expertise for rural development, which we can assume to mean biogas digesters such as those used at the Songhai Center, since large scale biofuels as we know them today hardly existed at that time outside of Brazil. One of the main points raised in the chapter was that Africa does not lack energy resources but that these have been poorly identified and distributed, a situation that calls for an interconnected African network for the creation of an integrated electrical energy market that could cover the energy needs of African countries through effective distribution and without waste. Furthermore, if such a network were coordinated by a continent-wide political organization that would organize complementary rather than competing economies, this could help prevent the total depletion of the dwindling resources of small nations. A compendium of Africa's clean energy resources was compiled and the establishment of a continental or sub-continental grid was proposed alongside decentralized power systems depending on the energy source. Finally, he counseled that Africa should use all the means at its disposal in order to succeed, while keeping in mind that its most pressing problems are self-sufficiency in food and health care - a timely piece of advice in light of the food versus fuel debates that have arisen due to biofuels development.

In order to describe the constitutional level of an ideal African energy or biofuels policy we can use as a case study, the consensus-based model of decision-making in traditional Asante society, which also appears to have been widespread in Africa. Even though the current nations of Africa are not designed on traditional models, these models had been in use for centuries and stable systems of governance had been developed that can still provide an internal logic, or an orientation that could guide the resolution of contemporary problems. Because the production of biofuels occurs mainly in sub-Saharan Africa, this logic could govern in the area at the local, national, regional or even in an eventual multinational union as proposed by scholars such as Professor Diop and more recently by Professor Muchie and others. In Asante society, the personal word of a chief or king or queen was not the law, but rather the consensus of his or her council. This council strongly represented the populace of the kingdom both in terms of the nature or demographics of its composition, and in the content of the decisions it made. The adherence to the principle of consensus was based on the "belief that ultimately, the interests of all members of society are the same, although their immediate perceptions of those interests may be different." The Asante believe that the means to the objective of attaining common basic interests is rational discussion. In their view, the quest for consensus went beyond simply achieving majority agreement because that would deprive the minority of their right of representation in the decision at hand, which they considered a fundamental human right. In other words, they resisted the urge to stop at a formal representation in the form of majority opinion in order to pursue substantive representation. The rationale for pursuing both was that while a majoritarian system is likely to induce disaffection by minority groups which are substantively poorly represented, democracy by consensus is better able to weather unrests that are generated by adversarial politics. Wiredu argues that although consensus may not always have been achieved, a system that is seriously committed to it would necessarily be different from a majoritarian one, even one that is buttressed by "checks and balances". Another main difference between the traditional system and the current one in place is that of the role of political parties. Under a majoritarian paradigm, parties are made up of people with similar views and aspirations with the unique goal of appropriating power to implement their policies. The Asante system of consensual democracy consisted of parties made up of lineages which were formed to participate in power and good governance with an underlying philosophy of cooperation rather than confrontation. While it may be unrealistic to expect an imminent transformation in this direction by the modern nation-states of Africa, it should be possible to suggest this principle of partnership and consensus as the underlying logic in the governance of environmental challenges such as the one posed by biofuels development. Combining this with the privileging of local perspectives in the IAD framework, we can expect a significant reduction in the deforestation rates, for example.

Conclusion

IAD has provided an analytic framework for studying the impact of biofuels on common pool resources, primarily forests in Africa. The case studies showed that with respect to biofuels, the most effective governance occurred when there was collaboration between the local inhabitants and the government officials, that is, in some of the Ugandan situations and to some extent in South Africa. Thus, legitimacy and cooperation are much more effective tools of governance than where there is secrecy or outright expropriation of land. In Cameroon, the collusion of the government with multinational corporations has led to an exploitative and predatory relationship with the forest communities. Lastly, even though the democratically elected government in Benin remains popular, the dominant top-down governance arrangement has resulted in ambitious programs for biofuels in the absence of an adequate regulatory framework and with little discussion with the Beninese people about their potential impact. Indeed, it has been suggested elsewhere that there should be mass education and awareness of the subject, thereby moving the discussion from the experts to the general public so as to prevent 'towering' or a perspective-limited conversation about the issue. There should also be increased support for localized and smaller-scale experimental projects that are in accordance with the cultural values of the affected people.

Some insights from IAD could provide more desirable outcomes even under the current African regimes. In particular, privileging the local perspective would likely make it easier to govern the common-pool resources as it requires that local, regional and national officials involve the affected people in decision-making. Furthermore, the role of the government would be more valuable to the people when it provides additional technical and material support to this end, rather than through the imposition of ordinances which are typically resisted. For instance, traditional tools such as informal communication and sanctioning can be expected to have a much better impact on community-based governance. The IAD framework can therefore provide tools to assist with governance challenges especially in those systems that are constructed exclusively from the top down. Its insights show that people can create arrangements more favorable to them through the processes of collective choice, constitutional choice and by ensuring effective governance that draws on a contextual understanding of biophysical and cultural realities.

However, its epistemological basis on the acquisitive rational individual and competition, in contrast with a focus on the community and consensus in traditional African societies, is a limiting factor in applying the framework to African concerns. The practical examples provided of the Asante principle of consensus as well as a more conceptually and pragmatically integrated African energy policy, therefore suggest a need for the development of more culturally and historically rooted models of institutional governance by African scholars that can effectively deal with modern environmental challenges such as those presented by biofuels.

Notes

1. Anthony Faiola, "The New Economics of Hunger," *The Washington Post*, April 27, 2008; Emma Okocha, "Bio-Fuel and World Food Shortages, Is Nigeria Ready to Avert Food Crisis?", *Vanguard*, May 6, 2008.
2. Silvia Secchi and Bruce A. Babcock, "Impact of High Crop Prices on Environmental Quality: A Case of Iowa and the Conservation Reserve Program," in *Iowa State University Center for Agricultural and Rural Development – Working Paper Series (07-WP 447)* (Center for Agricultural and Rural Development , Iowa State University, 2007).
3. Joseph Fargione et al., "Land Clearing and the Biofuel Carbon Debt," *Science* 319(2008).
4. Ibid.; Alexander E. Farrell et al., "Ethanol Can Contribute to Energy and Environmental Goals," *Science* 311(2006); Lee R. Lynd et al., "High Land Requirements and an Unfavorable Energy Balance Preclude Biomass Ethanol from Playing a Large Role in Providing Energy Services," in *Energy and American Society-Thirteen Myths*, ed. Benjamin K. Sovacool and Marilyn A. Brown (Dordrecht: Springer, 2007).
5. ABN, "Agrofuels in Africa-the Impacts on Land, Food and Forests." African Biodiversity Network, 2007. http://www.africanbiodiversity.org/media/1210585739.pdf.

6. Ibid.
7. Elinor Ostrom, Roy Gardner, and James Walker, "Rules, Games, and Common-Pool Resource Problems," in *Rules, Games, and Common-Pool Resources*, ed. Elinor Ostrom, Roy Gardner, and James Walker (Ann Arbor: The University of Michigan Press, 1994).
8. EAC – East African Community.
9. SADC – Southern African Development Community.
10. ECOWAS – Economic Community of West African States.
11. ECCAS – Economic Community of Central African States.
12. ABN, "Agrofuels in Africa-the Impacts on Land, Food and Forests."
13. SONGHAI, http://www.songhai.org/
14. Cheikh Anta Diop, *Black Africa: the Economic and Cultural Basis for a Federated State* (Lawrence Hill Books, 1987); "Le Problème Énergétique Africain," *Ankh*, no. 14/15(2005-2006). http://www.ankhonline.com/telecharger/cheikh_anta%20Diop_afrique_energie.pdf.
15. Kwasi Wiredu, *Cultural Universals and Particulars: An African Perspective* (Bloomington and Indianapolis: Indiana University Press, 1996).
16. Elinor Ostrom, "Institutional Rational Choice – an Assessment of the Institutional Analysis and Development Framework," in *Theories of the Policy Process*, ed. Paul Sabatier (Westview, 2007).
17. Clark C. Gibson, Margaret A. McKean, and Elinor Ostrom, "Explaining Deforestation: The Role of Local Institutions," in *People and Forests: Communities, Institutions and Governance*, ed. Clark C. Gibson, Margaret A. McKean, and Elinor Ostrom (Cambridge: The MIT Press, 2001).
18. Elinor Ostrom, "Institutional Rational Choice – an Assessment of the Institutional Analysis and Development Framework."
19. Sue Crawford and Elinor Ostrom, "A Grammar of Institutions," *American Political Science Review* 89, no. 3 (1995).
20. Ibid.
21. Ostrom, Gardner, and Walker, "Rules, Games, and Common-Pool Resource Problems."
22. Ostrom, "Institutional Rational Choice – an Assessment of the Institutional Analysis and Development Framework"; Elinor Ostrom, *Governing the Commons: The Evolution of Institutions for Collective Action* (Cambridge University Press, 1990).
23. Ostrom, "Institutional Rational Choice – an Assessment of the Institutional Analysis and Development Framework."
24. Ibid.
25. Ibid.
26. Elinor Ostrom, "Collective Action and the Evolution of Social Norms," *Journal of Economic Perspectives* 14, no. 3 (2000).
27. Ibid.

28. Ostrom, Gardner, and Walker, "Rules, Games, and Common-Pool Resource Problems."
29. Ostrom, *Governing the Commons: The Evolution of Institutions for Collective Action*.
30. Garrett Hardin, "The Tragedy of the Commons," *Science* 162, no. 3859 (1968).
31. Ibid.
32. Edella Schlager, "Rationality, Cooperation, and Common Pool Resources," *American Behavioral Scientist* 45, no. 5 (2002).
33. Ibid.
34. Thomas Dietz, Elinor Ostrom, and Paul C. Stern, "The Struggle to Govern the Commons," *Science* 302(2003).
35. José Sarukhán and Jorge Larson, "When the Commons Become Less Tragic: Land Tenure, Social Organization, and Fair Trade in Mexico," in *Protecting the Commons: A Framework for Resource Management in the Americas.*, ed. Joanna Burger, et al. (2001).
36. Ibid.
37. ABN, "Agrofuels in Africa-the Impacts on Land, Food and Forests."
38. William Weber et al., eds., *African Rainforest Ecology and Conservation – an Interdisciplinary Perspective* (New Haven and London: Yale University Press, 2001).
39. ABN, "Agrofuels in Africa-the Impacts on Land, Food and Forests."; Kalundi Serumaga, "Uganda: Oil Palm Plantations That Brought High Winds and Low Wages," *World Rainforest Movement Bulletin*, no. 109 (2006); WRM, "Uganda: Bidco Oil Palm Plantations at the Cost of Pristine Rainforests in Bugala," *World Rainforest Movement Bulletin*, no. 100 (2005); WRM, "Benin: Large Scale Oil Palm Plantations for Agrofuel," *World Rainforest Movement Bulletin*, no. 120 (2007).
40. Kalundi Serumega, "Uganda: Bidco Oil Palm Plantations at the Cost of Pristine Rainforests in Bugala."
41. Ibid.
42. A.Y. Banana and W. Gombya-Ssembajjwe, "Successful Forest Management: The Importance of Security of Tenure and Rule Enforcement in Ugandan Forests," in *People and Forests: Communities, Institutions, and Governance.*, ed. Clark C. Gibson, Margaret A. McKean, and Elinor Ostrom (2000).
43. Ibid.
44. RSA, "Biofuels Industrial Strategy of the Republic of South Africa," http://www.biotechsa.net/BiofuelIndustrialStrategy27November20071.pdf.
45. Neliswa Mtembu et al., "Rural Communities Express Dismay: "Land Grabs" Fuelled by Biofuel Strategy," http://www.biosafetyafrica.net/index.html/images/stories/dmdocuments/ruralcommunitiesexpressdismay.pdf.
46. Ibid.
47. Ibid.
48. ABN, "Agrofuels in Africa-the Impacts on Land, Food and Forests."

49. F. Gauthier et al., «Variabilité du regimé pluviométrique de l'Afrique de l'Ouest non sahélienne entre 1950 et 1989,» *Hydrological Sciences – Journal des Sciences Hydrologiques* 43, no. 6 (1998).
50. USDA, "Malaysia: Below-Normal Rainfall May Limit Palm Oil Yields Late 2004-Early 2005," USDA – Production Estimates and Crop Assessment Division, http://www.fas.usda.gov/pecad2/highlights/2004/08/maypalm/index.htm.
51. WRM, "Benin: Large Scale Oil Palm Plantations for Agrofuel."
52. Ibid.
53. Julien-François Gerber, "Cameroon: Oil Palm Plantations Fostered by New Biofuel Market Harm Local Livelihoods," *World Rainforest Movement Bulletin*, no. 112 (2006).
54. Ibid.
55. Amos Sawyer, "What African Analysts Can Learn from the Ostroms," *Journal of Economic Behavior & Organization* 57(2005).
56. Ostrom, "Institutional Rational Choice – an Assessment of the Institutional Analysis and Development Framework."
57. James F. Reynolds and David C. Paris, "The Economic Approach: Empirical and Theoretical Considerations," in *The Logic of Policy Inquiry* (Longman Publishing Group, 1983).
58. Ibid.
59. Ibid.
60. Ibid.
61. Ibid.
62. Alexander Rosenberg, *Economics – Mathematical Politics or Science of Diminishing Returns*, ed. David L. Hull, *Science and Its Conceptual Foundations* (Chicago and London: The University of Chicago Press, 1992).
63. Ibid.
64. Ostrom, "Institutional Rational Choice – an Assessment of the Institutional Analysis and Development Framework."
65. Ama Mazama, "Book Review: 'Eurocentrism' by Samir Amin," *Journal of Black Studies* (1995).
66. John Perlin, *A Forest Journey: The Role of Wood in the Development of Civilization* (Harvard University Press, 1991) ; C. Dustin Becker and Rosario León, "Indigenous Forest Management in the Bolivian Amazon: Lessons from the Yuracaré People," in *People and Forests: Communities, Institutions and Governance*, ed. Clark C. Gibson, Margaret A. McKean, and Elinor Ostrom (Cambridge: The MIT Press, 2000).
67. Ibid.
68. SONGHAI, http://www.songhai.org.
69. Ibid.

70. The paper, "Le problème énergétique africain" or "Africa and its future energy issues" was presented by Cheikh Anta Diop at the Kinsasha International Symposium in Kinsasha, Zaire, held from April 20-30, 1985. The theme of the symposium was "La science, la technique et le développement de l'Afrique – l'Afrique et son avenir," which is translated as "Science, Technology and the Development of Africa – Africa and its Future."
71. Diop, «Le Problème Énergétique Africain.»
72. Ibid.
73. Ibid.; *Black Africa – the Economic and Cultural Basis for a Federated State.*
74. Diop, «Le Problème Énergétique Africain.»
75. Ibid.
76. Kwesi Wiredu, *Cultural Universals and Particulars – an African Perspective*, ed. Charles S. Bird and Ivan Karp, African Systems of Thought (Bloomington and Indianapolis: Indiana University Press, 1996).
77. Diop, *Black Africa – the Economic and Cultural Basis for a Federated State*; Mammo Muchie, ed. *Making of the Africa-Nation: Pan-Africanism and the African Renaissance* (Adonis and Abbey Publishers, 2003).
78. Wiredu, *Cultural Universals and Particulars – an African Perspective.*
79. Ibid.
80. Ibid.
81. Ibid.
82. Ogundiran Soumonni and Susan Cozzens, "The Potential for Biofuel Production and Use in Africa: An Adaptive Management Approach," In *Globelics 6th International Conference.* Mexico City. http://globelics_conference2008.xoc.uam.mx/papers/Ogundiran_Soumonni_The_potential_for.pdf.
83. Ibid.
84. Sawyer, "What African Analysts Can Learn from the Ostroms."

• Chapter 18 •

THE ROLE OF FOLKLORE AND FOLK LIFE IN ENVIRONMENTAL EDUCATION AND SANITATION IN YORUBALAND

I. A. Ademiluyi and R.A. Asiyanbola

Introduction

This chapter examines the role of folklore and folk life in environmental education and sanitation in Yorubaland. It is argued in the chapter that the indigenous methods of educating and creating awareness on environmental issues in Yorubaland before the advent and widespread of western education was largely through folklore and taboos embedded in the folklore. The advent of the modern science with its perception of the environment has rendered many of these taboos irrelevant in environmental considerations. However, the indigenous ways of knowing are capable of raising awareness about environmental sanitation which is a part of the general environmental education process. A closer examination of the curriculum development goals for environmental education shows that no one learning system – formal, non-formal, and informal – can effectively cover the range of necessary awareness-raising. Thus, it is suggested that the methods of indigenous environmental learning (folklore, folksong, folk drama, etc.) be integrated into the national education curriculum to supplement formal Environmental Education System.

The term folklore was perhaps first used by W. J. Thomas in the middle of the 19th century as a substitute for popular antiquities.[1] There are various conceptions of the term which have made its definition difficult.[2] While Oduyale defines folklore as the traditions, customs and beliefs of a community,[3] the Folklore Society of London defines the term as "the comparison and identification of the survivals of archaic beliefs, customs, and traditions in modern ages."[4] Porter[5] observes a set of current discourses constructed around the term folklore which implicitly rejects the idea of folklore as solely a product of communal imagination or individual aesthetic sense, but rather the production of folklore is seen as also intimately related to power relations in the society, to economic factors, to public institutions and to academic disciplines.[6] Folkways (folk life) are the customary, normal, habitual ways a group does things. Folk life of a community is shaped largely by the folklore of that community.

From time immemorial in Yorubaland, as well as in other African societies, folklore has been one of the major local learning processes. It is the major source of the local knowledge, that is, knowledge that is unique to a given culture or society. This traditional way of knowing, or traditional education, is supported in Yorubaland because of its contributions to the growth, renewal and development of the society.[7]

Traditional education is the handing down of beliefs, customs, stories, etc. from generation to generation. It includes any long established customs or practice, which manifest in the forms of cultural expressions as songs, dance, stories, artworks, and crafts, as well as symbols, marks, and other recurring expressions of traditional concepts, any unwritten religions and teachings regarded as coming from the founder or earliest exponent of a religion and any codes of conduct handed over to succeeding generations.[8] Traditional education, therefore, is the avenue through which indigenous knowledge is passed from one generation to another. It is the process by which the society gradually absorbs or socializes its youth into its norms, religions, beliefs and moral values as well as the collective opinions of the whole society.

It is observed from contemporary research and documentary evidence that traditional education is functional. For instance, it is useful for immediate induction into the society; it is a preparation for adulthood and despite the erroneous view held about it, it also emphasizes spiritual and moral values, and social responsibility.[9] Traditional education is pragmatic in nature. For example, children must learn "to do by doing," and the experience gained goes a long way in enriching their knowledge. In most societies, both adults and children are informally involved in the traditional learning process through ceremonies, rituals, imitation, recitation and demonstration. Other inherent values include its elaborate nature and adaptation to local culture and environmental conditions of the society.

Atte,[10] in his article on indigenous knowledge as a key to local development, identified some reasons for the use of traditional or indigenous education. These include the enablement of the society to cope with the world of ecological processes and environmental events, classification of environmental data, structuring of environmental knowledge, and transmitting the knowledge so that as large a number of people as possible can use it for survival. For instance, terminologies and or classifications were used for understanding the physical environment such as soil characteristics, testing and analysis.[11] A thorough knowledge of plant species, growth environment, soil relationships, water requirement and uses are contained in the local or indigenous education. Knowledge of medicinal plants are also taught.[12] This includes plants for treating ailments like stomach ache, headache, malaria, as well as plants to keep snakes away from homes and those used to ward-off mosquitoes and so on.

The role of folklore in environmental education and sanitation

Abrahams has defined the environment as the sum total of all external conditions influencing the growth and development of an organism.[13] These factors could be physical, biological, social and cultural.[14] Furthermore, the environment is considered abused when an injurious or corrupting element is introduced which, by fouling it, reduces the satisfaction and utility derivable from growing and developing within it.[15] Apart from this, seasonal variability in the element of the environment particularly the climatic element (such as rainfall) can also cause environmental problem. Environmental degradation resulting from the interaction of human beings with the environment and in some particular cases reinforced by physical or natural forces (such as climate) ranges from flooding, soil erosion, deforestation, to air pollution. Environmental degradation arises from the dumping of solid wastes, littering of street with bio-degradable and non-biodegradable matter by food vendors, hawkers, dwellers, etc. In Africa, environmental degradation is often linked to management failure and political decline.[16] Before the arrival of industrialization and western education, most societies grappled with the vagaries of the environment through their traditions, customs and beliefs, that is, folklore. In dealing with the critical environmental problem such as flooding, there could be elaborate rituals. In Yorubaland, there were countless taboos and superstitions used to sustain the environment and keep it from disease vectors.

As observed in the literature, taboos and superstitions are often regarded as integral parts of indigenous knowledge.[17] Taboos are things that are embedded in the belief system of local societies and owing to the fact that they result from the accumulation of experience which are passed from one generation to the other. Taboos are simply the don'ts of a society; they are things that are considered as abomination.[18] Usually, punishment of one form or the other is spelt out for people that violate the society's rules and norms. Superstitions could be expressed as the beliefs or practices of people which may be deeply rooted in the culture and religious beliefs of the society. These are traditional beliefs that actions or events can cause or foretell an apparently unrelated event. Countless human activities are involved in superstitions and these include eating, sleeping, working, getting married, becoming ill, etc. For example, if a person dies, people are expected to open the door and windows of the room where the dead person lived, so that his or her spirit could leave the room. Also, if a black cat runs across one's way, or path, it is believed that one will have a bad luck that day.

From an investigation through questionnaire survey of Ibadan by Agbola and Mabawonku,[19] it was observed that the indigenous learning process could be through folklore, folk drama, folk songs, story, village meetings and taboos. About half (50.6%) of the sampled population learn folklore from the elders in their neighborhoods before school age (see Table 18.1). Folk-song or traditional music is described as chants embedded in the life style while folk-stories are the various tales popular with traditions, beliefs and customs passed on from generation to generation. Apart from entertainment function of the folk-song and folktales, they also are meant to educate as well.[20] Many of the Yoruba folk tale and folk songs contain useful themes which are environmentally sensitive. The categories of respondents who learnt environmental sanitation through folk stories were 15.4% in proportion. Folk song was used to educate 13.7% of the respondents. Folk songs can be through oratory skill composed as poem or chants. Some 8.6% of the respondents were taught environmental lessons through taboos, 7.10% watched folk drama related to the environment while 2.3% learnt about the environment through village meetings summoned by elders for the opening up of roads, felling and removal of aged trees among others (see Table 18.1).

Table 18.1: Methods of Indigenous Environmental Learning

Process	Frequency	Percentage
Folklore	200	50.6
Folk drama	28	7.10
Story telling	61	15.4
Folk songs	54	13.7
Village meetings	09	2.30
Taboos	34	8.61
Others	09	2.31
Total	365	100

Source: Agbola and Mabawonku (1996:85)[21]

Table 18.2 shows the respondents' most influential persons or adult on their knowledge or awareness of childhood environmental sanitation and problems. Majority of the respondents (64.6%) indicated their mothers as the most influential when they were young. The implication of this is that women are more influential in the training of children about the environment and other societal issues since they are close to the kids than men.

Table 18.2: Respondent's most influential persons or adult on their knowledge or Awareness of Childhood Environmental Sanitation

Process	Frequency	Percentage
Father	97	24.6
Mother	255	64.6
Senior sister	11	2.80
Senior brother	10	2.53
Grand father	06	1.52
Grand mother	11	2.80
Class teacher	05	1.30
Other	-	-
Total	365	100

Source: Agbola and Mabawonku (1996:85)[22]

As observed in Agbola and Mabawonku's investigation in Ibadan,[23] as in other Yoruba societies, the indigenous method of educating and creating awareness on environmental issues has been identified as through folklore. The method often adopted most of the time was that of taboos embedded in the folklore. As earlier stated, taboos are prohibition or strongly forbidden acts by social customs of the people. From the aforementioned survey, respondents from the core areas of the city like Beere, Elekuro, Mapo, Ogunpa, Bode and medium density districts like Orita-Aperin, Mokola, and Sango gave explanations concerning these taboos, and the reason(s) for them. Twenty-one taboos identified by the respondents were grouped into six as follows: taboos associated with refuse, taboos on cooking, taboos associated with eating, sweeping the floor, use of water and maintenance of building structure.

Prohibitions such as washing or kneeling on refuse heaps, stepping bare footed on melon waste peels; leaving melon peels until the following morning; carrying refuse with bare hands are some of the traditional ways of maintaining clean environment free of refuse. It is forbidden to carry out the above acts as it is said that germ infection on the knee called "kunle katan" could afflict anybody kneeling on refuse dump. Carrying of refuse with bare hands can also cause the shaking or unsteady nature of such hand. In actual fact, it is to prevent cholera if such hand is used for eating without cleaning Melon's waste, if left overnight, according to the taboo, will cause the housewife to have bad dreams in the night. The actual reason is to scare housewives from leaving them uncleared. It is forbidden to spit on the ground in order to avoid sore throats. The reason for this is that it is done to prevent the spread of diseases.

Dirty habits during food preparation are prohibited. Women are not supposed to talk while removing the hairs of a fowl being prepared for cooking. It is believed that such hair will grow again on the fowl's body but the real intention is to prevent saliva contamination. Housewives are forbidden from cooking with their heads uncovered because it can cause headache for such offenders. Obviously, it is to prevent hairs from the head from falling into the food being prepared in order not to cause digestive problem for the final consumer(s) of the food. Sitting on the mortar is also prohibited. The traditional belief is that it could cause offender to either die or excrete the whole content of his/her stomach. Realistically, this is forbidden in order to maintain the hygienic nature of the mortar since it is used in preparing food for the household.

In the indigenous Yoruba culture, to stand at the entrance of the house while eating is not encouraged. The traditional belief is that unseen spirit will empty the content of the offender's stomach; hence he will not be satisfied with the food. Similar thing is believed to apply to somebody eating in the open. Apparently, eating in the open is forbidden in order to prevent flies from coming in contact with the food.

The indigenous Yoruba people believe that it is forbidden to walk bare footed on an unswept floor early in the morning in order to maintain cleanliness of the floor. The rationale for this is that one can develop swollen feet. It is not allowed for someone to sweep the ground in the night so as not to annoy the gods. The underlying reason is that one will not be able to see properly what one is sweeping.

While chewing stick, it is highly forbidden to draw from the well. The chewing stick could drop into the well and contaminate the water for other users. It is also prohibited to use a cup placed on a water pot for drinking. It is expected that each person will bring his/her personal cup or bowl for dinning. Apparently, this is done to prevent the spread of disease through the mouth.

In order to maintain building structures, the indigenous people believed that stones are not allowed to be thrown on walls. This is to prevent the gradual cracking of the fence and generation of sand dust in the compound. To urinate in front of the house or beside the walls or fences is prohibited. This is to keep the house free of bad odor.

Countless other taboos which are environmentally sensitive and which are germane to various Yoruba people exist. It is amazing to note, as observed in the literature, that these informal education mechanisms have stood the test of time and have effectively nurtured the different environments until now. However, the advent of modern science with its perception of the environment, industrialization and western education has rendered traditional education almost irrelevant in environment considerations. For example, the elaborate rituals and especially taboos used to sustain the environment and keep it from all disease vectors have all been abandoned. Yet, these methods of indigenous environmental learning have stood the traditional societies so well in time past before the advent and widespread of the western education that such societies survived for centuries. From the foregoing, it is clear that the taboos are meant to inculcate some level of environmental sanitation discipline.

Recommendation and Conclusion

The advent and spread of modern science and western education notwithstanding, it is important to note that the indigenous methods of environmental education have the capabilities of raising awareness about environmental sanitation. Awareness about environmental sanitation is part of the general environmental education (EE) process which is that part of the general or specialized education which deals with the physical, biological, social cultural and economic aspects of the environment.[24] The four identified curriculum development goals for Environmental Education (EE) process are:

i. to focus on ecological concept, providing learners with the knowledge which can help make ecologically sound environmental decisions;
ii. to focus on information concerning many aspects of human environmental behavior;
iii. to focus on those skills needed for issue investigation, evaluation and value clarification; and
iv. to focus on those processes important to citizenship action (participation in problem solving and decision making).

A closer examination of the above goals shows that no one learning system – formal, non-formal and informal – can effectively cover the range of necessary awareness raising. The theoretical and policy recommendations in the literature center around, among others, recognition of the relevance of culture for quality education and quality life.[25] In Africa and in other developing countries, the importance of indigenous knowledge has been underscored in policy making. It is therefore suggested that the indigenous environmental education and sanitation methods be integrated into the national primary school education curriculum. Moreover, the modern society has a lot to learn from the past. Taboos are mythical means of controlling behavior to prevent disorder. The modern law has, for instance, not been able to deter men from criminality of all kinds. Where people have strong beliefs about them, taboos can prevent the commission of crimes Since the modern law seems to be incapable of preventing commission of crime, taboos should be reintroduced. Also, man naturally has immoral instinct that keeps working against the law, because the law is not potent enough in its application. Taboos can help to instill a sense of natural resistance to criminalities.

Notes

1. S. O. Eyong, "Folk Architecture: A challenge to contemporary urban architecture," *Nigerian Magazine*, Vol. 53 No. 3 (1987): 49-53.
2. James Porter, "The folklore of Northern Scotland: Five Discourses on Cultural Representation," *Folklore* Vol. 109 (1998): 1-14.
3. O. Oduyale, "Traditional Education in Nigeria," in O. Y. Oyeneye and O. M. Shoremi (eds.) *Nigerian Life and Culture*, Ogun State University 1985, pp. 230-244.
4. C. H. Holman, *A Handbook to Literature* (Indianapolis: Bobbs Merrill Company, 1972).
5. Porter, "The folklore".
6. Christopher Tilley, *Material Culture and Text: The Art of Ambiguity* (London: Routledge, 1991); Madhavi Sunder, "The Invention of Traditional Knowledge," *Law and Contemporary Problems,* Vol. 7, *Issue* 2 (2007): 97-124; C. Jane Becks, "Taking stock 1996 American Folklore Society Presidential Address," *Journal of American Folklore* Vol. 110(436) (1997): 123-139; John A. Burrison, "Transported Traditions: Transatlantic Foundations of Southern Folk Culture," *Studies in the Literary Imagination.* Vol. 36: Issue: 2 (2003).
7. Oduyale, "Traditional Education."
8. Oduyale, "Traditional Education;" Sunder, "The Invention."
9. Oduyale, "Traditional Education."
10. O. D. Atte, "Indigenous knowledge as a key to local level development: Possibilities, Constraints and Planning Issues in the context of Africa". A paper presented at Arusha on Challenges for Rural Development in Eastern and Southern Africa 12-14 Feb. 1989.
11. D. Michael Warren, "Indigenous Knowledge for Agricultural Development," In D. M. Warren, L. Egunjobi and B. Wahaab, (eds.) *Indigenous Knowledge in Education* Proceedings of a Regional Workshop on Integration of Indigenous Knowledge into Nigerian Education Curriculum. Published by Indigenous Knowledge Study Group c/o Centre for Urban and Regional Planning, University of Ibadan, Ibadan, Nigeria 1996, pp. 1-26; Tunji S. Titiola, "Application of Indigenous Knowledge in the Agricultural Sector Policy Formulation: An outline". In Warren, Egunjobi and Wahaab (eds.), *Indigenous Knowledge in Education*, pp. 27-37.
12. Layi Egunjobi and S. Osunwole, "Integration of Indigenous Health Systems into the Education Curriculum," In Warren, Egunjobi and Wahaab, eds. *Indigenous Knowledge in Education*, pp. 55-62.
13. C. Abrahams, *The Language of Cities*. Chicago: Equinox Books, 1971.
14. E. A. Keller, *Environment Geology*. New York: Bell and Howell Company, 1976.
15. G. E. D. Omuta, "Urban Solid Waste Generation and Management: Towards Environmental Sanitation Policy," In P. O. Sada and F. O. Odemerho (eds.) *Environmental Issues and Management in Nigerian Development*. Ibadan: Evans Ltd., 1988.

16. Pauline Von Hellermann, "Things Fall Apart?" *Africa*, 77, 3, 2007, p. 371.
17. Warren, Egunjobi, and Wahaab, eds., *Indigenous Knowledge in Education*, pp. 78-94.
18. Oduyale, "Traditional Education in Nigeria".
19. Agbola and Mabawonku, "Indigenous Knowledge," in Warren, Egunjobi and Wahaab, eds., *Indigenous Knowledge*, 78-94.
20. Olunlade Taiwo, "The Themes in Yoruba Children's Poetry," *Nigerian* Vol. 55 No. 4 (1987), pp. 83-87; and T. C., Nwosu, "The Structure, form and content of Nigerian folktales," *Nigerian Magazine* No. 136 (1981), pp. 64-71.
21. Agbola and Mabawonku, "Indigenous knowledge," 78-79.
22. Ibid.
23. Ibid.
24. T. Chelliah, "Environmental Education. Its Importance and Potentialities in Development and the Environmental Crisis".
25. Francesca Gobbo, "Globalization and education: Comparative perspectives," *International Sociology* Vol. 26 No. 2. 2011, 170-182.

• Chapter 19 •

FIGHTING "MODERN SLAVERY": SECURING LAND RIGHTS FOR THE SARAMAKA

Adam Carpinelli

Introduction

This essay is based on field interviews conducted among the Saramaka people of Suriname in August 2006 regarding their land rights case brought to the Organization of American States at the end of the 20th century. The Saramaka are people of African descent that live in the interior of the Amazon rainforest. Their ancestors, enslaved Africans, settled in the forest after fleeing from plantation slavery and formed their own autonomous communities. After over two hundred years of a fairly harmonious existence, they were again threatened by a new colonialism in the late twentieth century; neoliberal globalization. My research addresses how socio-economic and political changes impact human memory among the Saramaka since the era of the slave trade, with an emphasis on cultural links between Africa and the Americas. The way that neoliberal globalization is affecting the Saramaka people and their natural environment will be discussed.

Maroon Communities and Human Rights in Suriname

There is no room to explain the whole complex picture of Maroon communities in Suriname but several things should be explained.[1] There are six main groupings in Suriname: 1) Saramaka or Saramacca, 2) Ndyuka or Djuka, 3) Matawai, 4) Aluku or Boni 5) Kwinti, and 6) Paramaka or Paramacca.[2] Of the six Maroon communities in Suriname the Saramaka have the largest population. Roughly 25,000 Saramaka are dispersed in 70 villages.[3] This is not also accounting for those Saramaka residing within the urban outskirts of the interior rainforest, namely, the capital city of Suriname, Paramaribo. These groups can in no way be lumped together. What bind these societies are that 1) they all share the same history of slavery, colonialism and modern oppression; and 2) their cultural bases from Africa are similar. Other than this, everything from language, religion, food, music, art, and architecture are unique to each particular group. Another notable difference is the location and history of where they come from in Suriname.[4] According to international human rights law the Maroons in Suriname are considered "tribal peoples."[5]

As the plantation system developed in Suriname with the enslavement of both indigenous and African people, they escaped in groups farther south, far from the coastal plantations for their freedom and safety. Among the Saramaka, many more settlements developed in the interior. After the Maroons fought the colonial forces for a long time, and no end seemed in sight, peace treaties were signed in the last half of the 18th century to end hostilities. Consequently, with the cessation of hostilities, many of those early Maroon settlements were abandoned for new ones along the waterways. Therefore the Saramaka and other Maroon societies of Suriname are historically forest peoples.

Over the past three decades there has been a large body of research by lawyers, scholars and human and environmental rights organizations addressing human rights violations against the Saramaka.[6] These works have introduced the human rights of Maroons who have been historically unrecognized groups in the Americas.[7] It should be mentioned that there have been many instances of governmental oppression that have contributed to genocide waged on the Maroons as a result of a civil war.[8] There were reparations granted to several Saramaka families as a result of the murders when the case was raised before the inter-American Court of Human Rights, but resulted in the decision of a minimal and meager compensation.[9]

Several reports from the 1980s have described various human rights violations by the government of Suriname during the civil war.[10] What is important to understand about this civil war is that it was under the military dictatorship of Commander Desire Bouterse who purposely pitted the Maroon societies of Suriname against the Surinamese National Army thereby sparking the Saramaka spirit of guerilla warfare.[11] The war and human rights violations continue to undermine the Saramaka sense of pride and dignity. Land rights issues for the Saramaka are those of human rights.

The Natural Environment of Suriname

Suriname is located on the northeast Atlantic coast of South America.[12] It is bordered in the north by the Atlantic Ocean, in the west by Guyana, the east by French Guyana, and the south by Brazil. With a land area of 166,000 square kilometers and a population of about half a million, Suriname is one of the least densely populated countries in the world. The population center is in the capital city and only urban center Paramaribo and smaller towns in the coastal region.

The remaining 10% of the Surinamese population reside in the interior, which is predominantly comprised of scattered villages along some of the major rivers such as the Saramaka and Suriname rivers.

Suriname is part of the largest river basin in the world and is commonly referred to as the "heart of the Amazon," boasting the highest proportion of tropical forest cover than any other nation in the world. What this means is that Suriname is interconnected to a complex river basin sourcing from the Andes Mountains and reaching across the South American continent for about 4,000 miles, encompassing the national borders of eight modern countries. This systemic river basin is home to the planet's most significant surviving rain forest, hundreds of indigenous/forest peoples and endemic flora and fauna species.

The entire basin has experienced gross environmental exploitation and human displacement since the first days of European colonization in the late 15th century. Several hundred years later, in the 20th century, the period known as the rubber boom, promoted environmental exploitation prompting expansive transportation and communication networks. The rubber boom also diversified the extrapolation of other raw materials such as timber, gold and bauxite. The fact that the interior of Suriname, covered with undisturbed neotropical Amazonian forest, is largely uninhabited makes it one of the most important places anywhere in the world for tropical forest conservation.

Suriname's abundant and largely untapped natural resources, combined with its proximity to the Atlantic, has historically and contemporarily provided unique opportunities to foreign capitalist enterprises. Many of these natural resources include timber, gold, iron, ore, nickel, copper, platinum and hydroelectric power. Therefore, the root of Suriname's environmental problems today lies in the abundant natural resources that the country has in the rainforest. Mineral production is one of the most considerable economic exports in Suriname. There is very little or none in other territories or countries in the Caribbean. The Caribbean and most notably Suriname, has been one of the world's leading producing areas of bauxite. Operations are dependent on the state of the world aluminum market as Caribbean production is controlled by subsidiaries of North American aluminum manufactures and specifically the United States, such as ALCOA. Gold mining is also very prominent in Suriname and is predominantly by miners from Brazil.[13] The problem with these industries in Suriname is the unequal expense of severe environmental degradation and human and non-human animal exploitation.

Land Concessions and Environmental Exploitation

The negative environmental impact regarding the extrapolation of natural resources in Suriname is disgusting. The environmental history of land degradation in Suriname goes back all the way to plantation slavery. For purposes of this discussion the role of Asian and specifically Chinese logging corporations will be focused on in the late 20th and early 21st century.

Suriname is part of what is known as the Guyana Shield which constitutes about one hundred million acres of solid tropical woods. The management of this natural commodity on the part of the Surinamese government can be characterized by weak forest regulations and rampant corruption. Eighty percent of all Surinamese timber exports go to countries in South East Asia, notably China.[14] Chinese companies started in the last decade of the 20th century. Asian companies, predominantly those from China such as Tacoba Jin Lin Wood, control 90% of the world's ten billion dollar tropical timber trade.[15]

After centuries of colonial exploitation small-scale timber harvesting in Suriname, based on the use of indigenous and African slave labor beginning in the early years of colonial settlement right up to the turn of the 20th century. By 1904 the method of selective logging began.[16] This was a turning point making the human impact on the natural environment much worse compared to the previous several centuries of colonial occupation and sustainable indigenous methods. Industrialized logging started in the 1940s. The time span of land concessions in the first half of the 20th century were short durations (less than a decade) and very limiting compared to what would come latter.

For another 40 years Dutch-owned companies ran the show. This would change in the mid 1970s when Suriname gained independence from the Netherlands. During the 1970s CELOS, a polycyclic selective logging system, was introduced mitigating extractions rates over a two-decade harvesting schedule.[17] The profitability of this method is disputed and was only practiced for several years. By the last decade of the 20th century the forestry service developed a network of roads in the northern region of the country making it easier for forest access, penetration and corporate exploitation. During the same period the Forest Management Act stipulated the recognition of customary law.[18]

What can be assessed here is that foreign timber companies have not been aware of sustainable practices such as the CELOS. In addition, they have not demonstrated any knowledge or respect to the local terrain and specificities of the Amazonian rainforest of Suriname. In addition the plight of the forest peoples, their history and customary legal traditions have also blatantly been disregarded. The culture of Suriname is a rich one and the heritage of the Maroons is part of that culture and it would make sense to save it and the people who live along the waterways that feed, maintain and protect them. The area's hydroelectric power dam continues to threaten many of their settlements, including the archaeological sites and sacred burial sites, many of which it has already consumed.[19] The operations are displacing and destroying not only the modern settlements and the people, thus disrupting their already delicate survival, but also destroying the remaining archaeological sites that are the only remaining repository for the ancient cultural remains.

The lands upon which these unexplored archaeological sites exist are being exploited today for multinational interests by mining and timber corporations from Canada, China and the U.S. These land concessions, carelessly distributed by a corrupt Surinamese government, do not take into account any of the historical, environmental and archeological properties of these territories and sites. Most importantly there is no redress for the human rights violations that are incurred by the Maroon and indigenous societies of the country. As these sites and territories disappear, they do so with the cultural material left behind by the Maroons and indigenous people. The following section presents the Saramaka cultural memory of ancestral and natural connections with the land.

Land and the Saramaka

This section highlights the actual responses of Saramaka people to various questions about their land rights and sensibilities towards land within the context of their memories, participation and engagement with the land. Two main points should be understood: 1) awareness about the exploitation of their land has deeply affected the way that the Saramaka people view their land: and 2) these global shifts in the economy, exacerbated by the International Monetary Fund (IMF) and the World Bank, have created fundamental shifts for the Saramaka in economic ways that not only affect their worldview, but actual ways and quality of life.

The process of globalization has deeply affected many life ways for the Saramaka. Multinational logging corporations clear-cut the forest ruining natural habitats and further causing the extinction of indigenous wildlife and bauxite mines turn naturally clean rivers into murky swamps. Many Saramaka have relocated as a result. Freedom for the Saramaka people today is still very elusive. What has happened is that the Saramaka people have started to view land in a different away. Now the Saramaka people have literally become part of the global village. They are now seeing themselves as interconnected to a larger community of indigenous, forest, poor and oppressed people that are fighting for their human rights. The process of transformation that is affecting the Saramaka should be analyzed within the larger context of neo-liberal policies, capitalism, and globalization. Before breaking down these processes it is most important that the Saramaka people are recognized for their own traditional understanding of the land.

As my goal is not to legally analyze Saramaka people's land rights, it is my opinion that the people's own sense of land should be in no way overlooked and needs to be factored in the whole analysis of their land rights; and the issue of land rights for the Saramaka is not just a legal case, but also a personal struggle. Therefore, the perspective of land rights in this case and the whole world at large should be broadened from policy-oriented focus to the actual issues of agency and the sense of belonging that Saramaka people have with their land.

In this instance, the basic idea is the actual customary law of the people, which has more meaning to the Saramaka than international law. The Saramaka's connection with their land runs deep. The first question that I asked in my interviews with the Saramaka before their case was won was how people were connected to their land. Some responses referred to the actual resources and environment. Others spoke directly to the struggle that was fought on and for the land historically. One elder directly invoked stories about the brutalities of the plantation system that initially motivated the first Saramaka people to escape. In one such story, "A lady gave birth to a child and the child was crying nonstop and the plantation woman said, 'give me the child' to help her stop the crying...the lady gave the child to the plantation owner, thinking that she was helping" .[20] The story continues: "The lady took that child and turned his neck. She gave it back to the woman and told her to put it back on her backside. When the husband came back from the plantation and saw what happened he said that they should leave the plantation and go into the interior."[21]

This story did not just help to explain how Saramaka people escaped a brutal plantocracy for a better life, but how this particular elder initially responded to his land in a way that explained the whole people's history. His attitude about the land is rooted in slavery. He recognizes that the Saramaka people's land is hundreds of years old and that since "we are here more than three hundred years and the government still does not want to give us a right, but if people live for three hundred years in a place international law says it is yours and you do not have to fight for it, but the government does not want to accept it, they do not want to give us rights for the land. That is why we are here fighting for the lands."[22]

This point also contextualizes the realization that the land has always been in the people's possession. The Saramaka have a value about their land. It is not one about money; it is about a way of life. It is a reverence for an ancestral homeland, not a reverence for market forces. "The land is everything for us. We are living on it. There are plants we are planting. We are hunting. From the beginning we fought for the land so it is ours so if we [lose] it, then we [lose] everything."[23] The people holistically use the land to plant, to hunt and to revere ancestors. The Saramaka know that they fought for the land and that they deserve it. This is a consciousness of a land regarding its connections to farming, ancestral homage, and as a site for constant struggles. As Saramaka land has been a site for previous struggles it is important to understand exactly what the Saramaka people are up against today. Modern forms of slavery and colonialism are globalizing the contemporary world in catastrophic ways. The next section will show how European hegemony, through the process of colonialism, has transformed into a much larger concern that extends beyond western domination and imperialism.

Neoliberal Globalization and the Pan African World

For purposes of this discussion the form of globalization that is affecting the Saramaka is a global phenomenon of neoliberal capitalism. This neoliberal form of globalization has opened up the door for money hungry governments and multinational corporations to exploit environmental resources for big profit. Neoliberalism is a reemergence of policies that dates back to the past 500 years of European economic practices. Neoliberalism refers to an intellectual and political movement that espouses economic liberalism with purported intentions of promoting economic development and securing political liberty. This neoliberal movement is sometimes described as an effort to revert to the economic policies of the 18th and 19th century classical liberalism. In many ways these 500 years of European economic thought has shaped capitalism as it contributed to the establishment of the transatlantic slave trade.

This way of thinking is not only western but also inherently Eurocentric.[24] This means that if the world is to globalize under neoliberal policy then the world lacks a paradigmatic ability to cope with multiculturalism because it does not account for traditional epistemologies regarding diverse socio-cultural patterns and relationships between nature and other species. What this means to the Saramaka is that Eurocentric global capitalists, in collaboration with the Surinamese government, are making decisions on their behalf that can adversely affect their destiny. This past 500 years has crystallized into the neoliberal globalization that we discuss today.[25] In other words, if the roots of neoliberalism are in colonialism then modern neoliberal policies should be viewed as neocolonialism. Therefore, this helps to explain the present day modern slavery of the Saramaka people.

Modern slavery is reflected in laws and actions that are dictated without consulting the people or asking them what they think about things. It appears that the people feel neglected and do not see any interest on the part of the government to do anything for their benefit. One informant put the matter succinctly: "know that we are still in slavery, we are still slaves, we call it modern slavery."[26] This long enduring legacy of neo-colonial corruption speaks to a continuation of a neo-slavery where the so-called independence of Suriname from Dutch colonial rule is now just a case of sham independence.[27]

The contemporary reemergence of neoliberalism has had a tremendous influence among economic scholars and policy-makers, as it mainly crystallized during the Reagan years in the 1980s.[28] The term neoliberalism describes various economic theories, libertarian political philosophies and rhetoric influenced by neoclassical European thought. The neoclassical portrayal of government control over the economy considered it to be corrupt, inefficient and undesirable. This critique of government control suggests that a homogenous international system is implemented by the IMF and the World Bank to keep those governments in check and to mitigate corruption. In reality what has happened is that those international bodies for global governance are just as money hungry as any typical government.

Neoliberalism is not static or highly agreed upon economic and political ideology. It denotes a contemporary paradigmatic shift of political sentiments. Events such as the Stagflation Crisis of the 1970s, the Developing World Debt Crisis of the 1980s, and the Soviet Collapse in the early 1990s enabled the forward momentum for justifying the implementation of neoliberal policy. Basically the neoliberal project has been to globalize trade within a capitalist paradigmatic framework. What happens is that it allows people to think that homogenizing the world's infrastructure is the only solution as it blatantly disregards alternative approaches to doing business.

For the Saramaka their idea of trade and commerce as it relates to their land is totally different. First of all their understanding is not a linear western paradigm. As has been shown, they have a more cyclical and non-western approach to deciding land ownership and rights. For them, this is tantamount to a paradigmatic plague, a form of terrorism. This paradigm is a western epistemology that is so linear in orientation that it only sees the international standardization of commerce and protocol as the answer to the world's problems, even as it only makes it worse. This is because it gives no agency to people, like the Saramaka, when governments are to be so standardized and left to do the deciding for whole nations and are encouraged to do so. This neoliberal globalization is really just a form of neocolonialism as it is rooted in the classical European thought of the past 500 years. That is why this neoliberal orientation to globalization should be viewed as one of the worst forms of modern slavery today.

The term *globalization* means many different things to different people. And this point is very specific to what particular political lens one wishes to view this process. If globalization is an old process then it has contemporarily taken on a new character.[29] Whatever globalization was, it should contemporarily be viewed as a process of progress and movement in time and space in regards to the development and advancement of spatio-temporal capitalism, neocolonialism, modern slavery and imperialism. Scholars such as David Harvey suggest that the term (globalization) be replaced with the study of "uneven spatio-temporal development of capitalism".[30] Harvey's point gives us a better way of understanding present globalization because it puts oppression, and specifically poverty, at the center of understanding the process. This process of globalizing poverty leads to indebted countries and exploited labor that further leads to malnourishment and disease.

Susan George makes the point to understand globalization as "corporate-led…finance driven…neoliberal globalization"[31] She argues that transnational corporations (TNCs) have such market force in the world that they not only own so much but also control and influence international politics for capitalistic purposes.[32] When we speak of corporate power it must be understood that various socio-economic institutions have been set up on an international level to supposedly help countries in "crisis".[33]

What needs to be further critiqued about TNCs is that they have direct connections with policy makers, as this allows them to influence international politics for their own personal gains.[34] The personal gains for TNCs are things like the opening up and increased flexibility of trade borders. This inevitably results in the advancement of TNCs globally while poor people in localized areas suffer as TNCs move freely to whatever country they see as economically viable at the time of their interests.[35]

These institutions, and people controlling them behind the scenes, are only getting smaller in numbers. The global power of the meager 200 TNCs controls most of the world's foreign investment, employing the tiniest fraction of the world's population while they leave at least a billion people to abject poverty all over the world and specifically in Third World countries such as Suriname.[36] In a nutshell these international superpowers are dominating the world through their neocolonial practices.[37]

One example of such practices is the Structural Adjustment Program (SAP).[38] The most noticeable outcome of these programs on "crisis" countries is that there has been an increased debt, a large increase in unemployment, and destruction of the environment just to name a few things among a horrific universe of negative consequences.[39] The nations' rights to self-realization and agency have drastically declined as these international organizations are calling the shots for their neoliberal and capitalistic purposes. Therefore the most significant outcome for these "crisis" nations is severe dependency upon the outside world, especially first world countries, economically and politically.

The façade idea of SAPs is that they are supposed to do the opposite of what is really happening. Countries are left with no pride and dignity. This is characteristic of this neoliberal thinking that takes what might seem like a good idea and turns it into a nightmare. Pan African activist scholar Horace Campbell recognizes how countries with IMF programs experienced cuts in "health and support for education that increased poverty and misery."[40] In other words SAPs are more of a trap that locks countries into a cyclical process of oppression through economic indebtedness and dependency.

The Caribbean region today is following neoliberal reform policies that are mandated by the IMF and World Bank. These reforms liberalized their trading infrastructure via governments, which not only do the deciding for the people, such as the Saramaka, but also think that there is no other alternative. While the present liberalization of economies might seem very welcomed and legitimized by governments there is a sobering pitfall that appears to be far worse. The increased mobility of foreign direct investment and TNCs impose very real severe constraints on workers, communities and states. In the absence of effective controls and safeguards these small-state economies are susceptible to a whole range of new vulnerabilities.

The present socio-economic vulnerabilities of Suriname are being exploited by neoliberal policies. The kind of neoliberal polices that are affecting Suriname today in pernicious ways are ones such as free trade and the advent of the North American Free Trade Agreement (NAFTA).[41] The neoliberal foreign policies of the U.S. during Bill Clinton's administration misused free trade and NAFTA for their own domestic purposes without any exterior concerns for Caribbean matters.[42] Therefore the neoliberal tendencies of free trade policy demonstrate how first world countries exploit Third World countries for their own capitalistic benefits. During that same period (Clinton Administration) trade with the Caribbean was very important for the U.S. with a large increase in imports and exports. What happened during that time is that the Caribbean experienced higher prices, per capita, for U.S. exports.[43]

The policies of the Clinton Administration demonstrate how a neoliberal regional grouping, such as NAFTA, utilizes a neo-imperialist logic attempting to internationalize the world's people within the context of money.[44] Free trade increases the gap between rich and poor because it empowers individualistic countries and administrations, such as the U.S. and disempowers underprivileged countries who lack foreign policy that would bring mutual benefits for the entire region.

The fundamental problem with these policies and agreements is that they appear to be guided and motivated by strictly money/business interests. This approach is one that lacks certain ethics that would have a more serious commitment to people. This is completely contradictory to the pride and dignity of all human beings. These business interests have no interest in funding any program for health care or unions. They are more interested in exploiting the privilege of free-market capitalism where they can close down shop and set up quickly anywhere they chose. They have nothing invested in the people or the country because they are typically U.S., E.U. or Japanese-based corporations. In other words the neocolonial strategy of indirect rule is utilized for economic gains while poor people suffer. The Pan Africanist movement, which promotes the pride and dignity of all human beings, should stand up against global poverty, as it is the direct result of a global capitalism.

What the developing nations of the Caribbean have in common is poverty. Poverty is a systemic relationship of domination and oppression. It is perpetuated through lack of job security, fair and livable wages, affordable goods and services and most importantly access to quality education and health care. Without these things in regions such as the Caribbean, prostitution and sex work are inevitable and have become an economic and cultural reality in the lives of Caribbean peoples.[45] This next section takes one example of an emerging country that is taking advantage of neoliberal free trade that is directly threatening the Saramaka people's way of life.

Chinese Logging Corporations

Economic shifts in the global economy are deeply affecting people all over the world. The Saramaka are directly experiencing this as their government openly welcomes economic opportunities from TNCs. It appears that these governments and foreign companies have more rights to the land than the larger Surinamese population in general, especially those in the interior such as the Saramaka. Suriname has nurtured a new democratic regime following the neoliberal tendencies by IMF and the World Bank.[46] The result of this is that Suriname welcomes foreign companies to do whatever they want to get natural resources while severely exploiting the people and the environment. Contemporarily external forces are shaping the Surinamese economy.[47] None of this profit trickles down to the Saramaka people or the larger Surinamese population.

One specific economic development that is changing the global economy and has sparked the Saramakas' attention for land rights is the recent rapid industrialization of the People's Republic of China (PRC). China is presently emerging as one of the world's new superpowers.[48] China is presently the fourth largest economy in the world, controlling over ten percent of the world's trade.[49] Two major concerns of China's economic expansion are environmental degradation and increased demand leading to higher prices for raw materials.[50] The environmental point is very serious for the Saramaka because they holistically live in, off and with their natural environment. China has rapidly modernized its industries. This has led to an emerging middle class that has raised their incomes and purchasing power.[51] The result of this is a growing demand for natural resources and agricultural goods.

Basically China is now taking the lead among developing nations to dominate world commerce. This is quite opportunistic and also demonstrates a new feature of recent global economic events. This is because it marks a transition from the commercial domination of developed Western countries such as the U.S. to developing nations such as China.[52] China is at the forefront of breaking the pattern of northern economic domination from the past several decades.

Although the Saramaka lived a fairly harmonious life from the time of the Dutch peace treaties in 1762, they started to take notice of their continued exploitation. In one interview, I was told that when "a Chinese person came here first to go into....*Kumako,* I think two hours behind to go and ask about the concession that he had there, and he came to look for guys to bring him to go look for the concession. They told him that he did not talk to the captain so 'you cannot go there, you have to go talk to the captain you don't have nothing to do there'. He said 'no', that he 'doesn't need the captain he has his concession' there and he 'wants to go see it.' I don't know if he went or not--that is why they started the land rights organization, he was the first Chinese that came here because he had the concession."[53]

This example illuminates the ways in which Chinese corporations had never consulted anyone representing the Saramaka. One thing that seems to be stressed is that the people, community and specifically the *granman* expect a certain level of consultation regarding what goes on in their territory.[54] This relates back to the idea that their understanding of their land is a profoundly communal one. It is not available to the open market of free trade. The people have to discuss it first. For example, "If someone wants to come do something then you have to ask for permission because it belongs to them and no one can take it from them." This would seem like a basic courtesy for any human being to get permission, although it has not happened yet in Saramaka territory. "When people... comes and you know that you have to go to the captain to tell him that you are here to do things," one Saramakan explained. "But when the Chinese came here and said that 'I don't need the captain, I have my licenses to do whatever I want,' that is why we started the land rights [movement]."[55]

Both these points are reflective of the people's sense of customary law. The people deem it important. It is something that they take seriously as it is the basis of their retained African value system. These values have everything to do with the level of integrity that the Saramaka people have exemplified since slavery days. It is a sense of sovereignty, pride and dignity. They do not find their alliances with China just because they contest the capitalist hegemony of first world and developed nations. They are opposed to any form of exploitation by whoever or whatever inhibits their ability to live peacefully, undisturbed in a sustainable environment.

What China has attempted to do is create Third World solidarity with the Caribbean and Latin America by challenging hegemonic imperialism, such as that of the U.S.[56] China, albeit a non-capitalist economy, started to adopt market-oriented reforms in the 1990s that led to foreign investment. Therefore although China purports to be non-capitalist, under the rhetoric of socialism, they still engage in the capitalist mode of production.[57] The rhetoric of China should be thoroughly interrogated in this regard. This anti-imperialist/hegemonic rhetoric appears to be more of an excuse to increase foreign policy regarding their engagement with the developing world.[58]

As China appears to be in solidarity with the developing nations it does not mean that they are simply welcome everywhere and by everybody. The Saramaka and their territory are a case in point. One respondent expressed their indignation toward the Chinese loggers: "We don't want them here. If they come to work on the lands then we find nothing, they pick up your tree and bring it into town to make big money and then you don't have [anything]."[59] The Saramaka understand that there is no compensation for the business that happens on their land. They are isolated in the rainforest until businesses show up with concessions from the government. Therefore contact with the government is minimal. In addition the government does not appear to make a sincere effort to consult the Saramaka. After all it took a foreign company to attempt to start logging in their territory before they had any idea what the government was doing. In other words that government does not inform the Saramaka as to their national concerns with the interior and the people that live in the rainforest.

Conclusion

In the course of my fieldwork in 2007, the legal case for Saramaka land rights was pending. During those interviews we also discussed many aspects of their oppression as it related to governmental subordination. The government's apparent lack of interest in the Saramaka is a key factor in the continued legacy of poor and oppressed conditions that they experience day to day. It must also be mentioned that the Saramaka people constitute a very small percentage of the national population (roughly 5%). In addition, as they predominantly reside in the interior of the Amazon rainforest, very little of their history and culture is known by the 90% of the population that resides in the only urban center in the whole country. Therefore there are not only multinational forces that have caused barriers in the expedient resolution of the Saramaka's plight but sympathy and awareness are also lacking for domestic support.

This chapter has only focused on one imperialist power (PRC) to provide one concrete example of transnational imperialism that has affected the relative peace that the Saramaka have experienced since the peace treaties that they signed with the Dutch in the late 18th century up until the independence of Suriname in the last quarter of the 20th century. What is important to note is that as the Saramaka people are an African people, their understanding and value on their land is an African-based one. This means that the formal legal aspects of land and ownership are western-based concepts that have been imposed on them in order to fight for their rights in an international stage. The Saramaka charged the Surinamese government for violating their human rights and successfully won the case in the fall of 2007. Now it is a point of future investigation to examine the anticipated cooperation by the Surinamese government to accept their responsibility and to provide remedies to the Saramaka's case of human rights.

Notes

1. Maroon communities simply refer to enslaved Africans in the Americas and Africa who fled plantations and started autonomous communities. Commonly plantations were raided, destroyed, looted and enslaved Africans were also freed as a result of these rebel communities. The Maroons of Suriname were historically a huge threat to the colonial administration.
2. Respondent 9, Interview by Author, August 2007, Suriname, field notes. This respondent is an older male above the age of 60.
3. Respondent 9.
4. The Saramaka people specifically come from Jewish plantations in the northern region of the country. One specific plantation is that of Imanuel Machado who was a Jewish planter from Brazil. During the early 17th century many Jewish people were being kicked out of Brazil and were able to find a safe haven in Suriname. This is simply because the Dutch were more tolerant of Jewish people and their religion. This mainly was attributed to the general Dutch theological and economic ideals that were being spread in the Netherlands as a result of the Calvinist movement. These proto-Saramaka people who escaped plantations such as Machado's were in transit for several centuries until they established permanent settlements after signing peace treaties with Dutch colonists. Price, Richard. *Alabis World*. (Baltimore: The John Hopkins University Press, 1990).
5. Fergus MacKay, "Indigenous and Tribal Peoples in Suriname: A Human Rights Perspective," in *Indigenous Resurgence in the Contemporary Caribbean: Amerindian Survival and Revival*, ed. Maximilian C. Forte (New York: Peter Lang, 2006), 156.
6. Fergus MacKay and Ricardo Pane, "50 Years of Disrespect; Protected Areas in Suriname," *Cultural Survival Quarterly* 28 (Spring 2004): 43-46. Richard Price and Sally Price, "Maroons in the Americas: Heroic Pasts, Ambiguous Presents, Uncertain Futures," *Cultural Survival Quarterly* 25 (Winter 2001): 8-10. Richard Price and Sally Price, "Maroons under Assault in Suriname and French Guiana," *Cultural Survival Quarterly* 25 (Winter 2002): 38.
7. Fergus, "Indigenous and Tribal Peoples in Suriname: A Human Rights Perspective," In *Indigenous Resurgence in the Contemporary Caribbean: Amerindian Survival and Revival*, ed. Maximilian C. Forte (New York: Peter Lang, 2006), 156.

8. Richard Price, "Executing Ethnicity: The Killings in Suriname," *Cultural Anthropology* 10 (November 1995): 437-471.
9. David J, Padilla, "Reparations in Aloeboetoe v. Suriname," *Human Rights Quarterly* 17 (August 1995): 541-555.
10. *Report on the Human Rights in Suriname.* Washington D.C.: General Secretariat, Organization of American States, 1983. *Second Report on the Human Rights Situation in Suriname.* Washington D.C.: General Secretariat, Organization of American States, 1985.
11. Van Velzen, H. U. E. Thoden, "The Maroon Insurgency: Anthropological Reflections on the Civil War in Suriname," in *Resistance and Rebellion in Suriname: Old and New*, ed. Gary Brana-Shute (Williamsburg: William and Mary, 1990).
12. Also formerly known as Nederland's Guyana, Netherlands Guiana or Dutch Guiana.
13. Ellen-Rose Kambel and Fergus MacKay, *The Rights of Indigenous Peoples and Maroons in Suriname* (Copenhagen: IWGA, 1999). MacKay, Fergus and Ricardo Pane, "50 Years of Disrespect; Protected Areas in Suriname," *Cultural Survival Quarterly* 28 (Spring 2004): 43-46.
14. Respondent 7, Interviewed by Author, August 2006, Suriname, field notes. This respondent is an older male over the age of 40.
15. Respondent 7.
16. Marcus Colchester, *Forest Politics in Suriname* (Utrecht: International Books, 1995), 51.
17. Colchester, *Forest Politics in Suriname*, 53.
18. Respondent 7.
19. Respondent 7.
20. Respondent 9, Interview by Author, August 2006, Suriname, field notes. This respondent is an older male above the age of 60.
21. Respondent 9.
22. Respondent 9.
23. Respondent 10, Interview by Author, August 2006, Suriname, field notes. This respondent is an older male above the age of 40.
24. Egite S. Oyovbaire, "The Tyranny of Borrowed Paradigms and the Responsibility of Political Science: The Nigerian Experience," in *Political Science in Africa*, ed. Yolamu Barongo, (London: Zed Press, 1983), 239-254.
25. Norman Girvan, "Globalisation and Counter-Globalisation: The Caribbean in the Context of the South," in *Globalisation: A Calculus of Inequality*, ed., Denis Benn, Kenneth Hall and Ian Randle (Kingston: Kingston Publishers, 2000), 65-87.
26. Respondent 10.
27. This idea refers to a form of national and political post-colonial independence where corrupt local officials that are largely concerned with their own economic interests have only inherited the colonial oppression established by the European colonial system.

28. Girvan, "Globalization and Counter-Globalization," 65-87.
29. An apolitical idea of globalization is that it is simply a historic process, dating as far back as the Ancient world, that has to do with the transnational flow of goods, services, language, food and culture. This was initiated by ancient trade routes such as the Silk Road. Therefore if globalization is to be discussed contemporarily it must be connected to the political philosophy that it promotes today as will be discussed.
30. David Harvey, "Globalization in Question," *Rethinking MARXISM*. Volume 8, Number 4 (1995), 10-17.
31. Susan George, *Another World is Possible if* (London: Verso, 2004), 6.
32. Ibid., 7.
33. The reason that crisis is put here in quotes is because arguably every nation in the world is in some form of crisis today. Therefore it seems arbitrary to only have programs set up for those that are deemed in crisis when many global issues affect the whole world and not just a few countries that might appear to be more desperate or vulnerable.
34. Clarence Lusane, *Race in the Global Era: African Americans at the Millennium*, (Boston: South End Press, 1997), 218.
35. Harry G. Broadman, *Africa's Silk Road: China and India's New Economic Frontier*, (Washington: The World Bank, 2007), 356.
36. Lusane, Race *in the Global Era,* 113-114.
37. My argument is that the programs, initiated by commercial banks, governments, TNCs, World Bank and IMF, are neocolonial because they have a strong resemblance to the colonial form of indirect rule, whereby a state is controlled by exterior forces while appearing to be ruled by indigenous or local directors.
38. This is an economic policy that countries must follow in order to qualify for new loans from IMF and World Bank to supposedly help them repay debts owed to commercial banks, governments and the World Bank. This term pretends to emphasize the false idea of poverty reduction as their agenda.
39. Samir Amin, *Capitalism in the Age of Globalization* (London: Zed Books, 1997), 13.
40. Horace Campbell, Reclaiming *Zimbabwe: The Exhaustion of the Patriarchal Model of Liberation*, (Trenton: Africa World Press, 2003), 287.
41. Anthony T. Bryan, "The New Clinton Administration and the Caribbean: Trade, Security and Regional Politics," *Journal of Interamerican Studies and World Affairs* 39, (1997), 101-120.
42. Ibid,, 101-120.
43. Ibid., 101-120.
44. Amin, *Capitalism in the Age of Globalization*, 76.
45. Kamala Kempadoo, *Sexing the Caribbean: Gender, Race and Sexual Labor,* (New York: Routledge, 2004).
46. Bryan, "The New Clinton Administration and the Caribbean", 101-120.

47. Pitou van. Dijck, "Continuity and Change in a Small Open Economy", In *Twentieth-Century Suriname: Continuities and Discontinuities in a New World Society*. Edited by Rosemarijin Hoefte and Peter Meel, (Leiden: Leiden, 2001), 48-70.
48. Ian Taylor, *China and Africa: Engagement and Compromise*, (London: Routledge, 2006).
49. Horace Campbell, "China in Africa: Challenging US Global Hegemony," In *African Perspectives on China in Africa*. Edited by Frioze Manji and Stephen Marks, (Cape Town: Fahamu, 2007), 119-137.
50. Campbell, *China in Africa*, 119-137.
51. Harry G. Broadman, *Africa's Silk Road: China and India's New Economic Frontier*, (Washington, DC: The World Bank, 2007), 2.
52. Broadman, *Africa's Silk Road*, 41.
53. Respondent 8, Interviewed by Author, August 2006, Suriname, field notes. This respondent is an older male over the age of 40.
54. This is the Saramaka term for chief.
55. Respondent 9.
56. Campbell, *China in Africa*, 119-137.
57. Ibid., 119-137.
58. Taylor, *China and Africa*, 201.
59. Respondent 10.

• Chapter 20 •

MANAGING FILTH IN NIGERIA

Nnenna George-Kalu and George Kalu

Introduction

Oil spills are headline-catchers and rightly provoke indignation. But it is our ordinary everyday lifestyle that will, in the long term, do more to damage the environment. Counter-intuitive as it may at first seem, our lifestyles are likely to lead, albeit indirectly, to far more suffering than all the oil spills of the last 50 years put together.[1]

The advent of oil discovery and prospecting in Nigeria was heralded as a welcome development with great prospects of economic benefits. The boom meant that the country was firmly becoming established as a significant member of the international community. The potential ills of this boom were not envisaged and if they were, no regulatory mechanism was instituted to control the effects.

These ills have now gained prominence because their impact continues to rise, and they have seemingly defied solution. Civil Society Organizations, government agencies and even the oil-prospecting companies are, in varying degrees, involved in the attempt to solve the puzzle of bringing an end to oil-related environmental crisis. The corollary of these efforts is that the other "simpler" activities with future environmental implications are receiving less attention. In the fight to remedy the effects of oil-prospecting it is easy to forget that a precautionary approach to activities relating to the discovery of oil would have made for a different turn of events.

Environmental policies and legislation are on the increase in the nation. The MDGs are fully embraced as policy frameworks nation-wide and states reflect their willingness to pursue the realization of these goals of which the protection of the environment is one. The question is, to what extent are the lessons from oil reflected in our environmental control plans? On what do we base our policy choices and what do we intend to achieve in practical terms?

This chapter will analyze the current environmental protection efforts by governmental and non-governmental bodies in Nigeria in the light of current global awareness of ecological issues. This chapter suggests that a precautionary approach as opposed to a reactionary system would enable both the remediation of existent abuse and the prevention of future damage with sustainability as the goal. Further to this, it advocates a shift from the present emphasis on industrial and large-scale polluting activities to environmental control which focuses on the quality of life of Nigerian masses.

Growing Environmental Policies or Politics?

In this section the environmental protection strategy of Nigeria, with regard to policy-making and legislation, will be analyzed. The pertinent issues here are the determination of the relevance and practicality of our current policy and legislative stance on the environment for the Nigerian environment.

Nigeria made its major move towards environmental control when hazardous waste was dumped in Koko, Delta State (mid-West Nigeria) by an Italian vessel in 1988. The waste was imported through a transaction between two Italian waste broker firms and a Nigerian businessman. Drums of toxic substances were illegally shipped at the Koko Port by this contract in the guise that these were construction materials. Leakages from the barrels while they were in storage at the port exposed them as being toxic. Subsequent investigations exposed the unscrupulous deal leading to the withdrawal of the Italian Ambassador to Nigeria. A considerable number of workers from the Nigerian Ports Authority who had been given the task of removing the wastes were adversely affected in varying degrees from contact with the contents of the drums.[2] As a result of a concerted diplomatic pressure mounted by the Nigerian Government her Italian counterpart quickly mobilized two ships to clear up the wastes.[3]

The hue and cry following this discovery catalyzed the creation of the Federal Environmental Protection Agency (FEPA). The country had been caught unready because it had neither the policy framework nor the implementation outfit to tackle the issue.[4] The Federal Environmental Protection Agency Act (FEPA Act) has, therefore, existed in Nigeria since 1988. The interesting thing, though, is that the Federal Ministry of Environment (FME) was only created in 1999 by the Obasanjo administration. Environmental matters, prior to this time, were managed by the agency as a parastatal of the Federal Ministry of Works. The natural consequence of the creation of the FME was for FEPA to become absorbed into and its functions taken over by the new ministry.[5]

Several departments from pre-existing ministries were transferred to the new ministry. These include the forestry department of the Federal Ministry of Agriculture, the Environmental Health and Sanitation Unit of the Federal

Ministry of Health and the Oil and Gas Control Unit of the Federal Ministry of Petroleum Resources. The transfer of these departments to the FME shows a heightening awareness in the Federal government about the broad range of environmental issues. Significantly, the transfers grant greater impetus to the possibility of effective enforcement of the legislative measures under the FEPA Act. If the expanded scope of environmental legislation is to have any effect, it stands to reason that institutional re-adjustments must be made. It will be an interesting study to find out how these departments and parastatals have fared since these realignment in terms of effectiveness and efficiency, but the very fact that the Federal Government of Nigeria took the deliberate step to properly constitute a full-fledged ministry to oversee its environmental concerns underscores the importance which was accorded the issue and for which it should be commended.

The duties of the Ministry, as provided under the Act, can be summarized as ensuring that the physical environment is utilized in a sustainable manner in order to ensure development, improve the quality of life of the people and in line with international policies. The areas of focus for achieving this vision include: Municipal waste management and sanitation; industrial pollution control; conservation of bio-diversity and wildlife; and deforestation.

Under the FEPA Act states are mandated to promulgate their own environmental legislation which must adhere to national standards. Thus, most states have established their own environmental agencies and Ministries of environment. The trend has continued right down to the local government level.[6]

At the international level, Nigeria is not found wanting. It has ratified the Kyoto protocol on climate change and this convention has officially been in force in the nation since March 2005. This is a treaty which places little burden on developing countries while requiring developed nations to reduce their carbon emissions by specified rates within a time limit.[7] Ratification of this protocol by Nigeria, a developing nation, may therefore not really symbolize a major achievement.

The United Nations report on the implementation of the Agenda 21 in Nigeria lists the establishment of various bodies, policies and committees as a 'major milestone' towards the actualization of sustainable development in Nigeria. These include the FEPA Act, EIA Act, review of the National Environmental Policy Guidelines and Standards, establishment of a National Council for Environment and the implementation of Agenda 21 amongst others.[8] This report was written about seven years ago and if all of the above were then regarded as major milestones, it stands to reason that the outcome of these achievements should be evident by now.

In 2005 the Nigerian Environmental Sanitation Policy was approved by the Federal government. This was hailed as major step for the Nigerian environment because its targets include the provision of toilet facilities for both public places and households in homes.[9] Meeting these targets will certainly be of great benefit to the Nigerian environment but is there a feasible plan in place for this purpose? Environmental sanitation 'day' in most parts of the country are vigorously observed with people turning out en masse to clean the streets but this has little effect on the condition of sanitary facilities (the lack) in homes and workplaces.

Clearly, the nation does not lack for policy frameworks or strategies any longer. Every state has an environmental control agenda and various bodies have been set up for the purpose of implementing both national and state owned policies. The shortcoming lies in two areas the examination of which will form the central part of this chapter. First, the policies are many, broad and all-encompassing but they are targeted at the end-product of the damaging activities to the environment. Second, the usual problem with inadequate implementation of policies in Nigeria and misplaced priorities could affect the outcome. In the next section the two main issues stated here will be discussed culminating in an examination of the possibilities in adopting a precautionary approach as a solution.

Precaution versus Prevention

A vital gap remains in the system adopted by the government to tackle the environment challenges of the country. The administrative changes which have been made must translate to sustained action or nothing will be achieved. Protection of the environment should be treated with the urgency it deserves particularly when it is considered that some of the effects of environmental degradation cannot be remediated. Effects of environmental damage can be far-reaching as they range from direct impact to the health of humans and other living organisms to incidental impacts on the economic and social well-being of a community. A slow-paced approach to issues could later throw the country into a belated struggle to attempt to bring the situation under control as has been the case with the battle against the HIV/AIDS virus.

By the time the world awoke to the reality of the enormity of HIV/AIDS and its implications for the human race it appeared as though intervention was timely. The statistics today portray it differently. People have seen enough inexplicable illnesses ending in death that they have sought answers, and knowledge about the virus is gaining ground. But for many, the information is coming a little late in the day because of the many lives that have needlessly been lost that could have been saved had these discoveries emerged earlier. However, the information, buoyed by the aforementioned unpleasant experiences, has changed a few habits and saved lives. For instance nurses in our health facilities avoid re-using needles, and people are beginning to spend more on razors rather than sharing.

Uncertainty about the reality and effects of the virus led to delayed action and the consequences of this delay are yet manifest. Likewise two major factors that have affected environmental decision-making at various levels, including the international arena, are scientific and economic considerations.[10] Scientific uncertainty about environmental damage has consistently resulted either in delayed action or complete inaction. Related to this is the complacency occasioned by inconclusive and inconsistent scientific opinions on subjects such as climate change and global warming.

The above situation has influenced the majorly reactionary approach to environmental control such that most times legislation is being made to combat situations where remediation is impossible. One such legislation is the Environmental Damage (Prevention and Remediation) Regulations 2009 of the U.K. Its schedule 4 deals with remediation for damage to natural resources other than land and specifically Sections 4 to 7 state the types of remediation which apply to such damage. Mainly, they are categorized as complementary and compensatory remediation. As regards situations where remediation is not possible, the schedule provides that alternative natural resources with equivalent value to that which was damaged will be provided.[11] This legislation thus seeks to ensure that there is a replacement for resources which are irremediably damaged.

The fact that environmental damage is diverse and its effect is not totally ascertained has given rise to legislation of the like mentioned in the above paragraph. Such an expensive strategy can be avoided with the adoption of the precautionary principle. The European Union is renowned as the main proponent of the precautionary principle as a safeguard against the uncertainties of science. Principle 15 of the Rio Declaration states:

In order to protect the environment, the precautionary approach shall be widely applied by States according to their capability. Where there are threats of serious or irreversible damage, lack of full scientific certainty shall not be used as a reason for postponing cost-effective measures to prevent environmental degradation.[12]

In other words, it is a principle that advocates for protective action in situations where scientific evidence is uncertain either due to insufficient data or incomplete assessment. This principle has arguably attained the status of customary international law going by its inclusion in recent years in most international environmental treaty. The precautionary approach is the tool for realizing the sustainable development aim of Agenda 21.[13] It follows, therefore, that in explaining the objectives of the principle, sustainable development must first be understood. This has been succinctly defined by the 1997 Brundtland Commission as development that "meets the needs of the present without compromising the ability of future generations to meet their own needs."[14]

From the foregoing, sustainability requires that a considerable amount of foresight be exercised where probable risk exists in order that the present does not jeopardize the future. Here lies the difference between precaution and prevention. Preventive action is taken in reaction to risk which has been identified with certainty while precautionary action is employed in the face of likelihood that the risk may never be substantiated in the long run.[15] In the EC there is recognition that a situation of zero-risk is not conceivable, risk must therefore be more than hypothetical to warrant the application of the principle.[16]

The benefit in this approach to environmental protection is that environmental damage is usually irreversible or remediable at great cost. Preventative action may not always achieve the desired results and sometimes the debilitation increases even as restorative action is on-going. Prevention is the costlier stage of precaution as far the environment is concerned. The peculiar nature of each nation determines that precautionary measures be tailor-made to suit their capabilities and needs. This demands not only that environmental concerns and goals be identified but also that environmental values and priorities be determined and articulated as guides to focussed precautionary objective-setting.

Establishing Environmental Values in Nigeria

Hardin's illustration of the world as a pastureland where each farmer, in seeking his own optimal gain, keeps placing more and more cattle until the land is overcome by the unchecked increase in herd aptly describes the problems of ill-planned and ill-regulated development.[17] Any remedial action after the land has become overburdened would be at far greater cost than a careful approach to the use of the pastureland from start.

This situation is reminiscent of the development of Nigerian towns under the control of colonial rule. The colonial officials were creating an environment for themselves which did not visualize or plan for a Nigeria where indigenes will break out of the psychological fetters which were necessary to keep society in neat pockets of segregated existence. Administrative bodies after the colonial era are still working within the frame of infrastructural plans which are not all inclusive and lack the capacity to accommodate future growth.[18] This policy gave rise in the past to the establishment of Government Reservation Areas (GRAs) of old in most Nigerian cities across the geographical regions such as Ibadan, Kaduna and Enugu, where the 'cream' of society lived, including government officials and some privileged local elites. New post-colonial cities such as Abuja have even created a peculiar kind of segregation on a massive 'whole city' scale where people of lower economic opportunities will never have the ability to break through its ultra-expensive price scales. What would have been the poor man's remote hope of owning a property anywhere near the Abuja metropolis have even further been dimed by the destruction of the sprawling slums of Gwagwa and Karimo.

Urban to rural migration has been a worrying trend in Nigeria mainly because of the consequent over-burdening of city infrastructure due to over-population. In recent years, however, another trend is creeping in. Urban dwellers are moving to the rural areas in trickles. Is the government preparing for the future implications of this new development? Research shows that this phenomenon has been encouraged by various socio-economic and environmental factors ranging from over-crowding to poor sanitation in the towns. However, as the rural communities are benefiting from the re-population, it is opined that government should put incentives in place to further encourage it.[19] Our observation is that people tend more to move to rural areas with some form of basic amenities such as electricity. But the question is how long will it be before the ills that have led to the reverse in movement become fully manifest in the rural areas?

Many commentators such as Nwokocha have decried the rural-urban migration as a worrying trend which needs to be tackled through infrastructural development of the rural areas.[20] The now occurring reverse in choice of habitation has, however, not been propelled by any improvement in rural amenities. Nigerians, like any other people in the world today desire a decent quality of life and the situation in the cities has reached a breaking point.

In the light of the foregoing it becomes necessary to analyse the thrust of environmental policy-making and implementation in Nigeria. At the UNEP Global Ministerial Environmental Forum of 2005 the Federal Ministry of Environment presented a laudable report on policy strategies for the country.[21] The Sanitation and Human settlement sections of the report are particularly noteworthy in terms of having the potential of bringing about much needed change but this is contingent upon adequate implementation.

Three years later there is evidence of steps taken to achieve the goals stated in the report. In Abia State for instance a Mobile Sanitation Court has been established to prosecute those arrested for contravening the environmental laws, and also a waste recycling plant (one of 17 for different parts of the country) is already operational in Aba.[22] These measures, however, characterize the mode of environmental control in the country which intrinsically seeks to tackle the effects of damaging activities. Preventive measures when initiated are usually as a result of firmly established proof of damage.

A re-evaluation of environmental values in Nigeria by both government and non-governmental bodies is necessary such that goal-setting will be geared towards policy-making with a precautionary aim. Sustainability demands that the 'end-of-pipe' reaction to environmental harm be replaced by an anticipatory approach.[23] This essay does not advocate for what has been criticized as a radical application of the precautionary principle which argues for the total preservation of the environment for the benefit of the future generation, this seems to suggest that the present inhabitants of the earth are inconsequential.[24] The precautionary regime postulated here entails assessing our environmental situation as a nation, attempting to extrapolate on the future based on present findings and using this as a guide to balanced policy-making, legislation and compliance strategy.

Inclusion as a Precautionary Strategy

The Nigerian Environmental Action Study Team (NEST), in a recent online publication detailing is climate change projects, quite appropriately has planned for an outreach towards popular participation as one of four major components of its implementation plan.[25] Further information by NEST captures the seriousness and multifaceted nature of the effect of climate change in Nigeria.[26] Generating participation by the public may prove to be the most important strategy for attempting to control the effects of climate change. Here again, the scheme for intervention can only be reactionary.

The efforts of NEST, Green Alliance and other environmental NGO coalition groups in trying to bring coherence to the country's environmental control plans are laudable. The situation at hand, however, calls for greater realism which means that a basic grassroots form of participation is imperative. The ordinary people in Nigeria, whether urban or rural dwellers, must become sufficiently in touch with knowledge of the implications of many age-old activities on life and health. The people must understand that change is possible and this realization alone will propel change faster than the usually hortatory nature of government policies. According to Awori, "people should not be depicted as a problem but as a resource to be mobilized for positive development."[27]

The tragedy of haphazard planning cum lackadaisical response to public disaster by government is that even when they promise, after setting up panels of enquiry and receiving their reports, that such disaster will never happen again, they fail to put in place the recommended measures. Of course, people soon forget and then things quickly return to the status quo ante which inevitably poses the danger of the reoccurrence of the same tragedy. Take the instance of the January 2002 bomb blast at the Ikeja military barracks that led to the loss of thousands of lives in the Ejigbo canals. The government is yet to fulfil its promises to put in place policies and projects that will make the surrounding environment safer.[28] The 1854 epidemics of London[29] brought about change through more effective public health legislation and better sanitation-oriented considerations to housing structures. These were positive changes which had occurred after the loss of many lives, but the same cannot be said of similar tragedies of less magnitude that have occurred in Nigeria. Indeed, Nigeria does not have to experience epidemics of such epic scale to begin to take proactive steps to address our peculiar environmental situation which is mostly the result of poverty and worsening quality of life of the average citizen.

Schneidman and Collier in an interesting article write that poverty in Africa is no longer rising and has stabilized. They cite as an example the fact that African banks are not affected by the global banking crisis one of the reasons being that the banks have not engaged in sub-prime mortgages.[30] Relating this to Nigeria it is fact that the ordinary person has no access to mortgage facilities and therefore either owns no house or scrapes together whatever he can to build one. In a situation like this, compliance with standards on quality of building materials or sanitary facilities is conveniently jettisoned.

Many environmentally harmful lifestyle choices of Nigerians are being shaped by poverty but possible alternatives are not sought because of lack of information. For example the prohibitive cost of cooking gas has only driven the people to continue cutting wood for their domestic and commercial needs. The same goes for the lack of proper sewerage and sanitation system which has led to the indiscriminate deposit of human and other wastes, thus putting further pressure on land and water sources. These are some of the points of grievances of the Niger Delta people who are now taking the Federal Government and the oil corporations to task for not giving back to their impoverished communities from the wealth they derive from their area. The necessity here is not information on the merits of eco-centric protection of the environment. In the struggle for survival the average Nigerian wants information on how to improve his/her lot in life, the plight of some plant in the forest which is of little economic benefit would be meaningless. This is perhaps not because of any inherent disregard for non-economically useful ventures but is the direct consequence of abject lack.

Successful environmental control programs will be those which address real issues as affecting the reality of the day to day life of the people and thus can be 'owned' by the common man. These issues include: tackling the lack of dwindling productivity of farmlands; addressing the problem of the drying up of village springs and streams; explaining why selling communal land to land prospectors will enrich pockets briefly but mortgage the future of today's infants. Their needs determine what is considered as priority and their perceived worth of any scheme will be coloured by its relevance to meeting these needs.

Some western environmental attitudes may not be appropriate for the Nigeria of today. Such attitudes as held by environmentalists with eco-centric views will be quite unsuitable for a people whose struggle for survival is a daily battle. The average Nigerian farmer would consider any legislation which seeks to protect animals which plunder their produce as irrational. Indeed it is becoming really difficult for farmers who habitually practised the system of allowing land to lie fallow for a period to continue to do so. This is due to the problem of decreasing availability of farmland and poor yield of what little land they possess.

David King, president of the British Association for the Advancement of Science, asserts that Africa is being impoverished by ecologists who advice against the use of advanced techniques in farming in the continent for environmental reasons. He further states that the aversion to GM crops in some developed countries is 'a lifestyle choice for a community with surplus food'.[31] The effect of GMOs on human health is yet uncertain but the lesson in the opinion expressed here is that while being precautionary, Nigeria must choose its own priorities based on internally determined areas of need.

Simple environment-saving behavior which the ordinary person in developed countries is advised to practice include: switching off the electrical appliances rather than leaving them on stand-by, cycling to work rather than driving, recycling materials rather than discarding them. In Nigeria lack of constant power supply means that when electric power is available everyone uses it to maximum capacity. Cycling on our roads, particularly in the cities, is impractical for reasons of safety and the absence of proper road markings. Recycling of goods and household items is normal for mostly economical reasons. What then are the possibilities in a precautionary approach?

The Possibilities

Adopting a precautionary approach would mean that other more environment-friendly sources of economic development in the country will be explored and invested in. This will reduce over-dependence on oil and its attendant ills. Nigeria was a major exporter of agricultural produce like cocoa, groundnuts and palm-oil and food crop production was also very abundant. The country also showed great promise in the manufacture of cement, footwear and textiles amongst others. These and more avenues for economic growth still remain but with the concentration on oil in from the 1970s these others have suffered neglect.[32] Several projects are begun but abandoned at some point and the waste of resources caused by this is enormous.

The country now consistently imports goods that can be quite easily manufactured within its shores. Precautious measures would mean a re-examination of the probable future of the nation if this trend continues. It could also lead to an analysis of the future of the now oil-rich communities with a view of answering the question of how long the resource would continue to be abundant in those areas. This will definitely require the carrying of research. The possibility of decrease in oil if determined early should trigger investment in the resuscitation of other hitherto forgotten industries. This in turn will create more jobs for Nigerians, thus, improving the income of families. The link between poverty and environmentally harmful practices cannot be over-emphasized. When people are economically empowered the likelihood of accepting more expensive but healthier alternatives to socio-cultural behavior is greater.

Earlier in this chapter it has been mentioned that environmental damage can be irremediable. This is the underlying reason behind the U.K. legislation which requires the provision of an alternative site in such situations.[33] Because of the huge cost this measure will entail, the more viable plan would be to ensure preservation of what the environment and improvement where possible and necessary. It is bound to be easier to prevent or even slow the pace of deterioration than to produce alternatives.

Notes

1. Nick Spencer and Robert White, *Christianity, Climate Change And Sustainable Living*, (London: SPCK, 2007), 4.
2. "Nigeria Waste Imports from Italy (NIGERIA)," Trade and Environment Database (16/04/1999), http://www1.american.edu/TED/SUPER.HTM
3. See 'Italy Agrees It Will Retrieve Toxic Waste Sent to Nigeria,' an AP report published by the *New York Times*, July 24, 1988, http://www.nytimes.com/1988/07/24/world/italy-agrees-it-will-retrieve-toxic-waste-sent-to-nigeria.html.
4. Gbemiga Adelagan, 'The history of environmental policy and pollution of water sources in Nigeria (1960-2004): the way forward,' http://web.fuberlin.de/ffu/akumwelt/bc2004/download/adelegan_f.pdf
5. Oghogho Makinde and Bukola Ayanbule, *Environmental Impact Assessment Act, 1992, Nigeria: A Practical Insight to Environment Law*, Environmental Law, Global Legal Group Ltd., pp. 282-283, London, 2006. http://www.alukooyebode.com/publications/Chapter%2036%20-%20Nigeria.pdf.
6. Gbemiga Adelagan, "History of environmental policy," 3.
7. Patricia Birnie and Alan Boyle, *International Law And The Environment*, (Oxford University Press, 2002), 524.
8. Institutional Aspects of Sustainable development in Nigeria (Report of the United Nations Department for Economic and Social Affairs – Division for Sustainable Development, 1997).
9. J. I. Ezeji, 'Nigeria: government approves environmental sanitation policy,' (17/02/2005) IRC http://www.irc.nl/page/16315.
10. Phillipe Sands, *Principles of International Environmental Law*, (Cambridge: Cambridge University Press, 2003), 6.

11. Office of Public Sector Information, Statutory Instruments, 2009 No. 153, Environmental Protection, England, Damage (Prevention and Remediation) Regulations 2009. www.opsi.gov.uk/si/si2009/uksi_20090153_en_1 (Accessed on 24/04/2010).
12. The United Nations Conference on Environment and Development (UNCED) of 1992 held in Rio de Janeiro, Brazil. During this conference the Rio Declaration on Environment and Development was adopted, this consists of 27 principles including Principle 15 establishing the Precautionary Principle.
13. James Cameron, "The precautionary principle in international law" In Tim O'Riordan et al (eds.), *Reinterpreting The Precautionary Principle,* (London: Cameron May Ltd., 2002), 114-122.
14. World Commission on Environment and Development (Report of the Brundtland Commission), *Our Common Future* (1987), 8.
15. World Commission on Environment and Development (Report of the Brundtland Commission), *Our Common Future* (1987), 116-117.
16. J. Scott, "The precautionary principle before the European courts" In Richard Macrory et al, (eds.) *Principles of European Environmental Law*, (Europa Publishing, 2004), 60.
17. Garret Hardin, 'The tragedy of the commons,' (1968)162 Science, 1243-1248.
18. Geoffrey Nwaka, 'The urban poor, the informal city and environmental health policy in Nigeria,' http://www.engagingcommunities2005.0rg/abstracts/Nwaka-Geoffrey-final.pdf (Accessed on 22/03/09).
19. Gbemiga Adewale, 'Socio-economic factors associated with urban-rural migration in Nigeria: A case study of Oyo State Nigeria,' (2005) *J. Hum. Ecol.* 17(1): 13-16.
20. Ezebunwa Nwokocha, "Engaging the burden of rural-urban migration in a non-regulatory system: the case of Nigeria," 6, http://www.irmgard-coninx-stiftung.de/fileadmin/user_upload/pdf/urbanplanet/Nwokocha.pdf (Accessed on 22/03/09).
21. Innovative Practices For Consideration At The Ministerial Level Consultations At The Twenty-third Session Of The UNEP Governing Council/ Global Ministerial Environment Forum, Nairobi, Kenya, 21-25 February, 2005 (Report of the Federal Ministry of Environment, Nigeria).
22. "E-waste and sustainable development," 07/01/2008, The Tide Online, http://www.thetidenews.com/article.aspx?qrDate=01/07/2008.
23. Carolyn Abbot, "Environmental command regulation," In Richardson, B., and Wood, S., (eds.), *Environmental Law For Sustainability,* (Oxford: Hart Publishing, 2006) 64-65.
24. R. J. Berry, *When Enough Is Enough,* (England: Inter-Varsity Press, 2007), 115.
25. " Moving From Climate Change Capacity Development Project To Building Nigeria's Response To Climate Change," http://www.nestinteractive.org/projects.htm (Accessed on 21/03/09).
26. "Facts on climate change #1: repercussions for health and human settlements," see fn. 25.

27. Achoka Awori, (n.d.) 'An African perspective on environment and development': Voices From Africa: UN-NGLS, http://www.unsystem.org/ngls/documents/publications.en/voices.africa/number5/vfa5.04.htm.
28. http://punchng.com/Articl.aspx?theartic=Art20100128355329 (Accessed on 21/03/09)
29. Crossier, S., (n.d.) "John Snow: the London cholera epidemic of 1854", (2001-2009) Centre for Spatially Integrated Social Science, http://www.csiss.org/classics/ (Accessed on 21/03/09).
30. Witney Schneider and Paul Collier, "Africa: sustaining growth will challenge governments," AllAfricacom, http://www.allAfrica.com/stories200902200088.html (Accessed on 22/03/09).
31. Mark Henderson, "Green activists are keeping Africa poor," 08/09/08 Timesonline, http://www.timesonline.co.uk/tol/news/environment/article4699096.ece (Accessed on 23/03/09).
32. U.S. Department of State: Nigeria. http://www.state.gov/r/pa/ei/bgn/2836.htm (Accessed on 24/04/10).
33. See fn. 11.

• Chapter 21 •

GENDER, WORK AND TECHNOLOGY IN RURAL AFRICA

'Lai Olurode

Introduction

This chapter addresses an issue in the broad area of sociology of gender relations. Gender stereotypes and discrimination have been with society for a long time and have persisted in spite of social change dynamics. In no other area of social life have gender stereotypes been noticeable than in work activities. The work that is engaged in is not usually gender neutral. Technology has, however, been altering gender perception about work and in very complex ways, technology is abridging the worlds of work of men and women. The issues involved are diverse. This chapter weaved together the core dynamics involved through the study of how the introduction of technology is altering the gender balance in rural Africa and is hereby fostering collaborative activities between men and women.

In most of Africa, work is conceived of as a necessity and commitment to it is basic to a worthwhile existence. A representative view of an African conception of work can be gleaned from the worldview of the Yoruba of south-west of Nigeria which goes as follows: *"ise ni ogun ise, mura si ise ore mi, ise ni a fi ndi eni giga...iya re le lowo lowo, ki baba re lesin lekan, bi o ba gbojulewe won, oun to tan ni mo so fun o..."*; that is, "work is the known cure for poverty, it is through work that one can climb to the top, so, work hard my friend, your mother may be rich, your father may have horses at the stead, if you rely on these, they will surely lead you to ruins."[1]

Thus at every stage of socialization of the African child, emphasis is placed on the importance of work, so a life of leisure is frowned at. It was as if the Yoruba had anticipated the destruction that a life of ease would visit on the Spaniards in the 16th and 17th centuries and on which an historian had concluded that "wealth is not so good as work, nor riches so good as earnings."[2] Those who indulge their children are often portrayed as inflicting incalculable damage on such children who may later in life experience difficulties in growing up to be responsible. Many of African proverbs can readily be advanced in justification of this worldview. Admonitions which relate to the importance of work is gender neutral as both males and females are expected to respect the important place of work in everyday transaction. By being urged to shun a life of indolence, the individual is not expected to exert himself to a breaking point, commonsense is equally to be considered as a factor. Those who go about executing a task in the most strenuous way while neglecting *labor cum time-saving* procedures of accomplishing the same task are ridiculed as one who deploys little sense and good judgment into task performance and he is thus referred as "a companion or rather father of the lazy one" (in Yoruba *alagbara ma mero baba ole).*

Thus right from its origin, Africa had always embraced technology in its major activities of herding and gathering, hunting, horticulture and peasant farming. A variety of extant but traditional African technology in planting and food processing can be discerned. Thus the modern day technology, in as long as it helps to foster productivity and simplify tasks, cannot but be readily embraced by men and women. But it cannot be assumed that men and women have had equal opportunities and exposure to embrace modern day technology. The processes of globalization that were triggered first by the slave trade and later by colonialism had somehow worked more to erect barriers between the world of men and women in Africa. In no small way, the impact of the Victorian era in Africa were series of developments that removed women from the public domain and confine them to the sphere of domesticity. As had by now been well documented by many scholars,[3] women of Africa violently resisted attempts to re-define their place in the traditional society and had thus, with a measure of success, warded off forces of subjugation and domination under colonialism and even in the post-colonial phase. Thus when the colonial regime introduced cocoa as a cash crop, women successfully ignored efforts by the male-dominated colonial state to exclude them in its cultivation. Women fought to retain their right in land that had then become a prime possession by that time.[4]

There cannot be a better platform for assessing the gender differentials in access to technology in Africa than in rural production since this sector remains the most resilient of Africa's political economy by being a major provider of employment for its teeming population and indeed a sector in which women had been able to sustain their visibility and thus warded off marginalization. In both the actual production of agricultural goods and processing of same, African women's visibility and indispensability register powerfully. In most countries, women constitute the bulk of agricultural workers (over 50% in most cases) and the main food processors and preservers. Though the figures vary across Africa but in Benin where 70% of the female population live in rural areas, they carry out 60-80% of the agricultural work, they constitute 48% of the laborers in the agricultural sector in Burkina Faso; 73% of the economically active in agriculture in Congo where women produce 80% of the food crops; 57% of the agricultural force in Morocco were women; in Sudan women constitute 80% of traditional farmers and in Tanzania 98% of women defined as economically active were in rural agriculture.[5]

The focus of this chapter will be centered around agriculture generally especially on its different phases of weeding, planting, harvesting and processing as well as in other related activities. The main research question was how has the introduction of technology fettered or facilitated women's participation in rural peasant production and processing of agricultural goods. Has technology weakened or narrowed the gender balance in rural production? Has the use of technology in rural production been appropriated by African men to the disadvantage of the women folk?

Really, can data obtained in rural Africa sustain the hypothesis that the introduction of technology in rural production in Africa is blurring the gender threshold or are women being further pushed aside with every leap in rural production technology? In responding and articulating these issues, we relied on data gathered from rural sites in Nigeria and Kenya. Our notion of gender here is that which is malleable, elastic, changeable and indeed a dependent variable.

In addition to the introduction, the chapter is structured into four other sections. The second section is a presentation of an overview of Africa's political economy while the third looks generally at gender and the global technological divide. In section four, data from our field work in rural Nigeria and Kenya is presented. The section is preceded by some discussions of the research settings. Our conclusions are contained in section five.

Africa's Political Economy: An Overview

Africa, in many respects, remains an interesting phenomenon but a paradox of a sort. Its population is about 900 million. In respect of resource endowments, Africa remains very rich in mineral and forest resources. Africa's agricultural potentials are untapped. Its weather is moderate except for recent unsustainable environmental practices which are negatively impacting on its environment. Africa, nevertheless, has rich forest and mineral resources. Unfortunately, Africa's rich deposit of mineral resources is at the root of its political instability. In desperate attempts to capture state power, political elites engage themselves in prolonged and senseless political, military and ethno-religious warfare. The sites of these warfare almost coincide with the sites of rich and valuable diamond and oil deposit. Africa's resources are thus vulnerable to plundering by its states predators and their collaborators. It should not be surprising that the farming routines of most of its population remain perennially disrupted by ethno-religious warfare which mostly get provoked by struggles to gain privileged access to political and mineral resources. Once farming, the main occupation of over 70% of Africa's population, gets disrupted, it sounds the death knell of many and thus the deepening of their poverty profile. The frequent threat to state security has to be understood in this context.

Africa's health profile remains appalling, no thanks to the devastation and suffering that HIV/AIDS had inflicted on society as sub-Saharan Africa has 67% of global prevalence which comes to 28.5 million people. Though Somalia and Senegal have infection rates of below 1%, South Africa, Zambia and Zimbabwe figures hover around 15-20% of adults' population. However, three South African countries are worse off—Botswana (23.9%); Lesotho (23.2%) and Swaziland (26.10%).[6] The 2005 sentinel survey shows that Nigeria has 4.4% AIDS prevalence rate. The impact of HIV/AIDS pandemic on farming has been severe. It would not be fair to ignore the modest achievement that had been recorded with respect to democratic rule. Unlike the pathetic political situation some two decades ago, when most of Africa was under military dictatorship, today and in response to globalization, only a few countries in Africa could be counted as being under military rule. Political party tradition and competitive political practices are taking root even though the processes have not been smooth-sailing. Nigeria's 2007 general elections were marked by serious imperfections. In 2007 in Kenya, over 1000 people were massacred in elections in which the leading opposition candidate, Raila Odinga was believed to have won. Only a power sharing deal which was brokered by regional leaders averted further descent into chaos. Only recently did Robert Mugabẹ of Zimbabwe agreed to play by some rules. By and large, there are few specimens of good governance and leadership on the African continent with its 53 countries. The judiciary, in some countries, is asserting its independence and press freedom is becoming gradually entrenched. There are promising signs across the continent that pluralism and diversity are being tolerated ever than before.

These changes, some of which are due to internal dynamics while yet others are being engineered from outside, have not substantially impacted on agricultural practices which remains peasant-driven. With the unceasing collapse of the formal sector and most states' intolerance of the urban informal sector, more women are being pushed into rural production. Today, more women than men predominate in rural food production and processing, areas where technological revolution has been slow. The next section focuses on gender and the technology divide.

Gender and the Technological Divide

Gender, in everyday usage, refers to values, expectations and behaviors that are associated with the biological fact of being a male or female. But these values, expectations and behaviors are never settled or sacrosanct in that they are constantly being negotiated, re-negotiated, contested and open to manipulation. The meanings that are attached to the biological facts of maleness and femaleness are thus unstable and vary across societies. Moreover, they do not have the same meanings over time in the same society and even across societies. This explains why they are often viewed as not only relative cultural concepts but elastic as well. Gender role are rarely the same across societies.

In its simplest dictionary definition, technology refers to the practical use of scientific knowledge in industry and in everyday life. The essence of technology is to abridge time and reduce the intensity of labor in executing a task. In this age of globalization, the role of technology in everyday life is profound. We are simply overwhelmed by technology. The greatest revolution in history, that is, the industrial revolution from about the late 18th cent to early nineteenth centuries was accomplished with the introduction of technology in rural production. With agricultural revolution in Britain, a lot of rural producers were relieved of farm activities and were forced to seek relevance in cottage industries.

Though at different paces, the farthest corner of the world had been touched by technology. The modern means of transportation through land, air and water had brought the global society ever closer together and this had simultaneously facilitated social, cultural and technological exchanges. The built-up environment is a product of technology. It would be difficult for many to think of life without motor vehicles, without mobile phones, without heater during severe winter or without air conditioners or without reading glasses. Can we contemplate life without personal computers, internet and fax machines? But for technology, globalization which abridges time and space would have been unthinkable. Our births and deaths are thus constituted in the technological world. It has become so indispensable both to our living and dying as evidenced in assisted human reproduction, assisted death and human body preservation prior to burial. Our entire life is thus ruled by technology.

But in the usage and adoption of this technology, there are noticeable gender gaps which really have nothing to do with the character of this technology but rather with stereotypes and prejudices which flow from gender biases. These gender biases themselves flow from the assumption that activities that are carried out in the domestic sphere belong to female whereas those outside of it belong to male. Really, cooking, socialization functions and child minding activities are believed to be light jobs which are rarely remunerated nor do they reflect in the national official calculus and work measurement. Gender role division has come to be seen as fixed and unchangeable despite observations which point to the contrary.

The domestic sphere seems to be the last to be touched by technological revolution and probably because of women's dominance in this sphere, it is assumed to be undeserving of technological intervention until recently. There are two forms of technological divide that confront us: rural-urban and gender divide. But both processes push women to the fringe. As a matter of fact, ICT gulf between the rural and the urban dwellers and the increasing inequality between the rich and the poor are complicating efforts to reduce poverty. If individuals must access critical information that can assist in poverty alleviation, then familiarity with the internet culture is critical. Yet, the poor and women generally, especially in the third world are disadvantaged in this regard. The outcome is an unceasing process of marginalization in the context of globalization. Thus the intensification of technology has not robbed off satisfactorily on the poor and on women as a gender category whose profile has become worsened with every leap in technology.[7] It had been said that the income ratio of the richest 20% of the world's population to the poorest 20% has increased from 30:1 in 1960 to 75:1 in 2000. Worst hit in this category are women.[8] Women's workload has not reduced simply because of technological spread. Odd jobs in and beyond the home continued to be executed by women as evidenced in construction sites in Africa. Of course, it is generally acknowledged that women predominate among the poor. They have poor skill and had little or no access to education without which the shackles of oppression around their necks cannot be unchained.

Though the internet culture has been widespread among the global population but women remain disadvantaged for several reasons. First, there are more men internet users than women since there are more literate men than women in Africa. Secondly, women's are more abused and more misrepresented on the internet than men. Women are invariably often presented on the internet as objects for the pleasure of men. Pornographic portrayal of women on the internet is common[9] and women activists have been critical of this unequal treatment. This was a possibility because 'those who build applications and control the interfaces may gain a considerable say over the content of the information communicated and access by way of the net'.[10] So, the internet culture should be seen for what it is— a modality for cultural transmission in a globalizing world and does not really offer liberation to women as it equally extends the frontiers of male domination over females. In the argument marshaled by Murdock and Golding,[11] the anticipated triumph of virtual communities would certainly constitute another platform of expressing unequal power relationships between men and women. The duo of Murdock and Golding argued further that the notion that this technology portends liberation for users as they are able to participate not only as consumers but also as producers of cultural goods could be said to be simplistic since it fails to consider the social context of prior access to the internet and even after. Even where women have access to the internet technology, it is not only fettered for reasons of domestic and child care activities, women are often on the light side use of the technology—secretarial support services and chatting in the main.

When we look around our technology-driven environment, we notice that men exert greater leverage. Technology, though a gender neutral phenomenon, is regarded as a macho and as having a manly imprint. A machine could be driven or powered by men or women and the machine would respond to the press of the button regardless of gender. In the early years of my life, I noticed that women and girls rarely rode on bicycles and this practice has continued to the present time. Motor vehicles are mainly driven by men. Saudi women are still barred from driving motor vehicles. In the villages and towns of Africa, there are more male users of mobile phone than female. One might go away with the impression that women's contact with technology would contaminate technological devices. In the early colonial periods, African women were very active in the technology processes that drove local fabrics among other economic activities. The colonial experiences seemed to have changed all these and to women's disadvantage.

I have also noticed that in most places that I have been to in Africa, the mobile phone is more of an accessory for men than women. My mother, for instance, rejected my offer to buy one for her. But my father in fact demanded for one. And this has placed my father in an advantageous position to have monopoly of information flowing from us, his children who are scattered in far places in Nigeria. He decides what information flowing from us is to be shared with my mother. Of course, my father, for reasons of his bare literacy and numeracy is better placed to manipulate the mobile phone.

My central argument therefore is that the globalization of modern day technology rather than narrowing the gender gap is in fact suspected to be widening it as it seems to be empowering more male than female. This has been possible because technology has not been able to erode sufficiently into prejudice and stereotype which regard women simply as inferior and incapable of handling modern technology. In Africa's worldview however, women are not portrayed as inferior to men but their roles are to be interpreted as complementary. Women are always fighting all vestiges of marginalization even in the technological terrain. There is probably no other area of life where this hunch can be tested than in rural Africa where women engage in daily struggle against the odds of globalization and marginalization. The next section therefore discusses the interface between gender and technology in rural production.

Gender and Technology in Rural Africa

The diverse people of the continent of Africa are not strangers to technology. Technological innovations and inventions were also well received. Anthropological and historical accounts are rich with evidence of several instances of the use of technology in farming among other rural economic activities. The argument that Africa, through river Nile, was the cradle of civilization was an eloquent testimony to this assertion. In his work, Barnett[12] had given a robust account of how conversant Africa was with medicine, science and technology prior to the coming of European civilization. African women were not behind in these technological achievements. In the sphere of religion, the Yoruba associate a variety of textiles and cloths to different religious occasions. Women's religious textile is a complex phenomenon. In Yoruba technology, women's role is decisive.[13] Yoruba worship spirits and a plethora of gods. The worship of these gods have cloths and textiles that are associated with them. Women are active in the weaving and making of these dresses.

African pre-history contained evidence that had enhanced the reference to Africa as the cradle of civilization. For example, "Bipedal locomotion allowed the hominids who lived in Africa over two million years ago to carry objects with their forelimbs, and this made the use of tools possible. The ability to create tools made our species unique, and this achievement shows that the history of technology cannot be separated from human history."[14] Also, Nadel[15] had provided data on the Nupe people of Nigeria to buttress his argument about the black byzantium and the rich iron work industry among the Nupe people of Nigeria. That there is a widespread use of technology in Africa should not come as a surprise since Africans worldview through an analysis of proverbs, fairy tales and fables are supportive of the use of technology in everyday's life. Rural Africa, where a majority of the continent's near 900 million people live, remains a site of immense economic activities which revolve around farming and processing of agricultural products both for consumption and commercial purposes. Almost all the stages of farm crop production and processing entail the use of technological devices no matter how rudimentary.

The study by Ladipo[16] among women in Isoya communities of Ile-Ife in south-west Nigeria highlighted the significant changes that farming techniques had undergone. Women as wives were involved in maize shelling for their husbands either by beating the cobs with pestle or rubbing cobs over a grater-like mould. This method was labor-intensive and thus time-consuming. Large-scale processing was thus difficult. Nowadays, in the communities around Isoya, pedal operated shelling technology is in use and some diesel driving engine as well. The impact of these on productivity has been immense apart from affording women more time to spend on other activities. African women are also active in fishing. Along settlements that are found along river courses are men and women who are active in fishing. Alamu[17] came across two categories of fisherwomen in Kainji dam area of Nigeria: (i) the active ones who own their fishing implements (nets, boats, paddle etc) and are involved in fishing directly; and (ii) the passive ones who do not go out physically to fish but nevertheless own fishing equipments which are operated for them by their husbands, grown-up children (either male or female) and or hired labour.[18] Fishing by women was not, however, a common experience in Tombo village near Sierra Leone.[19] Among the common and simple technology in use were the hook and line, bow and arrow, trapping, poisoning and spearing techniques. With the spread of globalization and its technology, nets are now used in conjunction with boats. Women are fighting exclusion from these new processes of fishing which give extra income.

In these countries however, the techniques of fish processing and preservation are an entirely female activities. These include smoking, solar drying, cooking (grilling and frying) and salting. Fish processing has however undergone several technology changes according to Steady.[20] Wire mesh was more durable. Smoking has equally undergone innovations. Smoking structure is made of mud or bricks that are supported by iron or wood pools with grill layers made of wire mesh. Income losses that are normally experienced through poor preservation techniques are thus being warded off. Women's visibility in this industry has persisted over the years. Though men take active part in fishing but not in its processing and marketing. Along the creeks of Nigeria and in the riverine areas, Ilaje women and their children dominate the fishing industry. Women therefore are taking the benefits that are flowing from the introduction of new technology into the fishing industry.

As a way of enriching our discussion on gender and its relationships to the mode of production in rural Africa, we gathered data from rural parts of Nigeria and Kenya. The data from Nigeria were obtained between July and August 2008 while the Kenyan data were sourced in December 2008. I will first analyze and discuss data from Nigeria.

Analysis and Discussions of Data from Nigeria

Nigeria has an official population of about 140 million according to the census returns from its census of 2005 whereas Kenya's total population is about 30 million.[21] In both countries, the rural dwellers constitute the majority where peasant farming and processing of farm crops remain a major economic activity. Women and children are very active in all the processes of agricultural production and food processing which activities entail extensive use of all sorts of technology. In some cases, women could constitute more than a half of farmers. As farmers, they engage in clearing of bushes, preparing moulds, planting, weeding, harvesting, transportation, packaging, processing, marketing and storage among others. Women are equally involved in house-building.

In Nigeria, we visited three communities for the purpose of data gathering. These communities were located around Iwo, the most densely populated local government in Osun State, Nigeria, with a population of over 191,000. The settlements were Telemu, Asamu and Ilemowu. Telemu has a population of about 15,000 made up of largely women and children whereas Asamu has a population of about 1,500 and Ilemowu has about 1,000 people. In all these rural communities, subsistence and commercial cash crop agriculture are practiced. Cassava cultivation is common in all the villages and its cultivation is done by men and women. In the past, processing of cassava tubers into flour was an exclusive female occupation but with the introduction of technology, men are now involved. But men and women own the technology. Land clearing could be by men and women as well with the aid of cutlasses and hoes. Planting is usually manually executed. Weeding is done with the aid of chemicals such as Atrazine, diuron, pendinesnalin, s-metalachlor, igrancombi, gramazone and clearweed among others. The application of these chemicals could be undertaken by men and women. Some of these chemicals are applied before and after planting. Harvesting could be by both men and women. Children's labor could be employed in farm work. Grinding is done with the aid of technology and this is mainly male-dominated whereas when it was labor-intensive, it was mainly a woman's job. Processing of cassava tubers entail the following activities – peeling (mainly by women and usually manual); grating (usually with the aid of technology and by men); pressing (by men); sieving by women; frying by women and packaging by women. Whenever mechanization had been introduced in any of the stages, men show interest. But both men and women own the technology. The same observation was made in the processing of palm fruit into oil. Previously female-dominated, men have now registered their presence as processing is now with the aid of technology. The processing of palm products involve boiling of the fruits (mainly by women); grating with grinding machine or squeezing by feet (mainly women); oil extraction (mainly women); water evaporation through boiling (women); and storage in containers by women. Previously, other than in planting and harvesting, men played marginal role in the processing of palm products. The same trend was observable in the processing of groundnuts. Cocoa is a major commercial crop. There were more men involved than women. The processing and harvesting of cocoa involve both men and women. Both family and hired labor could be used in these processes. Cocoa processing include the following: seed extraction from pods (manually by men and women); fermentation of seeds for 3-5 days, drying under the sun (men and women); sorting (men and women); packaging and marketing could be by men and women. In the past, the last stages of packaging and marketing were mainly the preserve of men. Today, women entrepreneurs have challenged men's monopoly of this profitable trade in cocoa cash crop. Women who are not merely supportive of their husbands but who actively participate in farming and farm crop processing fair better than the category of women whose role merely complement their husbands during farm work. In all these communities, cooperative societies, particularly the traditional thrift system are active in supporting rural dwellers economic aspirations. There was a marked absence of government patronage. The roads were in derelict conditions and there was a plethora of abandoned projects by government. In Telemu village, these included a maternity health center, a small scale palm oil processing factory at Agbede's compound, a bore-hole water scheme at Ode-Oba, a palm oil processing technology and a primary school building. In Asamu, there was a cassava processing venture that had since been abandoned by government. At Ilemowu too, abandoned projects included an uncompleted building at Ilupeju community Grammar School, a borehole in the same school and a universal basic education (U.B.E.) project at the U.A.C. primary school.

Apart from poor road connection to the main market in Iwo and the problem of abandoned projects which could have aided economic activities and productivity, other problems which hamper rural economic activities in Nigeria are bush burning, frequent malaria attack, lack of storage facilities, poor patronage, labor shortages and theft of agricultural crops.

Analysis and Discussions of Data from Rural Kenya

Data were gathered from the Coast Province. The villages studied included Tiwi and Msambweni in the Kwale district. Tiwi has a population estimated at 20,000 while Msambweni's population was estimated at about 75,000. There were more women and children than men. Men and women in the villages engage in sand digging and granite quarrying apart from peasant farming. Some of the implements used in breaking rocky stones into granite include hammers, axe and metal bars. Women lease or buy land in which stones and rock are buried and then engage men in the digging. Women also engage in weaving of mats and thatch roof materials. Men are mainly farmers and also engage in fishing. But in Vanga and Shimoni, women also engage in fishing. But they do not venture into the high seas. In fishing activities, canoe, nets and gun are used. Maize is the most common staple in Kenya and along the coast in particular. There are three maize milling machines in Tiwi. These are owned by men who hire men to operate the machines. Women are only employed as sweepers and cleaners in the mill. Red dye tree grows well around Tiwi village. There is a factory which processes the dye seed. The factory is owned by government in partnership with some expatriates. Men work in the technical section of the factory whereas women carry out light jobs. There are complaints by villagers that many rural folks had been down with tuberculosis arising from pollution from the factory. The technology in use in processing bouba seed into sweet is simple. Once the seed is brought out manually, the thread is removed, and then grinding is the next stage inside a mortal (chinu) with a pestle (mutsi). This is followed by sifting. A kilogram of sugar and a kilogram of the seed is added in a cup of water plus food coloring. The final product is mabuyi or maoyo. This sort of processing is widely practiced by women and young girls. The products are taken to the urban centers for sale.

In Msambweni, men mainly fish while women engage in subsistence farming in rice, cassava and maize. Women are also active in house building. They prepare the materials for roofing while the task of roofing is men's work. Women also fetch water and mix with sand laterite during house construction. There is widespread joblessness in the village as it is difficult to get land for farming because of the activities of land grabbers. Land mainly belongs to government and most peasant farmers are squatters or tenants. So, land tenure is insecure. A large number of men had deserted the area in response to the growing phenomenon of landlessness. Poverty is evident on the faces of the ruralites along the coast of Kenya. Rural women are struggling to improve their life chances. Though compared to Nigeria, rural women along the coast of Kenya are yet to be visible in the use of technology in rural production. Rural technology seems to have had a wider spread in Nigeria than in Kenya. In Nigeria, rural women are visible and aggressive in the adoption of technology in rural production and in the processing of agricultural goods. Rural-based women in Nigeria have urban-based families who are probably supporting them with money to finance access to new technology. Of course, Nigeria's economy is bigger than Kenya's. In the conclusion which follows, we present a summary of efforts and struggle by rural African women to take advantage of technological globalization in rural productive activities as a driver of empowerment generally and thus the pursuit of a more abundant life.

Conclusion

Africa's rural societies are indeed complex and a straight-jacket analysis would be unhelpful. Some observers may go away with the impression of an unchanging and static social structure. To some, however, the rural scene has undergone profound social change processes with regard to its population structure, sources of income and the vast encroachment of the forces of globalization as manifested in the presence of new values and new technology. Others may yet be right to express the view that the gender divide, especially regarding gender roles in rural productive activities, has begun to be narrowed for reasons of the introduction of new technology.

These varying pictures present different aspects of the same reality. And none of the positions could be regarded as dead wrong. In the sphere of rural production in Africa, women initially register more impact in the sphere of supportive and thus in roles that complement those of their male spouses while simultaneously pursuing their own independent farming practices. But for a long time and in true appreciation of the worldviews of the people of Africa on gender roles division, women in rural Africa contest men's monopoly of certain economic activities. The result today is that in East and West Africa, women farmers are a common phenomenon. They play active roles in clearing bushes for planting, weeding, harvesting, marketing, processing and preservation of farm crops. Women are equally taking advantage of new technologies as these are introduced into farming and processing of agricultural products. Women appreciate and understand the workings of these machines. Sometimes, they also invest in them, and if need be engage men to operate them. Women also take advantage of improved seedling and insecticides. Indeed, the introduction of new modes of production is helping to re-define men's – women's relationship as well as the balance of power between them. In this process of re-negotiating gender relationship, new dimensions of power and empowerment are being thrown up.

African women have refused to be consigned to the periphery of social action by the forces of globalization as they are disclosed through the spread of technological innovations in rural agricultural practices. In a way, our hypothesis that gender role division is becoming blurred by the introduction of new technology in rural productive activities has some empirical validation. Men and women are collaborating in the use of new technology. Taboos, which in the past, restrict women from venturing into investments in agricultural cash crops, have since begun to be honored more in the breach. I also noticed, during fieldwork in Nigeria, that men are assisting in domestic tasks and in childcare, a role regarded as mainly women's. This change process has partly been triggered by some positive impact that the adoption of new technology in rural economic activities has facilitated. Labor exerting activities in which women partake are now less stressful and opportunities for greater income now abound. There really cannot be a better tool for women's empowerment than in the introduction of new technologies and the trickling down of new innovations. The people of Africa are falling into poverty at an alarming rate. People are demobilized and thus become unproductive by simple ailments such as malaria and missiles. A sustainable approach to poverty reduction must place rural producers and particularly women at the center for reasons of their strategic role in rural production. A widespread adoption of technology in rural production would help to reduce women's double load in productive and reproductive engagements and create some room for leisure as it reduces the drudgery associated with farming. Fortunately, African women could not be said to display much phobia in the adoption of rural technology in farming. In addition, the use of technology in rural production would help in moderating the potential conflict in work-life settings. Whatever time that is saved from long hours of work would help to moderate stress-related encounters in domestic work for rural men and women. This was unlike the findings in North America where changing gender roles have been found to be shifting conflict in work-life balance from women to men now that traditional gender roles which see women as homemakers are losing favors with two-earner couples (*USA TODAY*, March 26 2009).

But even with this, there remains three impediments. First, poor access roads to the villages remain a major barrier in the ability of rural producers to take full advantage of new development in science and technology. Secondly, the absence of crop processing and crop preservation technology in large numbers accounts for a lot of waste of farm crops and thus the loss of income and poverty. Thirdly, the global trade terms remain unfavorable to small producers in Africa within the context of barriers to trade. The African Union should consider these issues as more relevant to the needs of its rural majority than the pursuit of an elusive United States of Africa.

Notes

1. See L. Olurode, "The conception of work among the Yoruba of South-West Nigeria," *Journal of Enterprising Communities*, Vol. 2, No 1, 2007.
2. See D. Landes, *The Wealth and Poverty of Nations*, p. 172.
3. For example see N. E. Mba, *Nigerian Women Mobilized: Women's Political Activity in Southern Nigeria, 1990-1965* (Ibadan: Intec Printers Ltd., 1982).
4. See O. Ojo, "More than Farmers' wives: Yoruba women and cash crop production (1920-1957)" in Adebayo Oyebade, ed., *The Transformation of Nigeria: Essays in Honor of Toyin Falola* (Trenton: Africa World Press 2002), pp. 383-404.
5. See http://www.fao.org/docrep as at April 24 2009.
6. See http//www.avert.org/aafrica.htm as at April 24 2009.
7. See L. Olurode, "Gender, Globalization and Marginalisation in Africa," *Africa Development,* Vol. xxviii, Nos. 3 & 4.
8. See S. Hill, "Development Needs of Women," Paper Presented at the UNESCO's Asia Pacific Gender Equity in Science and Technology Programme (APGEST) (Philippines 19th June 2003).
9. For details, see R. Schaefer, *Sociology*. (New York: McGraw Hill 2005).
10. See J. Slevin, *The Internet and Society* (Cambridge: Polity Press 2001), p 39.

11. See G. Murdock and P. Golding, *Culture, Communication and Political Economy* in J. Curran and M. Gurevitch, eds., *Mass Media and Society*. (London: Hodder Arnold 2005).
12. Barnett, 2000: 210-214.
13. Asakitikpi, 2005: 71-87.
14. Barnett, 2000: 195.
15. Nadel, 1942.
16. See W. C. Barnett, "Medicine, Science and Technology," in Toyin Falola ed. *Africa: Culture and Societies before 1885*. Vol. 2 (Durham, NC: Carolina Academic Press, 2000), 210-214.
17. For further reading see A. O. Asakitikpi, "Owo Women's Religious Textiles," in P. E. Elisha and A. W. Babatunde, eds., *Yoruba Religious Textiles* (Ibadan: Book Builders 2005), 71-87.
18. See Barnett, "Medicine, Science and Technology," 195.
19. See S. F. Nadel, *The Black Byzantium* (London, 1942). For further reading in P. Ladipo, "Unexpected Gender Issues in the introduction of a Maize Sheller," in A. Afonja and B. Aina, ed., *Nigerian Women and Social Change* (Ile-Ife: OAU Press, 1995).
20. See S. O. Alamu, "Women in Artisanal Fish Production: A Comparison of Kainji and Jebba Lake Areas, Nigeria," A research paper submitted to the Department of Sociology, University of Lagos as part of the requirement for the award of Master of Science (Sociology) degree of the University of Lagos, Nigeria, 1998, 98.
21. For further reading see Alamu, "Women in Artisanal Fish Production: A Comparison of Kainji and Jebba Lake Areas, Nigeria," 7.
22. See F. C. Steady, "Women's Work in Rural Cash Food Systems: The Tombo and Gloucester Development Projects, Sierra Leone," In S. Muntemba, ed., *Rural Development and Women: Lesson from the Field* (Geneva: ILO 1985).
23. See Steady, "Women's Work."

• Notes on Contributors •

Timothy Olubisi Adejumo is an Associate Professor of Agricultural/Environmental Microbiology and Acting Director of the Centre for Research and Development at Adekunle Ajasin University, Akungba-Akoko, Nigeria. He is a Georg Forster Research Fellow of the Alexander von Humboldt Foundation, Germany since 2005. He collaborates with scientists on many projects in Nigeria and abroad, especially at the Department of Molecular Phytopathology, Georg-August University, Göttingen and Max Rubner Institut, Detmold, Germany, where he carried out his postdoctoral research. He has worked for many years in several research institutes, including the International Institute of Tropical Agriculture (IITA), Ibadan, Nigeria, where he was a Research Associate and Graduate Research Fellow. He has many publications on survey and detection of mycotoxins in human food supplies, botanicals and microorganisms as biopesticides for controlling mycotoxigenic fungi and pests of important economic crops. His current research is on microorganisms as nature's recyclers, especially using fungi in the bioremediation of soil pollutants.

Ajewole Ademiluyi Israel is an Associate Professor with the Department of Geography and Regional Planning, Olabisi Onabanjo University, Ago-Iwoye, Nigeria. He obtained a PhD in Geography from the University of Ibadan. He also holds a Master's degree in Urban & Regional Planning from the University of Wales, United Kingdom. His research interests include: Regional Development, Housing, and Transportation. He has over thirty publications to his credit. He has attended many local and international conferences where he made scholarly presentations.

Ahmad Bawa Abdul-Qadir is a Senior Lecturer in the Department of Business Administration, Kaduna State University, Kaduna, Nigeria. He holds a PhD in Management Studies of the Usmanu Dan-Fodiyo University Sokoto, Nigeria. He also holds a certificate in Economic Policy Analysis of the International Law Institute, Washington D. C. His teaching and research interests include Development Studies, Corporate Governance and Microfinance and small business development. He is a member of both the Nigerian Institute of Management and The Academy of Management Nigeria. He is currently Head of Department, Department of Business Administration, Kaduna State University, Nigeria.

Ann Albuyeh is Professor of English Linguistics at the University of Puerto Rico. She received her PhD from the University of Wisconsin-Madison. She has taught in Tehran, Iran, and Jalingo, Nigeria, as well as at Harvard University and the University of Wisconsin-Madison. Professor Albuyeh has also been a faculty associate at the International School of Theory at the University of Santiago de Compostela in Spain. She writes and has worked as a magazine and newspaper reporter and editor for English-language publications in the Middle East. Although her academic research has included theoretical issues in language acquisition, her linguistics publications have most often focused on the evolution of English from Old English to the current varieties of English pidgins, creoles, and standard dialects worldwide, with a particular emphasis on Africa and the Caribbean.

Maurice Amutabi is an Associate Professor in Social Sciences at the Catholic University of Eastern Africa. He has previously taught at Central Washington University, USA (2005-2010). Amutabi holds a PhD in History (Africa) from the University of Illinois at Urbana-Champaign, USA. He received his B.A (Hons) and M.A degrees from the University of Nairobi, Kenya. Amutabi is the author of *The NGO Factor in Africa: The Case of Arrested Development in Kenya* (New York: Routledge, 2006). Amutabi is co-author of *Nationalism and Democracy for People-Centered Development in Africa* (Moi University Press, 2000). He has also co-authored *Foundations of Adult Education in Africa* (Cape Town/Hamburg: Pearson/UNESCO, 2005). He has written two novels, *Because of Honor* (a novel on Islam in Africa) and *These Good People* (a novel on corruption in Africa). He is also the author of *Nakhamuma Stories* (a collection of short stories from the Abaluyia community of western Kenya). Amutabi's book *Islam and Underdevelopment of Africa* is forthcoming. His chapters have appeared in over two dozen books. His articles have appeared in several refereed and reputable journals such as *African Studies Review, Canadian Journal of African Studies, International Journal of Educational Development; and Jenda: A Journal of Culture and African Women Studies*. Amutabi has made presentations at over one hundred national and international conferences. He is the Vice-President of the Kenya Studies and Scholars' Association (KESSA), Kenya's premier research and academic organization.

R. Abidemi Asiyanbola holds B.Sc. (Hons) in Geography from Obafemi Awolowo University, Nigeria; Master of Urban and Regional Planning (MURP) degree and PhD degree in Geography from the University of Ibadan, Nigeria. He is currently a Senior Lecturer in the Department of Geography and Regional Planning, Olabisi Onabanjo University, Ago-Iwoye, Ogun State, Nigeria. He is a member of a number of professional bodies which include the Union for African Population Studies (UEPA/UAPS), International Union for the Scientific Study of Population (IUSSP/UIESP), International Sociological Association (ISA), Urban and Regional Development Research Committee (ISA-RC21), Population Association of America (PAA), Nigerian Institute of Town Planners, and Association of Nigerian Geographers (ANG). He is a past recipient of the Council for the Development of Economic and Social Research in Africa (CODESRIA) small grant for thesis writing and the Centre for Research on Inequality, Human Security and Ethnicity (CRISE), University of Oxford, United Kingdom – grant on gender and conflict research. Asiyanbola has made several presentations at local, national and international conferences. His researches have focused more on the social aspects of geography and urban and regional planning. He has particular interest in the interrelationship between society, practice and space (place, people and practice).

Adam Carpinelli is a doctoral student and teaching assistant in the department of history at Washington State University, focusing on African cultural retentions and memory systems of the African Diaspora in the Americas during the time of enslavement. His historical analysis of trans-Atlantic linkages between Africa and the Americas primarily looks at cultural expressions in folklore, art, music, drumming and dance, resistance and religion. He is interested in how neoliberal globalization, socio-economic and political changes impact human memory. His intention is to contribute to this area of historical scholarship by providing trans-Atlantic thematic approaches that will enhance our understanding of the era of the slave trade and the tremendous impact it has on the world today. Publications include: "The Cowrie Shell", *Commodities, Culture, and History: The Products That Have Changed the World*, Edited by Heather Streets-Salter, (New York: Facts on File, forthcoming); "Political Developments in Central Africa: 1750-1880," *Encyclopedia of World History* (Santa Barbara: ABC-CLIO, forthcoming 2010); and "Taino Survival and Caribbean Archaeology" co-written with Dr. Pedro Ferbel-Azcarate, *Proceedings for the International Association for Caribbean Archaeology*, Santo Domingo, Dominican Republic, 2003.

Ademola O. Dasylva, Professor of African literature and Oral literature, a poet and biographer, literary theorist and critic. He is co-coordinator, Ibadan Cultural Studies Group, a study group for the promotion of excellence in African and African American cultural studies. He was convener 2008 Ibadan International Conference on African Literature (IICAL). He is a Fellow, Salzburg Seminar, Session 374, Austria, 2000. He is a native of Ado-Ekiti (Ekiti State), of Aduloju family. He attended Our Lady & St. Kizito's Catholic Seminary, Ede, Osun State. He has his B.A. Hons, (English) and M.A. (African Literature), University of Ife (now Obafemi Awolowo University) Ile-Ife. He obtained his Ph.D in African Literature, from Nigeria's premier University of Ibadan, Nigeria. Dasylva teaches Drama, Poetry, the African Novel, and Oral Literature/Folklore Studies at the Department of English, University of Ibadan, Nigeria. Dasylva's scholarly publications include, *Understanding Wole Soyinka: Death and the King's Horseman*; *Studies in Drama*; *Classificatory Paradigms in African Oral Narrative* (a monograph), and co-edited with Prof. Kola Owolabi *Forms and Functions of English and Indigenous Languages in Nigeria*; co-author with Dr. Toyin Jegede, *Studies in Poetry*. His collection of poems, *Songs of Odamolugbe* won the 2006 ANA/Cadbury National Award for Poetry, and one of the 9 poems that made the final list in the 2009 NLNG Literature Award. In March 2009, at the Africa Conference, Professor Ademola Dasylva won the 2009 Distinguished Africanist Award for Research Excellence, University of Texas at Austin, USA. He is widely published in local and international scholarly journals.

Olukayode Segun Eesuola specializes in political behavior, with bias for protest /social movement; and he has been a lecturer in the Department of Political Science, University of Lagos, Nigeria, since 2005. He is also an intellectual aficionado of African political thought and culture; an interest which extends his current research to the political philosophy in Ifa, an ancient Yoruba system of divination, as well as the roles it played in the politics and administration of the Yoruba people before colonialism. Eesuola is an alumnus of the University of Lagos, Nigeria, and the Institute of American Politics and Political Thought, USA.

Toyin Falola teaches at the University of Texas at Austin. He has authored and edited over a hundred books on various aspects of African studies.

Nelson Olubanji Fashina obtained his MA and PhD in African/African Diaspora Literature, Theory & Criticism from the University of Ibadan, Ibadan, Nigeria, where he is currently a Senior Lecturer. He teaches African literature in all genres, African American, Caribbean and Third World Literatures, Literature and Gender, Sacred Texts, Folklore, Semiotics, Structuralism, Psychoanalysis and Creative Writing with conjunctive interest in Stylistics, Sociology and Cultural Studies. He was recipient of the RI Foundation Exchange Scholarship to England in 1992 and the US Department of State Fulbright Scholar Award on Postmodern American Literature at the University of Louisville, Kentucky in year 2000. He is Honorable Colonel of the Commonwealth of Kentucky and Honorable Citizen of Louisville, awards which were conferred by the Governor of Kentucky and the Mayor of Louisville, respectively. Nelson Fashina is not only a scholar-critic but also a poet-playwright, director, novelist and oral performance artist whose published plays, novel and poetry including Gods at the harvest, *A Season of Harvest*, *Foliage in the Woods*, and *Lyrics of the River* are enjoying the reading appreciation of higher educational literature departments. With about thirty articles in learned journals, his new direction of research borders on how to distil African specific critical language and theory from African and African Diaspora mega cultures and religion such as Ifa divination corpus.

Nnenna George-Kalu is a lawyer who has worked both in private practice and the civil service. She left the Ministry of Justice in Abia State, Nigeria late 2008 as a Principal State Counsel for further studies at the Newcastle University in the United Kingdom. This was a result of being elected fellow of the Ford Foundation (USA) International Fellowships through which she has studied for a Master's degree in Environmental Law and Policy (LLM). Nnenna is a beneficiary of the Emerging Leaders Development Program (ELDP), a project of the MacArthur Foundation USA whose aim was to identify, select and nurture up-coming leaders across Africa as agents of change in their respective countries. She is also a 2007 nominee of the Archbishop Desmond Tutu Leadership Program. She co-founded a non-governmental organization (NGO) known as Mission Africa Projects with her husband Rev George O. Kalu which focused mainly on empowering young people to achieve their potentials in life.

George O. Kalu graduated with BSc (combined honors) in Education and Economics from Abia State University, Uturu, Nigeria, and later obtained MA from the University of Gloucestershire, United Kingdom. He is an ordained Minister of the Presbyterian Church of Nigeria, and presently a Minister of the United Reformed Church, serving the joint pastorate of Weoley Hill United Reformed Church and Weoley Castle Community Church in Birmingham United Kingdom where he lives with his family. George has a passion for people empowerment and for the promotion of a sustainable environment. These led to his founding with his wife Nnenna a non-governmental organization, the Mission Africa Projects. He is currently a PhD candidate at the Oxford Centre for Mission Studies, Oxford UK.

Maria Leus is a professor in the Department of Architecture, PHL & University of Hasselt, Belgium, since 1989. She is responsible for the architectural design studio and the seminar of building structure design in the first year of the masters degree program in architecture. In 1997 she accepted a position as professor in the department of conservation and restoration of monuments and sites at the Artesis University College of Design Sciences in Antwerp, Belgium, where she currently holds a professorship in 'historic city center studies'. She also leads a cross-over interdisciplinary workshop on the re-use of cultural urban heritage sites. Over the years she has been thesis chair to several masters students, doctoral student as well as acted on the committee of numerous other post- graduate programs. Her working experience involves projects relating both to architectural design and the conservation and restoration of monuments and sites, with research interests around the themes of revitalization of towns, cultural identity and sustainable tourism. Her current scientific research is focused on educational projects relating to the re-use and revalorization of monuments and sites. This activity is translated in publications and lectures in international congresses. She furthermore has been a tutor and member of the jury at several recent international workshops: In 2004 Leus was a visiting professor at the Department of Architectural Technology at the Cape Peninsula University of Technology in Cape Town, South Africa and in 2010 at the Department of Architecture of the Anton de Kom University in Suriname.

Jolanda Morkel is a qualified architect, senior lecturer and doctoral candidate at the Department of Architectural Technology, Cape Peninsula University of Technology, Cape Town, South Africa. She is an Associate member of the South African Planning Institute (SAPI) and Member of the Association of Heritage Assessment Practitioners (AHAP). Prior to joining CPUT in 2001, she held a research and teaching position in the masters program in the Department of Town and Regional Planning at the University of Stellenbosch. Her teaching and research foci revolve around architectural design, townscape management and local distinctiveness (urban design), instructional technology, and designing sustainable learning spaces supporting multi grade pedagogy. She acted as head of department (2003) and faculty manager (2004 –2006). Ms Morkel currently holds the portfolio of coordinator for the design stream and service learning components of the undergraduate program in architectural technology, a Royal Institute of British Architects Part 1 accredited qualification. She coordinates service-learning projects in disadvantaged rural communities in the Western Cape. After successful completion of phase 2 of the Meulenhof Primary School upgrade in Philadelphia, she is currently involved in work at a number of rural multi grade schools. This project is run in collaboration with the Centre for Multigrade Education at Wellington. In both 2002 and 2005, she was invited as visiting academic at the Department of Architecture at the University of Hasselt in Belgium.

Olalekan Monsuru Muritala is a Doctoral candidate and an Assistant lecturer in the department of History, University of Ibadan. His research and teaching interests include: economic history of Africa, migration, urban and environment history. He has attended conferences both local and international. He has worked for a period of five years in the financial sector of Nigeria before returning to academics.

Mike Odugbo Odey is Professor of Economic History. He obtained his Ph.D. from the University of Jos, Nigeria (2002), M.A (1994) PGDE (1987) and B.A. (Hons) History (1980). His teaching and research interests include issues in Comparative Economic Development, Poverty Analysis/Policies and Food Security Systems, Inter- Group Relations as well as the Nigerian and African Development Question. He has over 50 well researched publications in these areas in International/Local journals. He is co editor of *Historical Research and Methodology in Africa*, (Makurdi: Aboki, 2007). He is the author of *The Development of Cash Crop Economy in Nigeria's Lower Benue Province, 1910-1960* in 2009. His latest work is *Food Crop Production, Hunger, and Rural Poverty in Nigeria's Benue Area, 1920-1995* by Carolina Academic Press in 2010. Odey is the current Head of Department of History, Benue State University, Makurdi, Nigeria and Member of Council, Historical Society of Nigeria.

Soji Oyeranmi is currently a Lecturer in the Department of History and Diplomatic Studies, Olabisi Onabanjo University (formerly Ogun State University), Ago Iwoye, Ogun State, Nigeria, and is also a PhD candidate in the Department of History, University of Ibadan, Nigeria. His research interests include environmental history and urban history. Oyeranmi's PhD thesis is on "Environmental Management and Development in Ibadan, 1900-2000." He has attended a number of international conferences and has a number of local and international publications.

Bukola Adeyemi Oyeniyi teaches History and International Relations in the Department of History and International Relations at the Joseph Ayo Babalola University, Nigeria. He holds a Bachelor's degree in Classical Studies and also a Master's degree in History from the University of Ibadan. He is currently at the threshold of completing his doctoral degree at the Institute for History, Universiteit Leiden, the Netherlands. He has attended a number of international conferences and has a number of publications in peer-reviewed journals such as *Social History, African Affairs, Outliers,* and *Quest: An African Journal of Philosophy/Revue Africaine de Philosophie.*

'Lai Olurode was trained at the University of Lagos for his first and master's degrees and obtained his doctorate degree in political sociology from the University of Sussex in Britain in 1984. He was a recipient of the Commonwealth scholarship for his doctorate, and was a Fulbright senior fellow. In 2003, he was a visiting professor at the University of California at Irvine and has given lectures at the Kennesaw State University, Armstrong Atlantic University, and Savannah State University, among others. In 2008, he was a visiting fellow at the Oxford Centre for Islamic Studies. He has enjoyed generous research funding from the Ford Foundation and from the Friedrich Ebert Foundation. His research works have appeared in *Rural Africana*, *Africa Development,* and the *Journal of Enterprising Societies*, among others. He has authored and edited several books. He specializes in gender, social inequality, and social change.

Shadrack Wanjala Nasong'o is an Associate Professor of International Studies and J.S. Seidman Research Fellow at Rhodes College in Memphis, Tennessee. He teaches courses in Comparative Politics, International Relations, African Politics, and Research Methods. His research interests are in the politics of democratization, conflict studies, nationalism, and political economy. Dr. Nasong'o is the author of *The Human Rights Sector in Kenya: Challenges and Keny Issues* (Nairobi: 2009); *Contending Political Paradigms in Africa: Rationality and the Politics of Democratization in Kenya and Zambia* (Routlegde, 2005); co-editor of *Kenya: The Struggle for Democracy* (Zed Books, 2007); and editor of *The African Search for Stable Forms of Statehood: Essays in Political Criticism* (Edwin Mellen, 2008). He has published over twenty peer-reviewed chapters and refereed journal articles as well as three encyclopedia entries. His scholarly articles have appeared in the *African Studies Review; Journal of Third World Studies; Journal of Contemporary African Studies; African and Asian Studies; Nigerian Journal of International Affairs; Taiwan Journal of Democracy;* and *Estudios de Asia y Africa.*

Ogundiiran Soumonni is a doctoral student in the School of Public Policy at the Georgia Institute of Technology in Atlanta, Georgia. His area of focus is Energy Policy with additional interests in Science, Technology and Innovation (STI) Policy more broadly. His research and publications include work on biofuels development, electricity policy, and nanotechnology for poverty alleviation. He is currently working on a dissertation on electricity planning in West Africa, using an environmental policy framework. Prior to entering the PhD program, he worked as a materials engineer in the area of advanced materials for energy-efficient lighting and display technologies.

Elinami Veraeli Swai holds a doctorate degree in Adult Education and Women Studies from Pennsylvania State University, U.S.A. Her major research was on learning and the development of identity from a socio-cultural and semiotic approach. Currently, Elinami is a Senior Lecturer and Dean in the School of Curriculum and Teacher Education, University of Dodoma, Tanzania. Her scholarly interests are combined in the analysis of macro political and social and cultural transformations and their effects on human development, including social inequality, family and class relations. Elinami has made several presentations at national and international conferences and has written and contributed papers in journals such as *Africa Today* and the *African Files* at Issue Enzine. She has also contributed chapters in books such as Toyin Falola and Matthew Heaton, (Eds.), *Endangered Bodies: Women, Children and Health in Africa* (Africa World Press, 2006); Toyin Falola and Niyi Afolabi (Eds.), *The Human Cost of African Migrations* (2007); Moses Oketch and Maurice Amutabi (Eds.), *The Distance Future: Perspectives on Lifelong Learning in Africa* (Kluwer); Chima J. Korieh and Philomena Okeke (Eds.), *Gendering Transformations: Gender, Culture, Race, and Identity in a Global Context* (Ihejirika). Elinami is the author of *Beyond Women's Empowerment in Africa Exploring Dislocation and Agency* published by Palgrave MacMillan Press.

Aribidesi Usman is an associate professor of Anthropology and African & African American Studies at Arizona State University. His research and publications focus on African history and archaeology, especially in the precolonial and contact period, African urbanism, regional political and economic interaction, frontier dynamics, Africans and transatlantic contacts, indigenous technology, and cultural resource management. His publications include: "State-Periphery Relations and Sociopolitical Development in Igbominaland, North central Yoruba, Nigeria," a BAR monograph (Archeopress, London, 2001); *Movement, Borders, and Identities in Africa*, co-edited with Toyin Falola, (University of Rochester Press, 2009), and several articles in anthologies and journals. He is currently working on a book, *The Yoruba Frontier: A Regional History of Community Formation, Experience, and Changes in West Africa* (ca. 1200-1900 AD)."

Hauwa Evelyn Yusuf holds a PhD in Sociology and Anthropology from the Obafemi Awolowo University Ile- Ife, Nigeria. She is currently a lecturer in the Department of Sociology, Kaduna State University, Kaduna, Nigeria. Her academic interest borders on issues of domestic violence, mainstreaming gender, and poverty and its possible eradication. She is involved in counseling on issues of human existence. She served as the Head of Department of Sociology Kaduna State University from 2007 to 2010. She is currently the Deputy Dean of Student Affairs and the Gender Focal Person of the Kaduna State University. Hauwa is also the CEO of the Center for study and resolution of domestic violence, a registered NGO in Kaduna State Nigeria.

• Index •

151211-200-1-60W